Infrastructure Possibilities and Human-Centered Approaches With Industry 5.0

Mohammad Ayoub Khan
University of Bisha, Saudi Arabia

Rijwan Khan
ABES Institute of Technology, India

Pushkar Praveen
Govind Ballabh Pant Institute of Engineering and Technology, India

Agya Ram Verma
Govind Ballabh Pant Institute of Engineering and Technology, India

Manoj Kumar Panda
Govind Ballabh Pant Institute of Engineering and Technology, India

A volume in the Advances in Web Technologies and Engineering (AWTE) Book Series

Published in the United States of America by
IGI Global
Engineering Science Reference (an imprint of IGI Global)
701 E. Chocolate Avenue
Hershey PA, USA 17033
Tel: 717-533-8845
Fax: 717-533-8661
E-mail: cust@igi-global.com
Web site: http://www.igi-global.com

Library of Congress Cataloging-in-Publication Data

Names: Khan, Mohammad Ayoub, 1980- editor.
Title: Infrastructure possibilities and human-centered approaches with industry 5.0 / edited by Mohammad Ayoub Khan, Rijwan Khan, Pushkar Praveen, Agya Verma, Manoj Panda.
Description: Hershey, PA : Engineering Science Reference, [2024] | Includes bibliographical references and index. | Summary: "This book collects the ideas on Smart society and human-centered approaches for research opportunities and Improvement from the different authors. This book will be a combination of Agriculture process improvement, water supply system, transportation, health and many other factors"-- Provided by publisher.
Identifiers: LCCN 2023037468 (print) | LCCN 2023037469 (ebook) | ISBN 9798369307823 (h/c) | ISBN 9798369307830 (eISBN)
Subjects: LCSH: Industry 4.0. | Infrastructure (Economics) | Human-machine systems.
Classification: LCC T59.6 .I53 2024 (print) | LCC T59.6 (ebook) | DDC 303.48/34--dc23/eng/20240126
LC record available at https://lccn.loc.gov/2023037468
LC ebook record available at https://lccn.loc.gov/2023037469

This book is published in the IGI Global book series Advances in Web Technologies and Engineering (AWTE) (ISSN: 2328-2762; eISSN: 2328-2754)

British Cataloguing in Publication Data
A Cataloguing in Publication record for this book is available from the British Library.

For electronic access to this publication, please contact: eresources@igi-global.com.

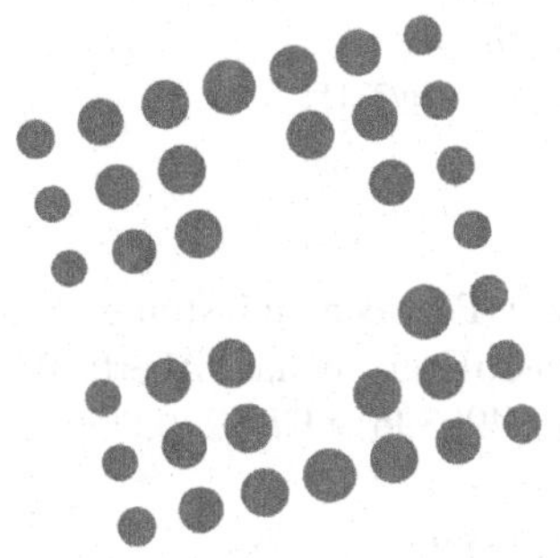

Advances in Web Technologies and Engineering (AWTE) Book Series

Ghazi I. Alkhatib
The Hashemite University, Jordan
David C. Rine
George Mason University, USA

ISSN:2328-2762
EISSN:2328-2754

Mission

The **Advances in Web Technologies and Engineering (AWTE) Book Series** aims to provide a platform for research in the area of Information Technology (IT) concepts, tools, methodologies, and ethnography, in the contexts of global communication systems and Web engineered applications. Organizations are continuously overwhelmed by a variety of new information technologies, many are Web based. These new technologies are capitalizing on the widespread use of network and communication technologies for seamless integration of various issues in information and knowledge sharing within and among organizations. This emphasis on integrated approaches is unique to this book series and dictates cross platform and multidisciplinary strategy to research and practice.

The **Advances in Web Technologies and Engineering (AWTE) Book Series** seeks to create a stage where comprehensive publications are distributed for the objective of bettering and expanding the field of web systems, knowledge capture, and communication technologies. The series will provide researchers and practitioners with solutions for improving how technology is utilized for the purpose of a growing awareness of the importance of web applications and engineering.

Coverage

- Integrated user profile, provisioning, and context-based processing
- Information filtering and display adaptation techniques for wireless devices
- Knowledge structure, classification, and search algorithms or engines
- Case studies validating Web-based IT solutions
- Radio Frequency Identification (RFID) research and applications in Web engineered systems
- IT education and training

IGI Global is currently accepting manuscripts for publication within this series. To submit a proposal for a volume in this series, please contact our Acquisition Editors at Acquisitions@igi-global.com or visit: http://www.igi-global.com/publish/.

The Advances in Web Technologies and Engineering (AWTE) Book Series (ISSN 2328-2762) is published by IGI Global, 701 E. Chocolate Avenue, Hershey, PA 17033-1240, USA, www.igi-global.com. This series is composed of titles available for purchase individually; each title is edited to be contextually exclusive from any other title within the series. For pricing and ordering information please visit http://www.igi-global.com/book-series/advances-web-technologies-engineering/37158. Postmaster: Send all address changes to above address.

Titles in this Series

For a list of additional titles in this series, please visit:
http://www.igi-global.com/book-series/advances-web-technologies-engineering/37158

Internet of Behaviors Implementation in Organizational Contexts
Luísa Cagica Carvalho (Instituto Politécnico de Setúbal, Portugal) Clara Silveira (Polytechnic Institute of Guarda, Portugal) Leonilde Reis (Instituto Politecnico de Setubal, Portugal) and Nelson Russo (Universidade Aberta, Portugal)
Engineering Science Reference • copyright 2023 • 471pp • H/C (ISBN: 9781668490396) • US $270.00 (our price)

Supporting Technologies and the Impact of Blockchain on Organizations and Society
Luís Ferreira (Polytechnic Institute of Cávado and Ave, Portugal) Miguel Rosado Cruz (Polytechnic Institute of Viana do Castelo, Portugal) Estrela Ferreira Cruz (Polytechnic Institute of Viana do Castelo, Portugal) Hélder Quintela (Polytechnic Institute of Cavado and Ave, Portugal) and Manuela Cruz Cunha (Polytechnic Institute of Cavado and Ave, Portugal)
Engineering Science Reference • copyright 2023 • 337pp • H/C (ISBN: 9781668457474) • US $270.00 (our price)

Concepts, Technologies, Challenges, and the Future of Web 3
Pooja Lekhi (University Canada West, Canada) and Guneet Kaur (University of Stirling, UK & Cointelegraph, USA)
Engineering Science Reference • copyright 2023 • 602pp • H/C (ISBN: 9781668499191) • US $360.00 (our price)

Perspectives on Social Welfare Applications' Optimization and Enhanced Computer Applications
Ponnusamy Sivaram (G.H. Raisoni College of Engineering, Nagpur, India) S. Senthilkumar (University College of Engineering, BIT Campus, Anna University, Tiruchirappalli, India) Lipika Gupta (Department of Electronics and Communication Engineering, Chitkara University Institute of Engineering and Technology, Chitkara University, India) and Nelligere S. Lokesh (Department of CSE-AIML, AMC Engineering College, Bengaluru, India)
Engineering Science Reference • copyright 2023 • 336pp • H/C (ISBN: 9781668483060) • US $270.00 (our price)

Advancements in the New World of Web 3 A Look Toward the Decentralized Future
Jane Thomason (UCL London Blockchain Centre, UK) and Elizabeth Ivwurie (British Blockchain and Frontier Technology Association, UK)
Engineering Science Reference • copyright 2023 • 323pp • H/C (ISBN: 9781668466582) • US $240.00 (our price)

Architectural Framework for Web Development and Micro Distributed Applications
Guillermo Rodriguez (QuantiLogic, USA)
Engineering Science Reference • copyright 2023 • 268pp • H/C (ISBN: 9781668448496) • US $250.00 (our price)

701 East Chocolate Avenue, Hershey, PA 17033, USA
Tel: 717-533-8845 x100 • Fax: 717-533-8661
E-Mail: cust@igi-global.com • www.igi-global.com

Table of Contents

Detailed Table of Contents

chapter, some applications of DL in healthcare have been envisaged, and it has been concluded that this technique is very successful in healthcare.

Sheelesh Kumar Sharma, ABES Institute of Technology, Ghaziabad, India
Ram Jee Dixit, ABES Institute of Technology, Ghaziabad, India
Deepshikha Rai, ABES Institute of Technology, Ghaziabad, India
Shalu Mall, ABES Institute of Technology, Ghaziabad, India

Artificial intelligence and machine learning have the potential to transform education by making it more effective, personalized, and engaging. In the context of smarter education, these technologies can be used in many ways to enhance learning for students and teachers: Intelligent Content Creation: AI can create learning content such as quizzes, exercises, and interactive experiments based on the specific needs of the course; Education Analysis: AI can analyze data from a variety of sources, such as student interactions with lessons, attendance records, and assessment results to understand learning patterns and trends. Teachers can use this information to identify struggling students, adjust instruction, and create interventions to improve overall learning; Automatic Grading and Feedback: Machine learning algorithms can grade assignments and tests, providing instant feedback to students. These technologies allow students to explore complex concepts in a practical and engaging way, making learning more effective and memorable.

Sheelesh Kumar Sharma, ABES Institute of Technology, Ghaziabad, India

In this chapter, the authors explore the role of AI in monitoring agricultural essentials, such as soil health, water management, and crop diseases. They delve into the various applications and benefits of AI in monitoring agricultural essentials. Through the integration of weather forecasts, soil moisture sensors, and machine learning algorithms, AI systems can determine the precise water requirements of crops in real-time. AI-based monitoring systems equipped with sensors and cameras can detect pest activity and provide real-time alerts to farmers. By harnessing the power of AI-driven monitoring systems, farmers can access real-time data, make informed decisions, and optimize resource allocation for sustainable agriculture. The applications discussed in this chapter, including soil health monitoring, precision water management, crop disease detection, pest monitoring, and yield prediction, highlight the transformative potential of AI in the agricultural sector.

Pratibha Verma, Constituent Government Degree College, Puranpur, India
Pratyush Bibhakar, Galgotias University, India

In this chapter, agriculture stands as the foundational pillar of the Indian economy, serving as a vital catalyst for growth and yielding substantial dividends in terms of revenue. Despite the commendable growth rate of the Indian economy, the state of agriculture remains less than satisfactory, marred by a multitude of disregarded challenges. India holds a prominent position as a leading supplier of essential commodities such as spices, pulses, saffron, and milk, significantly contributing to the nation's GDP and profoundly impacting the lives of the majority of its populace. Regrettably, these contributions have witnessed a decline. This study aims to elucidate the primary concerns and obstacles confronting Indian agriculture while also proposing potential remedies.

Because of increased pollution and population, the agriculture system is being affected drastically. New technologies should be implemented in this sector. The author discussed irrigation techniques and patterns using artificial intelligence that can change our irrigation process and minimize water consumption. Artificial intelligence approaches for soil fertility properties, the connection between soil quality, effect of fertiliser on soil possession and crop fecundity, season's growth, modelling of some soil properties, and estimation of polluted soil are being discussed. One more essential factor that reduces crop growth, quality, and yield is called weeds. Different types of weed control applications using ANN and ANFIS model have been discussed. To minimize monetary losses for farmers, disease control becomes a necessary parameter for better crop production. Image-based strategies using machine learning and deep learning for exact location, order of the disease, for precise and accurate identification have been considered.

Agriculture has significant challenges in the 21st century, which is why cutting-edge technologies like artificial intelligence (AI) and image processing are being employed to alter current farming operations. The primary study concern is the increased availability of visual data from a variety of sources, such as drones, satellites, and ground-based imaging equipment, which give critical information on crop and soil health as well as animal monitoring. The authors look at how AI algorithms can analyze this rich data and detect illnesses, pests, and nutritional deficits early on to help precision agriculture and reduce chemical use. In addition, they explore how image processing is used in automated irrigation management, where artificial intelligence (AI) uses meteorological data and satellite photos to optimize water use and protect resources. They also explore mechanized harvesting with AI-controlled robots that not only improve efficiency but also increase productivity.

Historical societies have been propelled into new eras of progress by the growth of industrial paradigms, each of which was characterised by distinctive technological developments and sociological changes. These revolutions, starting from the Industry 1.0's mechanisation to Industry 4.0's digitalisation, have radically changed the way that various industries and economies around the world operate. Industry 5.0 is now emerging as a disruptive force on the verge of a new phase. Innovative technologies like artificial intelligence (AI), robotics, and the internet of things (IoT) will be combined with human ingenuity, according to its promises.

Moving from Industry 4.0 to Industry 5.0 represents a significant shift in manufacturing, focusing on human collaboration, customization, and sustainability instead of just automation and digitization. Industry 5.0 brings substantial advancements that reshape the industrial landscape. This chapter examines the transition, emphasizing the importance of infrastructure, ethical concerns, and future implication. Industry 5.0 relies on a robust infrastructure, blending physical and digital elements to enable advanced technology integration. It places human-centric strategies at its core, emphasizing human interaction, well-being, and skill development. The synergy between infrastructure and human-centric approaches is pivotal for Industry 5.0's success, enabling ergonomic workplaces, seamless human-machine collaboration, data access, and customization. Simultaneously, human-centric strategies prioritize collaboration, skill development, and ethical considerations, fostering a harmonious synergy.

This chapter examines the transformative potential of machine learning in shaping smart health services within the framework of Industry 5.0. Through a comprehensive exploration of applications, methodologies, and real-world case studies, this chapter illustrates how machine learning algorithms are revolutionizing healthcare services. From real-time data analytics to personalized treatment pathways, the integration of machine learning empowers healthcare practitioners to make informed decisions that drive efficiency, accuracy, and patient-centred care. The chapter highlights the symbiotic relationship between machine learning and Industry 5.0, showcasing how data-driven insights and real-time collaboration are fostering the evolution of smart health services. As healthcare transitions from reactive to proactive, this chapter envisions a future where machine learning-driven smart health services not only optimize processes but also enhance patient well-being, marking a transformative step toward a patient-centric, technologically empowered future.

significantly contributes to the transformation of agriculture by facilitating data-driven decision-making, enhancing productivity, and optimizing numerous processes. There are so many areas where machine learning can be applied in agriculture processes like crop and soil monitoring, yield prediction, pest and disease management, etc. Pests and diseases are major obstacles to achieving higher productivity. The creation of effective techniques for the automatic detection and forecasting of pests and illnesses in crops is very much important. In this chapter, the authors use different machine-learning algorithms to detect early signs of pests, diseases, and weed infestations in crops.

Preface

As the editors of this reference book, *Infrastructure Possibilities and Human-Centered Approaches With Industry 5.0,* we are thrilled to present this comprehensive compilation of knowledge and research in the realm of Industry 5.0 and its profound implications for society. The emergence of Industry 5.0 represents a pivotal moment in the history of industrial revolutions, one that brings the promise of a harmonious fusion between advanced technology and human ingenuity.

Industry 5.0, unlike its predecessors, is defined by its human-centric approach. In contrast to the traditional focus on mass production and efficiency, Industry 5.0 aspires to empower individuals, enabling them to harness cutting-edge technologies to craft tailor-made products and services that cater to their distinct needs. This philosophy asserts that technology should serve humanity, rather than the other way around. By uniting the capabilities of advanced technologies with human expertise, Industry 5.0 holds the potential to address some of the most pressing global challenges, such as poverty, inequality, and environmental degradation.

A prominent domain where Industry 5.0 is poised to make a significant impact is personalized and flexible production. By amalgamating the Internet of Things (IoT), Artificial Intelligence (AI), and other advanced technologies, it becomes feasible to create products that are uniquely customized to the demands of individual consumers. This is a profound departure from the traditional methods of mass production, which churn out identical goods in large quantities. Through personalized and flexible production, Industry 5.0 offers the promise of a more inclusive and sustainable economy that is finely attuned to the needs of each individual.

Another field that stands to benefit immensely from Industry 5.0 is healthcare. The incorporation of advanced technologies like wearables and sensors enables the real-time collection of vital health data and the delivery of personalized healthcare services. This could revolutionize healthcare delivery, leading to improved outcomes for patients. Furthermore, the integration of robotics and AI has the potential to spawn groundbreaking medical technologies and treatments previously thought to be unattainable.

However, alongside the transformative potential of Industry 5.0, there are also challenges that must be confronted. Job displacement due to automation is a pressing concern, as many roles may become automated, leading to unemployment and social upheaval. Addressing this challenge necessitates the development of policies and programs to facilitate workforce transitions and skill acquisition for the evolving job landscape.

ORGANIZATION OF THE BOOK

Chapter 1

This chapter delves into the advent of Industry 5.0, highlighting the profound changes brought about by advanced automation, interconnected systems, and data-driven decision-making. It explores the necessity of a highly skilled and adaptable workforce in this new era and focuses on the integration of adaptive and personalized learning strategies within Industry 5.0 education. By leveraging cutting-edge technologies such as artificial intelligence and machine learning, this chapter elucidates how adaptive learning can tailor educational content, pacing, and assessments to individual learners, fostering engagement and knowledge retention. Additionally, it investigates how personalized learning approaches enhance Industry 5.0 education by creating immersive learning environments through simulations, virtual reality, and real-world case studies. The chapter conducts an in-depth literature review, presents compelling case studies, and offers an analysis of existing implementations in the realm of adaptive and personalized learning.

Chapter 2

Agriculture, being the foundation of food security and economic stability, transcends borders, playing a pivotal role in sustenance, employment, and environmental sustainability. This chapter addresses a crucial concern among farmers - the lack of knowledge regarding crop selection and fertilizer management. It emphasizes the misconception that more fertilizer leads to higher productivity, which often results in overutilization, leaching, and soil degradation. The chapter discusses the implications of cultivating the same crop repeatedly and its impact on soil fertility. It explores the need for informed and sustainable agricultural practices and offers insights into the challenges faced by the agricultural sector.

Chapter 3

This chapter focuses on the integration of Artificial Intelligence (AI) and Machine Learning (ML) in smart education. It explores the applications, benefits, and challenges of these technologies in enhancing learning experiences, personalizing education, and improving learning outcomes. The chapter also addresses ethical and privacy concerns, emphasizing the need for robust policies and guidelines to protect students' rights. By enabling personalized learning experiences, tailoring content, delivery, and assessment to individual needs, and supporting competency-based education, AI and ML have the potential to revolutionize education. However, the chapter acknowledges the challenges of privacy, security, algorithmic biases, teacher training, and ethical implications and offers recommendations to harness the full potential of these technologies in creating a smarter and more effective educational environment.

Chapter 4

This chapter explores the transformative potential of AI-driven skill development in the context of Industry 5.0. As Industry 5.0 demands a skilled and adaptable workforce, the chapter delves into how AI can analyze data to personalize skill development, bridge the skills gap, and enhance workforce preparedness. It emphasizes the positive outcomes of investing in AI-driven skill development for the evolving job market and its pivotal role in addressing the workforce needs of Industry 5.0. The chapter underscores the importance of equipping individuals with the skills required to thrive in this rapidly evolving industrial landscape.

Chapter 5

The emergence of deep learning and its applicability in healthcare are explored in this chapter. With a focus on the extensive data requirements in medical technology, the chapter investigates the applications of deep learning, including neural networks, convolutional neural networks, generative adversarial networks, and recurrent neural networks, in fields such as pattern recognition, natural language processing, image processing, and speech recognition. The chapter emphasizes the success of deep learning in healthcare, particularly in image processing for the proper diagnosis of patients through various medical imaging techniques. It offers insights into how deep learning techniques can revolutionize healthcare.

Chapter 6

This chapter sheds light on how Artificial Intelligence and Machine Learning can transform education by making it more effective, personalized, and engaging. It delves into several aspects, including intelligent content creation, education analysis, automatic grading and feedback, and the practical applications of these technologies for both students and teachers. The chapter highlights how AI and ML enable students to explore complex concepts practically and engage in more effective and memorable learning experiences. It underscores the transformative potential of these technologies in the education sector.

Chapter 7

Chapter 7 explores the role of AI in monitoring critical agricultural factors such as soil health, water management, and crop diseases. By integrating weather forecasts, soil moisture sensors, and machine learning algorithms, this chapter demonstrates how AI systems can precisely determine the water requirements of crops in real-time. Additionally, AI-based monitoring systems equipped with sensors and cameras can detect pest activity and provide real-time alerts to farmers. The chapter showcases the transformative potential of AI in agriculture by discussing soil health monitoring, precision water management, crop disease detection, pest monitoring, and yield prediction.

Chapter 8

This chapter centers on the pivotal role of agriculture in the Indian economy and the challenges it faces. It highlights India's contributions to the production of essential commodities and its impact on the nation's GDP and the livelihoods of its population. However, the chapter points out the declining state of Indian agriculture and aims to elucidate the primary concerns and obstacles facing the sector while proposing potential remedies. The chapter focuses on the challenges Indian agriculture confronts and offers insights into possible solutions.

Chapter 9

This chapter discusses the impact of pollution and population growth on the agricultural sector. It emphasizes the need for new technologies and explores irrigation techniques using artificial intelligence to optimize water consumption. Additionally, the chapter delves into AI approaches for soil fertility properties, the effects of fertilizers on soil quality, disease control, and weed control applications using artificial neural networks (ANN) and adaptive neuro-fuzzy inference systems (ANFIS). The chapter underscores the importance of innovative technologies in addressing challenges and enhancing crop production.

Chapter 10

Chapter 10 highlights the significance of employing cutting-edge technologies like artificial intelligence (AI) and image processing to transform current farming operations. It focuses on the availability of visual data from various sources, such as drones, satellites, and ground-based imaging equipment. The chapter explores how AI algorithms can analyze this data to detect crop diseases, pests, and nutritional deficits early, leading to precision agriculture and reduced chemical use. It also delves into automated irrigation management using image processing and AI, which optimizes water use and resource conservation. Furthermore, the chapter discusses the use of AI-controlled robots in mechanized harvesting to improve efficiency and productivity.

Chapter 11

This chapter provides an insightful historical perspective on the various industrial revolutions, leading up to the emergence of Industry 5.0. It highlights the disruptive force of Industry 5.0, driven by innovative technologies such as artificial intelligence, robotics, and the Internet of Things. The chapter underscores the promise of Industry 5.0, which combines these advanced technologies with human ingenuity, paving the way for a new phase of industrial progress. It provides context for the subsequent chapters by elucidating the ongoing transformation of industries and economies worldwide.

Chapter 12

Chapter 12 discusses the shift from Industry 4.0 to Industry 5.0, focusing on human collaboration, customization, and sustainability rather than just automation and digitization. It emphasizes the importance of infrastructure in Industry 5.0, blending physical and digital elements to enable advanced technology integration. The chapter places human-centric strategies at the core of Industry 5.0, prioritizing human interaction, well-being, and skill development. It highlights the synergy between infrastructure and human-centric approaches, vital for the success of Industry 5.0, enabling ergonomic workplaces, seamless human-machine collaboration, data access, and customization. The chapter also underscores the significance of human-centric strategies in fostering collaboration, skill development, and ethical considerations.

Chapter 13

Chapter 13 delves into the transformative potential of machine learning in shaping smart health services within the framework of Industry 5.0. It comprehensively explores the applications, methodologies, and real-world case studies of machine learning algorithms in revolutionizing healthcare services. From real-time data analytics to personalized treatment pathways, the integration of machine learning empowers healthcare practitioners to make informed decisions that drive efficiency, accuracy, and patient-centered care. The chapter highlights the symbiotic relationship between machine learning and Industry 5.0, showcasing how data-driven insights and real-time collaboration are fostering the evolution of smart health services. As healthcare transitions from reactive to proactive, this chapter envisions a future where machine learning-driven smart health services optimize processes and enhance patient well-being.

Chapter 14

This chapter offers a historical review of the industrial revolutions, from Industry 1.0 to Industry 4.0, and culminates in the emergence of Industry 5.0. It emphasizes the unique features of Industry 5.0, which include a harmonious balance between human and machine interaction, sustainability, and an innovative blend of technologies such as artificial intelligence, robotics, and the Internet of Things. This chapter serves as a foundational context for understanding the evolution of industry and the technological innovations that have brought us to the cusp of a new industrial phase.

Chapter 15

Chapter 15 discusses the importance of computational thinking as a problem-solving method. It explores how computational thinking can enhance students' ability to think critically and systematically. The chapter highlights the significance of computational thinking in problem decomposition, pattern recognition, and algorithm creation. It underscores the role of computational thinking in fostering higher thinking capacities among students and strengthening their cognitive processes. This chapter offers insights into the implications of computational thinking on student thinking, emphasizing the significance of this problem-solving skill in education.

Chapter 16

This chapter focuses on the integration of machine learning in agriculture, particularly in the areas of crop and soil monitoring, yield prediction, pest and disease management. It highlights the pivotal role of machine learning in data-driven decision-making, productivity enhancement, and process optimization in the agricultural sector. The chapter emphasizes the challenges of pests and diseases as major obstacles to achieving higher productivity and discusses techniques for early detection and forecasting of these issues using various machine learning algorithms. It provides valuable insights into the application of machine learning to address critical challenges in agriculture.The increased reliance on IoT and advanced technologies also raises the specter of cybersecurity threats, including cyberattacks and data breaches. Robust cybersecurity measures are essential to safeguard sensitive data and ensure individual privacy and security.

Furthermore, ethical and social considerations are paramount as Industry 5.0 technologies become deeply integrated into society. Issues such as privacy, data protection, and social inequality must be rigorously addressed through dialogue and collaboration with stakeholders from diverse backgrounds and perspectives.

The primary aim of this book is to collate the latest ideas and research opportunities in smart society and human-centered approaches. The chapters encompass a wide array of topics, from revolutionizing agriculture with AI and machine learning to enhancing food quality, optimizing fertilizer management, and delving into the applications of AI and machine learning in smart health services, transportation, and education. This book is tailored for research scholars, graduate engineering students, and postgraduate students specializing in computer science, agriculture, and health engineering.

We hope that the diverse range of topics covered in this book, as outlined in the tentative table of contents, will serve as a valuable resource for readers seeking to explore the possibilities and challenges presented by Industry 5.0. We extend our gratitude to the authors who contributed their expertise to this compilation, and we are confident that their insights will inspire and inform scholars, students, and professionals alike.

Sincerely,

Mohammad Ayoub Khan
University of Bisha, Saudi Arabia

Rijwan Khan
ABES Institute of Technology, India

Pushkar Praveen
Govind Ballabh Pant Institute of Engineering and Technology, India

Agya Verma
Govind Ballabh Pant Institute of Engineering and Technology, India

Manoj Panda
Govind Ballabh Pant Institute of Engineering and Technology, India

Chapter 1
Adaptive and Personalized Learning in Industry 5.0 Education

Harshit Singh
Galgotias University, India

Usha Chauhan
Galgotias University, India

S. P. S. Chauhan
Sharda University, India

Agrima Saxena
https://orcid.org/0009-0009-0562-1088
Galgotias University, India

Priti Kumari
National Institute of Technology, Patna, India

ABSTRACT

The advent of Industry 5.0, characterized by advanced automation, interconnected systems, and data-driven decision-making, has revolutionized traditional industries, necessitating a highly skilled and adaptable workforce. This chapter explores the integration of adaptive and personalized learning strategies within the framework of Industry 5.0 education. By leveraging technological advancements such as artificial intelligence and machine learning, adaptive learning tailors educational content, pacing, and assessments to individual learners' needs, fostering engagement and knowledge retention. Personalized learning approaches enhance Industry 5.0 education by creating contextually relevant and immersive learning environments through simulations, virtual reality, and real-world case studies.

DOI: 10.4018/979-8-3693-0782-3.ch001

I. INTRODUCTION

The emergence of Industry 5.0 signifies a profound shift in the industrial landscape, characterized by the convergence of advanced automation, interconnected systems, and data-driven decision-making. This transformative paradigm is reshaping traditional industries and necessitating a workforce that is not only highly skilled but also adaptable to the dynamic challenges of the modern era (Alvarez-Aros EL, 2021). As Industry 5.0 drives innovation and redefines job roles, there is a growing recognition of the pivotal role that education plays in preparing individuals to thrive in this evolving environment.

Moreover, the concept of personalized learning further enriches Industry 5.0 education by creating immersive and contextually relevant learning experiences. Through techniques such as simulations, virtual reality, and real-world case studies, learners are not merely recipients of information but active participants in their educational journeys. This approach aligns with the ethos of Industry 5.0, which emphasizes collaboration, innovation, and real-world problem-solving.

This research paper delves into a critical aspect of Industry 5.0 education – the integration of adaptive and personalized learning strategies. Leveraging the unprecedented advancements in technology, particularly artificial intelligence and machine learning, adaptive learning has the potential to reshape how knowledge is imparted. By tailoring educational content, pacing, and assessments to the specific needs and capabilities of individual learners, adaptive learning can significantly enhance engagement and foster the retention of crucial knowledge and skills.

A comprehensive exploration of the benefits associated with adaptive and personalized learning within the framework of Industry 5.0 has been conducted here. It draws insights from an in-depth literature review, examines case studies of successful implementations, and critically analyzes existing strategies (A. Akundi, 2022). The findings underscore the potential of these pedagogical approaches to foster improved skills alignment, enhance workforce readiness, and promote a culture of lifelong learning. However, as with any transformative initiative, challenges such as data privacy, learner autonomy, and the necessary technological infrastructure warrant careful consideration.

The ultimate aim of this research is to contribute to a deeper understanding of how adaptive and personalized learning can effectively equip individuals to meet the dynamic demands of Industry 5.0. By providing recommendations for educators, policymakers, and industry leaders, this paper seeks to navigate the transformative educational landscape that Industry 5.0 has ushered in. In an era where Industry 5.0 is reshaping industries and workplaces, the integration of adaptive and personalized learning emerges as a crucial enabler for cultivating a workforce equipped not only to adapt but to thrive in the ever-evolving technological landscape.

II. EDUCATION AND INDUSTRY 5.0

The concept of "Industry 5.0," also referred to as the "Human-Centric Industry" or "Society 5.0," builds upon the foundations established by the preceding industrial revolutions (Industry 1.0 to Industry 4.0), accentuating the seamless integration of individuals and cutting-edge technology. The primary objective of Industry 5.0 is to amalgamate the finest attributes of both human ingenuity and machine capabilities, thereby fostering collaborative and human-centered solutions (E. Coronado, 2022). This stands in contrast to earlier industrial revolutions, which predominantly focused on facets such as automation, digitization, and networking (A. Akundi, 2022).

Key characteristics that define Industry 5.0 encompass:

1. Human-Machine Collaboration: Industry 5.0 underscores the significance of synergistic collaboration between human entities and machines, recognizing the distinct proficiencies each entity brings forth. While robots contribute swiftness, precision, data processing capabilities, and automation, humans provide a dimension of creativity, empathy, intricate problem-solving abilities, and adaptability (E. Coronado, 2022).
2. Personalization: A pivotal aim of Industry 5.0 is the delivery of bespoke goods and services tailored to meet specific customer requisites. This notion is universally applicable, spanning diverse sectors including manufacturing, healthcare, retail, education and beyond (Duggal AS, 2021).
3. Localization: The concept of localized manufacturing and consumption takes precedence, aimed at mitigating environmental impacts and fostering community engagement. State-of-the-art manufacturing techniques like 3D printing and localized production contribute to heightened sustainability and efficiency (Gürdür Broo D, 2021).
4. Integration of Physical and Digital Realms: Industry 5.0 blurs the demarcation between the physical and digital domains through the amalgamation of real-time data sourced from Internet of Things (IoT) devices, sensors, and digital platforms with tangible processes. This amalgamation facilitates informed decision-making and swift responsive actions.
5. Enhanced Intelligence: Industry 5.0 strives to harness the capabilities of AI and automation to augment human aptitude, rather than being predominantly fixated on AI-driven automation. The focus shifts towards AI systems assisting humans in decision-making, intricate problem-solving, and demanding tasks.
6. Ethical Considerations: An integral facet of Industry 5.0 is its unwavering commitment to ethical considerations, encompassing aspects such as privacy, security, and the responsible utilization of technology (Praveen Kumar Reddy Maddikunta,2022).
7. Socio-Wellbeing: Industry 5.0 is resolutely dedicated to crafting solutions that prioritize societal welfare and sustainable advancement. This stands in stark contrast to the outcomes of previous industrial revolutions, which often engendered social and environmental predicaments.

However, Multiple industries are wholeheartedly embracing the principles of Industry 5.0. In education, for instance, AI-driven technologies and innovative tools are being harnessed to revolutionize the learning experience, creating tailored curricula, providing real-time feedback, and promoting interactive engagement within learning environments (C. Bigan, 2022). This shift echoes the essence of Industry 5.0, which accentuates the fusion of human-machine collaboration to craft solutions that enhance educational outcomes, address societal challenges, and foster sustainable advancement, signifying a profound transformation in the approach to technological integration (E. Coronado, 2022).

A. Evolution of Industry 5.0

Industry 5.0 is a nascent concept, still in its formative stages. It represents a theoretical construct inspired by the antecedent industrial revolutions (Grabowska S, 2022). A chronological overview of the industrial progression predating the emergence of Industry 5.0 is elucidated below:

- The inaugural industrial revolution, designated as Industry 1.0, transpired during the late 18th to early 19th centuries. This era witnessed the shift from agrarian societies to industrialized economies. Notably, it marked the automation of manufacturing processes utilizing steam and water power, catalyzing the establishment of factories and mass production.
- The Second Industrial Revolution, spanning the late 19th to early 20th centuries, saw the inception of the internal combustion engine and electricity. This epoch precipitated noteworthy advancements in communication, transportation, and manufacturing, bolstering productivity and enabling global interconnectedness.
- The "Industry 3.0" or Third Industrial Revolution materialized in the latter half of the 20th century, characterized by the rise of computers and automation. The integration of computers into operational management augmented process control and precision. Concurrently, this era initiated the digitalization movement and the utilization of production management software.
- The dawn of the 21st century introduced "Industry 4.0," characterized by the amalgamation of digital technology, the Internet of Things (IoT), data analytics, and artificial intelligence into manufacturing processes (Mostafa Al-Emran1, 2022).
- The nascent concept of the "Human-Centric Industry" within Industry 5.0 builds upon the foundational tenets of Industry 4.0, yet accentuates the imperative of human-machine synergy. Envisioned is an era where robotics and AI collaboratively interface with humans, enhancing efficiency, customization, and societal well-being (E. Coronado, 2022).

Throughout the trajectory from Industry 1.0 to Industry 4.0, hallmarks such as increasing automation, digitalization, and interconnectivity have persistently surfaced. Industry 5.0 signifies a paradigmatic shift towards equilibrium, valuing and harnessing the unique contributions of both human intellect and machine capabilities to navigate intricate challenges and formulate innovative solutions (see Figure 1).

Figure 1. Evolution of Industry 5.0

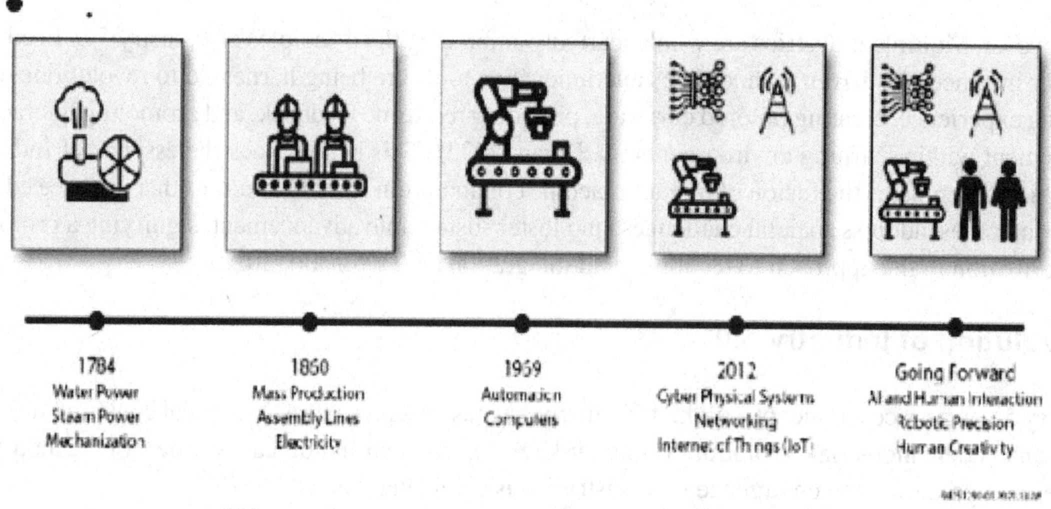

B. Evolution of Education 5.0

Education 5.0 embodies a contemporary concept that is still evolving, inspired by the progression of educational paradigms. A chronological overview of the evolution of education leading up to the emergence of Education 5.0 is expounded below.

Education 1.0: The nascent stage of formal education systems. It emphasized traditional teacher-centered instruction, reliant on oral communication and basic materials.

- Education 2.0: Evolving in the 19th and early 20th centuries, this phase introduced printed materials and textbooks, enabling more structured dissemination of knowledge. It marked a shift towards formalized curricula.
- Education 3.0: The 20th century witnessed the integration of technology, such as projectors and audio recordings, into classrooms. This phase aimed to enhance teaching methods and engage students in more interactive ways.
- Education 4.0: The digital age ushered in a transformative era, where technology took center stage. Online learning platforms, multimedia tools, and virtual classrooms became integral. Learning transcended physical boundaries, enabling remote education and personalized learning experiences.
- Education 5.0 (Emerging Concept): Building upon the foundations of Education 4.0, Education 5.0 envisions a future where technology and human interaction synergize seamlessly. It emphasizes the importance of emotional intelligence, critical thinking, and adaptability in a rapidly changing world (Javed AR, 2022).

Education 5.0 departs from previous stages by promoting holistic development, personalized mentorship, and a deep understanding of diverse subjects, fostering harmony between advanced educational technology and human educators. Like Industry 5.0's human-machine collaboration, Education 5.0 nurtures versatile, socially conscious individuals for modern complexities (S. Nahavandi, 2019).

Figure 2. Evolution of Education 5.0

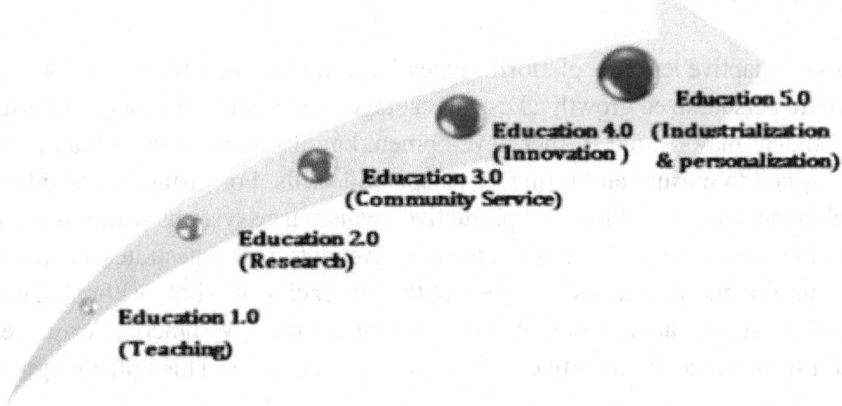

III. RAMIFICATIONS OF INDUSTRY 5.0 IN EDUCATION

Industry 5.0 represents the next phase of industrial development, focusing on the integration of advanced technologies like AI, robotics, the Internet of Things (IoT), and more into the manufacturing processes. When considering the ramifications of Industry 5.0 on adaptive and personalized learning, several significant impacts emerge (Mostafa Al-Emran, 2022).

A. Customized Learning Pathways

Adaptive learning platforms are at the forefront of educational innovation, utilizing advanced algorithms to meticulously dissect a student's learning behavior. These algorithms delve into a treasure trove of data points, unraveling the intricacies of how a student absorbs information, which content resonates most profoundly, and where challenges may arise (B. Andres, 2022). Armed with this comprehensive understanding, the platform crafts a personalized learning expedition for each student. It deftly modulates the complexity of content, recommends supplementary resources to fortify weak points, and even orchestrates the sequence of topics to synchronize with the student's pace. In essence, adaptive learning platforms are the architects of bespoke learning journeys, intricately tailored to enrich and elevate each student's educational odyssey.

B. Real-World Relevance

In the realm of adaptive and personalized learning, the pursuit of real-world relevance goes beyond mere theory, encompassing a profound integration of learning materials into the tangible realm of Industry 5.0 applications. This immersive approach entails a dynamic fusion of case studies, vividly explanatory videos, and interactive simulations meticulously designed to vividly illustrate the seamless integration of Industry 5.0 technologies within authentic manufacturing and production scenarios.

By melding these educational tools, students can forge a tangible bridge between the theoretical underpinnings they acquire and the practical intricacies of real-world contexts. This transformative strategy not only elevates the overall engagement quotient of the learning experience but also instills a deep-seated sense of applicability within the minds of learners, empowering them with insights that resonate profoundly with their envisioned careers in the dynamic landscape of Industry 5.0.

C. Skill Development

The capabilities of adaptive learning platforms extend beyond the identification of skill gaps, venturing into a realm where personalized growth takes center stage. These platforms not only discern the unique areas where a student may require further development but also curate meticulously tailored learning experiences designed to nurture and refine those specific skills. For instance, consider a student who aspires to excel in the domain of AI-driven predictive maintenance systems within the expansive canvas of Industry 5.0. In response to this aspiration, the adaptive platform orchestrates a comprehensive blend of cutting-edge programming courses that unravel the intricacies of AI algorithms, data analysis modules that unravel the secrets held by complex datasets, and immersive practical exercises meticulously crafted to mirror the essence of predictive maintenance technologies. This holistic approach empowers students to transcend conventional boundaries, gaining mastery over a spectrum of competencies crucial for thriving within the intricate tapestry of Industry 5.0's transformative landscape (B. Andres, 2022).

D. Continuous Learning

The ever-accelerating evolution of Industry 5.0 propels professionals into an era where perpetual skill refinement is not a choice but a necessity. In this dynamic context, adaptive learning emerges as a steadfast ally, proactively catering to the imperatives of ongoing skill enhancement. By seamlessly interweaving a tapestry of micro-learning modules, succinct yet impactful short courses, and incisive skill-based assessments, adaptive learning offers a flexible and accessible avenue for learners to continually embrace new knowledge and cultivate competencies that remain consonant with the fluid demands of Industry 5.0 (S. Nahavandi, 2019). This nimble approach empowers individuals to proactively navigate the twists and turns of this transformative industrial paradigm, ensuring they remain at the vanguard of innovation and remain responsive to the ever-evolving landscape of Industry 5.0 industries.

E. Data-Driven Insights

Mirroring the core tenet of Industry 5.0's data-driven essence, adaptive learning platforms deftly gather and dissect a wealth of data about students' intricate learning trajectories. This comprehensive process equips educators and administrators with a powerful analytical prism through which they can glean invaluable insights into the tapestry of individual and collective performance. By delving into the intricate nuances of how students interact with learning materials, and identifying the peaks of proficiency and the valleys of challenges, these data-driven insights bestow upon educators the acumen to orchestrate strategic decisions that elevate the learning experience to unparalleled heights (Javed AR, 2022). With a keen understanding of where students excel and where hurdles arise, educators are empowered to tailor interventions, sculpt resources, and create an environment that not only fosters meaningful comprehension but also nurtures a profound connection between learners and the absorbing world of Industry 5.0 concepts.

F. Collaborative Learning

In the realm of adaptive and personalized learning, collaborative learning takes on a dynamic dimension by establishing virtual environments that facilitate joint ventures among students. These shared learning experiences not only emulate the cooperative dynamics prevalent in Industry 5.0 teamwork scenarios but also mirror the symbiotic interactions witnessed between human professionals and cutting-edge machines within the Industry 5.0 framework.

Within these specially crafted virtual spaces, students are allowed to seamlessly collaborate on multifaceted projects and engage in collective problem-solving endeavors. By bridging diverse disciplinary backgrounds, these collaborative efforts reflect the quintessential essence of Industry 5.0, where individuals with varying expertise harmoniously unite to conceptualize, refine, and troubleshoot intricate systems (Gürdür Broo, 2021).

Much like the harmonious interplay between humans and machines characterizing Industry 5.0 settings, students in these adaptive and personalized learning environments discover the power of synergy as they harness their collective intelligence to address challenges that extend beyond traditional academic boundaries (Tyagi AK, 2023). This pedagogical approach not only nurtures valuable skills in teamwork and cooperation but also readies the next generation for a future where interdisciplinary collaboration is paramount.

G. Innovative Teaching Methods

Incorporating inventive pedagogical techniques into the fabric of adaptive learning platforms extends to the integration of transformative technologies such as virtual reality (VR) and augmented reality (AR). This visionary approach redefines the educational landscape, exemplified by a scenario where a student delving into the intricacies of industrial automation seamlessly harnesses Augmented reality headset to engross themselves within a realm of enriched learning.

Within this augmented realm, students are empowered to transcend the confines of traditional learning methodologies. This convergence empowers them to not only comprehend but also tangibly engage with the nuances of automated production lines through interactive virtual simulations.

The fusion of educational goals with technological advancements becomes apparent as the student progresses through intricately designed automated processes, addresses potential challenges, and enhances intricate sequences for optimal outcomes. Moreover, this experiential journey unfolds within a secure and meticulously controlled environment, effectively mitigating real-world risks while fostering a profound understanding of industrial automation (Alvarez-Aros EL, 2021).

This visionary amalgamation of AR technology and adaptive learning not only catalyzes conceptual understanding but also instills a transformative sense of experiential learning. In doing so, it ushers in an era where students transcend traditional boundaries, gaining proficiency and confidence through dynamic interactions with virtual constructs (Tamás Ruppert, 2022). As education converges with technology, the potential to nurture adept and agile professionals primed for Industry 5.0 dynamics becomes an imminent reality.

H. Global Connectivity

Within the realm of adaptive learning, the concept of global connectivity ushers in a transformative era where geographical boundaries dissolve, facilitating a seamless convergence of students and educators from every corner of the globe. This interconnectedness nurtures a rich tapestry of collaboration, enabling a vibrant exchange of ideas and a shared journey of learning that transcends cultural nuances.

This harmonious interplay between individuals of diverse backgrounds fosters a symphony of perspectives, each contributing a unique note to the collective learning symposium. By cultivating an environment where global insights converge, the adaptive learning landscape evolves into a fertile ground for cross-cultural pollination, where students glean not only subject matter knowledge but also a profound appreciation for the multifaceted tapestry of human experience.

An extraordinary dimension comes to the fore as students actively partake in international projects, discussions, and forums. These dynamic engagements serve as incubators for the development of critical skills such as intercultural communication, adaptability, and a global outlook. As students deliberate and collaborate on a global scale, they are primed to embrace the intricate dynamics of Industry 5.0, where seamless cross-border cooperation and understanding are integral.

In the transformative wake of this globalized educational paradigm, students emerge not only as proficient professionals but also as ambassadors of cultural empathy and global harmony. Armed with an enriched learning experience that spans continents, they are poised to seamlessly navigate the interconnected realm of Industry 5.0, where diversity and collaboration are the bedrock of success (Tamás Ruppert, 2022).

I. Challenges of Automation

In the face of automation's sweeping transformations, adaptive learning emerges as a strategic solution, deftly tailoring its offerings to tackle the evolving challenges head-on. This dynamic approach orchestrates a symphony of targeted courses and meticulously designed learning paths, meticulously calibrated to cultivate the very human-centric proficiencies that thrive in the tapestry of an Industry 5.0 landscape.

Central to this pedagogical evolution are skillsets that encapsulate the essence of human ingenuity. As automation's march persists, adaptive learning takes on the role of a skilled sculptor, chiseling out competencies that transcend the mechanized domain (Praveen Kumar Reddy Maddikunta, 2022). Among these prized aptitudes is creative problem-solving, a quintessential ability that endows learners with the acumen to unravel complex puzzles through innovative and imaginative approaches.

As Industry 5.0 unfolds, adaptive learning stands as a beacon, not just imparting knowledge, but sculpting adept and agile individuals equipped with the quintessential human skills that endure in the ever-evolving realm of automation. This seamless fusion of technology and human wisdom ensures that, despite the march of machines, the essence of humanity remains a steadfast pillar of progress (E. Coronado, 2022).

J. Ethical and Social Considerations

The integration of ethical and social considerations into the fabric of adaptive learning transcends mere instruction, embarking on a transformative journey that imbues learners with a profound sense of conscientiousness (A. Bakir, 2023). This pedagogical evolution manifests through the infusion of dynamic discussions, intricate case studies, and immersive scenarios meticulously curated to illuminate the intricate ethical tapestry woven by the advent of Industry 5.0 technologies.

In this immersive realm, students embark on a captivating exploration of the ethical echelons that Industry 5.0 unfurls. Engaging discussions become crucibles of introspection, where learners traverse the ethical dimensions of data privacy, a cornerstone concern in an era of prolific information exchange. The intricacies of job displacement unfold as case studies invite learners to navigate the delicate equilibrium between automation and human livelihoods, prompting a deeper reckoning of societal responsibilities.

As learners unravel the complex threads of ethical discourse, they confront the profound societal impact of automation. Immersed within scenarios that mirror real-world dilemmas, students are challenged to grapple with the intricacies of technology's role in shaping cultures and societies. This experiential journey extends far beyond theoretical musings, sculpting individuals primed to shoulder the ethical compass as they march towards Industry 5.0 professionalism (A. Akundi, 2022).

The culmination of this transformative process is a generation of Industry 5.0 professionals armed not only with technical prowess but also with an unwavering ethical resolve. Equipped with a nuanced understanding of their responsibilities, they become architects of progress, harmonizing technological advancement with a resolute commitment to the betterment of humanity. Thus, adaptive learning emerges as a crucible of enlightenment, where ethical consciousness becomes the cornerstone of Industry 5.0's future (Gürdür Broo, 2021).

Figure 3. Ramifications of Education 5.0

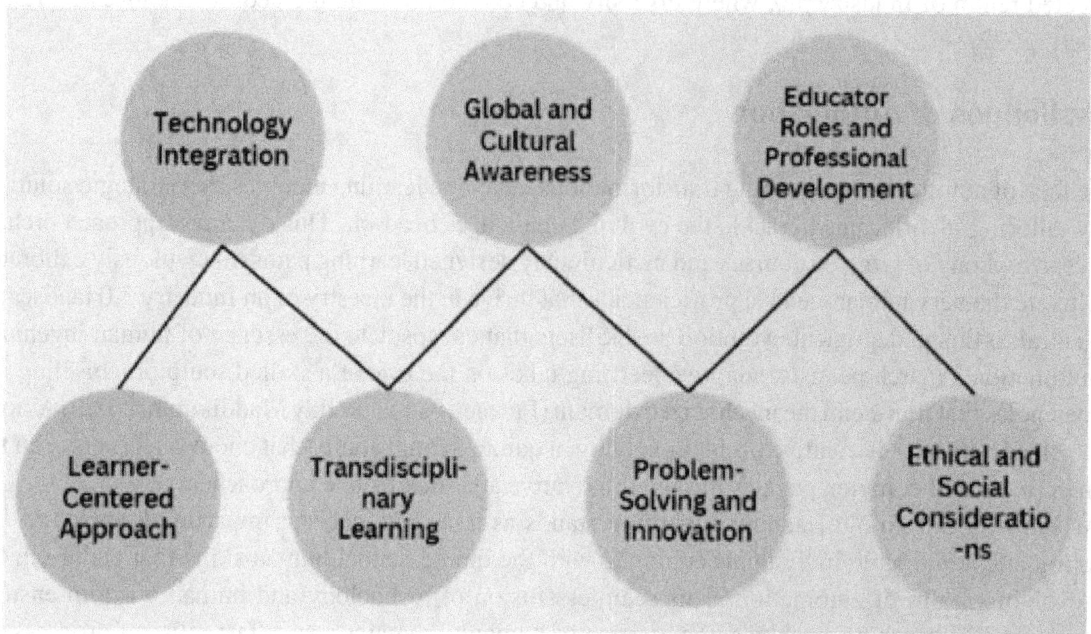

IV. CHALLENGES FACED BY EDUCATION 5.0

Adaptive and personalized learning in the context of Industry 5.0 education offers innovative approaches to prepare learners for the dynamic and technologically advanced workforce. However, it is not without its challenges. Here are some of the challenges faced by adaptive and personalized learning in Industry 5.0 education:

A. Complex Technological Infrastructure

Implementing the dynamic realm of adaptive and personalized learning necessitates the establishment of a robust and intricately woven technological infrastructure, underpinned by a symphony of advanced components (Mostafa Al-Emran, 2022). Among these, the orchestration of AI algorithms, data analytics frameworks, and cutting-edge learning management systems forms the backbone of this pedagogical evolution (Akshara Pramod, 2022). Through the application of technology and innovation, smart infrastructure can potentially extend access to fundamental services, stimulate economic growth, promote environmental sustainability, and elevate the overall quality of life within local communities. However, overcoming challenges related to funding, skilled workforce, and community involvement demands a holistic approach that entails collaboration among numerous stakeholders (Praveen P. et al., 2023). However, the endeavor of erecting and sustaining such an intricate ecosystem poses challenges that span the domains of resources and technical intricacies.

Constructing this technological landscape demands substantial resources, encompassing financial investments, skilled personnel, and time-intensive efforts. The intricate interplay of AI algorithms necessitates a team of adept data scientists, AI engineers, and educators, collectively steering the ship toward effective personalized learning experiences (C. Bigan, 2022). These experts sculpt the algorithms that unravel the unique contours of each learner's journey, shaping the individualized pathways that propel education into the Industry 5.0 era.

Yet, this journey is not without its technical tribulations. The sophisticated machinery of AI algorithms, with their intricate neural networks and deep learning architectures, requires a foundation of computational power and storage capacity. Procuring and maintaining the hardware and software infrastructure to support this AI-driven orchestration presents a formidable technical challenge, necessitating optimization and scalability.

Furthermore, the harmonious convergence of data analytics must be choreographed with precision. The vast reservoirs of learner data, spanning preferences, progress, and performance metrics, demand intelligent analytics engines to distill meaningful insights. This entails the development of intricate data pipelines, capable of seamlessly processing, analyzing, and transforming raw data into actionable intelligence that informs personalized learning interventions.

Amidst these intricacies, the learning management system emerges as the conduit that binds this technological tapestry. It serves as the interactive canvas upon which learners and educators coalesce, facilitating the seamless flow of personalized content, assessments, and engagement. The construction and perpetual evolution of this system stand as a testament to the relentless pursuit of technical excellence, as it strives to provide an intuitive and adaptive interface for the diverse spectrum of stakeholders (Praveen Kumar Reddy Maddikunta, 2022).

In the grand tapestry of adaptive and personalized learning, the construction and sustenance of this technological infrastructure emerge as a formidable yet essential endeavor. It requires a judicious allocation of resources, a synergy of technical expertise, and an unwavering commitment to the transformative potential of education in the Industry 5.0 landscape.

B. Data Management and Privacy

The cornerstone of effective personalized learning resides in the meticulous collection and analysis of substantial volumes of learner data. This process stands as a linchpin, enabling the tailoring of educational experiences that resonate uniquely with each individual. However, this data-driven approach unfurls a tapestry of challenges, centering on the critical dimensions of data privacy, security, and ethical considerations.

As educational institutions and platforms harness an intricate web of data points to shape personalized learning pathways, the imperative of safeguarding individuals data privacy assumes paramount importance. Learners entrust their personal information, learning patterns, and progress metrics to the educational ecosystem, necessitating stringent measures to prevent unauthorized access or breaches .

In tandem with data privacy, data security stands as a sentinel guarding against potential vulnerabilities. The sheer magnitude of learner data amassed in the pursuit of personalized learning underscores the necessity for fortified security protocols (R Varsha, 2020). Safeguarding this wealth of information from cyber threats and unauthorized exposure becomes a foundational obligation, underpinning the integrity of the entire personalized learning endeavor.

Furthermore, ethical considerations come to the fore as personalized learning delves into the intricate landscape of learner data. The responsible and transparent use of this data becomes an ethical imperative, demanding adherence to stringent guidelines that prevent misuse, bias, or unwarranted profiling.

Striking a harmonious balance between data-driven insights and ethical boundaries is a delicate yet essential tightrope walks.

In this data-rich terrain, navigating the complex landscape of data protection regulations becomes a focal point. Ensuring full compliance with data privacy laws and regulations safeguards the rights and interests of learners, reinforcing the ethical fabric of personalized learning. Additionally, engendering and preserving learners' trust serves as an invaluable currency, reinforcing the educational ecosystem's commitment to transparency, accountability, and responsible data stewardship (B. Andres, 2022).

In short, the collection and analysis of learner data, while foundational to effective personalized learning, unfurls a complex tapestry of challenges demanding vigilant attention. As the educational landscape evolves, the harmonious interplay between data privacy, security, ethical considerations, and regulatory compliance becomes an integral part of the personalized learning narrative, fostering an ecosystem that prioritizes both individual advancement and collective integrity.

C. Customization at Scale

Expanding personalized learning to cater to a substantial volume of learners presents a multifaceted challenge. The creation of tailored content, assessments, and individualized learning trajectories demands a substantial investment of time and resources. Each learner's unique needs and preferences must be considered, necessitating a level of attention that escalates the complexity of implementation (B. Andres, 2022).

D. Educator Training and Buy-In

A pivotal aspect of successful implementation lies in the imperative to equip educators with the acumen and proficiency necessary to navigate the terrain of data-driven insights seamlessly. The orchestration of personalized learning mandates not just technological prowess, but a profound understanding of how to decipher and wield the rich tapestry of learner data that underpins this approach. As such, a concerted effort towards comprehensive training initiatives becomes a linchpin in ensuring that educators are not only adept at utilizing the technological tools at their disposal but are also astute in discerning patterns, identifying areas for intervention, and tailoring instructional strategies to meet the diverse needs of learners (Javed AR, 2022).

Simultaneously, it is an endeavor that extends beyond technical proficiency. The endeavor to cultivate a culture that wholeheartedly embraces the infusion of technology into pedagogy necessitates an unwavering resolve to surmount the often formidable resistance to change. This endeavor entails a dual-faced challenge: the need to cultivate a deep-seated enthusiasm for technology-driven pedagogies among educators and concurrently, to assuage apprehensions that might arise in response to the perceived displacement of traditional teaching methodologies.

In essence, the transformational journey towards a technology-infused pedagogical landscape is not solely a technological transition; it is a nuanced exercise in cultural revolution. It demands not only technical training but the cultivation of a mindset that perceives technology as an enabler, an instrument to amplify the efficacy of education rather than a disruptor of established practices. This, in turn, compels educational institutions and stakeholders to invest in comprehensive professional development initiatives that encompass not just the technical but the intricate interplay of technology, pedagogy, and the art of pedagogical adaptation (B. Andres, 2022).

The endeavor to foster a culture that celebrates technology-driven pedagogies thus assumes the role of a dynamic journey of persuasion, a narrative that unfolds through the storytelling of success stories, the showcasing of tangible benefits, and the establishment of a supportive ecosystem that bolsters educators as they navigate uncharted technological waters (Praveen Kumar Reddy Maddikunta, 2022).

In essence, the challenge of training educators and nurturing a culture of technological integration is a multifaceted endeavor that intertwines the technical and the cultural. It necessitates not just proficiency in harnessing data-driven insights but a collective shift in perspective, an evolution from viewing technology as a challenge to embracing it as a powerful ally in the orchestration of effective, adaptive, and personalized learning experiences (Akshara Pramod, 2022).

E. Content Diversity and Quality

Creating learning content that is of high quality, diverse, and directly relevant to the unique needs of each learner presents a considerable challenge. It requires meticulous attention to detail, significant time investment, and a deep understanding of individual learning styles and preferences. Moreover, this content must not only meet the rigorous standards set by established curricula but also seamlessly align with the rapidly evolving demands of industries in the real world. Striking this delicate balance ensures that learners receive not only personalized education but also the essential knowledge and skills demanded by the modern workforce (Javed AR, 2022).

F. Overcoming Inherent Biases

The utilization of algorithms within adaptive learning systems introduces a potential concern: the inadvertent perpetuation of biases that might already exist within the data. This can result in the unintended amplification of disparities in learning opportunities among different groups of learners, thereby exacerbating inequalities. Additionally, there is a risk that these algorithms may unintentionally reinforce societal stereotypes, further hindering the goal of providing a fair and unbiased educational experience. As such, careful attention must be given to the design and training of these algorithms, coupled with ongoing monitoring and adjustment, to ensure that personalized learning remains a truly equitable and inclusive endeavor (R Varsha, 2020).

G. Learner Autonomy and Self-Direction

Recognizing the diverse learning profiles among students is crucial, as not all learners possess the same level of proficiency in self-directed learning—an integral component for the success of personalized educational approaches. It's important to acknowledge that some students might encounter challenges in effectively navigating their learning journeys without the guidance and structure provided by traditional classroom settings. These students may require additional support and scaffolding to develop the necessary self-regulation and time-management skills. Striking a balance between fostering autonomy and offering targeted guidance is essential to ensure that all learners, regardless of their self-directed learning capabilities, can fully benefit from and excel within a personalized learning framework. For these individuals, adapting to a more self-directed approach might necessitate additional support mechanisms (Javed AR, 2022).

Educators could play a pivotal role in helping students cultivate effective self-regulation strategies, time-management techniques, and personalized goal-setting practices. Moreover, fostering a supportive and collaborative learning environment, where students can share insights and learn from one another, could aid those who find self-directed learning less intuitive.

Strategically integrating moments of structured guidance or mentorship into the personalized learning process can act as stepping stones for learners who require more assistance. Balancing the encouragement of autonomy with targeted interventions ensures that personalized learning remains accessible and beneficial for every student, regardless of their initial level of comfort or proficiency in self-directed learning. By nurturing these skills within a personalized framework, educators empower students to not only adapt to this transformative mode of learning but also to thrive within it.

H. Continuous Monitoring and Adaptation

The dynamic nature of adaptive learning systems necessitates an ongoing cycle of vigilance, assessment, and refinement to uphold their efficacy in response to the evolving needs and progress of learners. As students advance along their personalized pathways, their aptitudes, preferences, and goals undergo continuous development. Consequently, the adaptive mechanisms guiding their education must be regularly observed and fine-tuned to ensure alignment with these shifting dynamics (Tyagi AK, 2023).

This demands a proactive approach to monitoring, where educators and technologists collaboratively engage in real-time analysis of student interactions, performance data, and feedback. By discerning patterns and identifying potential gaps or areas of improvement, these insights can inform timely adjustments to the adaptive algorithms, content selection, and learning trajectories. Furthermore, staying attuned to learners' changing requirements permits the seamless integration of new resources, technologies, or learning methodologies that may emerge over time.

The continuous monitoring and adaptation process not only sustains the efficacy of personalized learning but also fosters a sense of responsiveness and relevance within the educational experience (S. Nahavandi, 2019). By acknowledging and addressing the evolving needs of learners, adaptive systems can remain agile, enhancing engagement, learning outcomes, and overall satisfaction. This commitment to perpetual improvement serves as a testament to the dynamic potential of adaptive learning in shaping a truly adaptable and effective educational landscape.

I. Resource Allocation and Equity

Effectively allocating resources to support the implementation of adaptive and personalized learning in Industry 5.0 education presents a multifaceted challenge, one that can potentially introduce disparities between institutions with varying levels of funding and resources. The intricate landscape of technology integration demands substantial investments in robust IT infrastructure, software development, and data analytics capabilities (A. Akundi, 2022). Simultaneously, crafting tailored learning content that resonates with individual students' needs necessitates a dedicated commitment to content development, encompassing diverse formats and approaches.

Furthermore, ensuring educators possess the requisite skills to navigate this technology-driven educational paradigm requires comprehensive training initiatives. These programs not only familiarize teachers with the intricacies of adaptive systems but also cultivate a nuanced understanding of interpreting data-driven insights to enrich the learning experience. While well-funded institutions might readily embrace these resource demands, the potential exists for under-resourced counterparts to encounter obstacles in accessing and implementing these cutting-edge tools and practices. Such disparities could inadvertently lead to an uneven playing field, perpetuating educational inequalities rather than mitigating them (Goyal, Deepti Tyagi, 2019).

Addressing these resource allocation challenges mandates a balanced approach, wherein collaborative efforts across educational sectors, public and private partnerships, and innovative funding models converge. Equitable access to adaptive and personalized learning demands a concerted endeavor to bridge the resource gap, ensuring that all learners, regardless of their institutional backdrop, can harness the benefits of Industry 5.0 education in a manner that propels them toward success in the dynamic and technologically advanced workforce.

J. Alignment With Industry Needs

Securing the alignment of personalized learning pathways with the dynamic shifts in industry demands and the evolving contours of the job market is an imperative of paramount significance. As Industry 5.0 ushers in transformative advancements at an unprecedented pace, the criticality of preparing learners to navigate this ever-changing landscape cannot be overstated.

In this context, personalized learning emerges as a potent tool, capable of tailoring education to not only address current industry requisites but also instill in learners the resilience and agility required to anticipate and embrace future shifts. This calls for a proactive approach that employs real-time data insights and predictive analytics to anticipate emerging skill sets and professional trajectories. By weaving these anticipatory insights into the fabric of personalized curricula, educators can empower learners to acquire competencies that resonate with the pulse of the rapidly evolving job market (B. Andres, 2022).

Moreover, a symbiotic relationship with industry stakeholders is pivotal in this endeavor. Collaboration with businesses, technology leaders, and other key players enables the identification of emerging trends and the fine-tuning of personalized pathways to seamlessly integrate with the demands of the future workplace. This synergy not only equips learners with the tools they need to remain competitive but also engenders a continuous learning mindset that enables them to adapt and thrive in the face of uncertainty.

In essence, the harmonization of personalized learning with dynamic industry demands is not merely a desirable goal—it is an educational imperative that empowers learners to stay ahead in an ever-shifting professional landscape, ensuring their sustained relevance and competitiveness in the unfolding narrative of Industry 5.0.

K. Balancing Personalization and Standardization

Harmonizing the realms of personalized learning experiences and the attainment of standardized learning outcomes and competencies entails a nuanced and intricate undertaking. The delicate nature of this task lies in orchestrating an educational environment that tailors itself to the unique strengths, interests, and paces of individual learners, while simultaneously upholding the benchmarks of knowledge and skills established by standardized curricula.

In this intricate interplay, educators must deftly navigate the fine line between fostering a personalized journey that nurtures creativity and critical thinking and adhering to the foundational requirements that enable learners to engage effectively with broader educational and professional landscapes. Striving for personalization without sacrificing fundamental competencies necessitates a thoughtful design

Figure 4. Challenges for Education 5.0

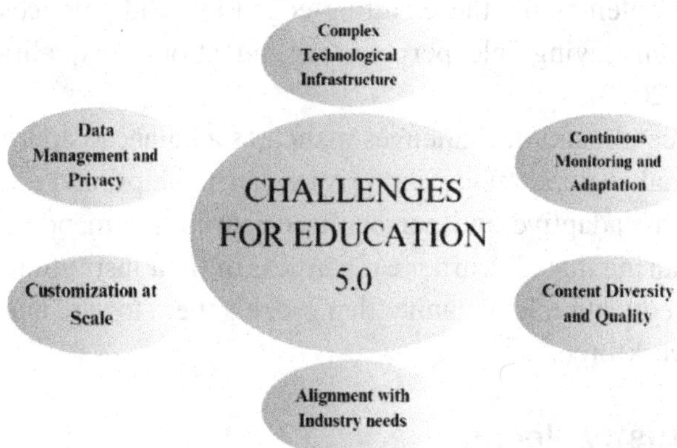

that integrates adaptive technologies, diverse learning modalities, and adaptable assessment methods (Mostafa Al-Emran, 2022).

Moreover, this balance is further complexified by the ever-evolving demands of Industry 5.0, which mandates a workforce proficient in interdisciplinary skills and rapid adaptation. As such, aligning personalized learning pathways with both traditional academic goals and the dynamic needs of emerging industries demands an astute orchestration that empowers learners to not only master standardized content but also develop the agility and innovative thinking that are indispensable in the modern world.

Ultimately, achieving this equilibrium requires a multifaceted approach that engages educators, learners, policymakers, and industry stakeholders in a collaborative effort to reimagine education as a holistic endeavor, one that bridges the personalized and the standardized to produce graduates who are both versatile and well-prepared for the challenges of Industry 5.0 and beyond (A. Bakir, 2023).

Addressing these challenges requires a comprehensive approach that involves collaboration among educators, technologists, policymakers, and learners themselves. Solutions may include targeted training for educators, robust data governance frameworks, diversified content creation strategies, transparent algorithmic decision-making, and ongoing research into the effectiveness and ethics of personalized learning systems (Javed AR, 2022).

V. CONCLUSION

This chapter discussed the advent of Industry 5.0 in a new era of industrial transformation, marked by unprecedented levels of automation, interconnected systems, and data-driven decision-making. As traditional industries evolve at an astonishing pace, the necessity for a highly skilled and agile workforce has become paramount.

Moreover the exploration of adaptive learning, empowered by advancements in artificial intelligence and machine learning, has illuminated the potential to revolutionize education. By tailoring educational experiences to the unique needs and capabilities of individual learners, adaptive learning enhances engagement, accelerates knowledge retention, and optimizes skills alignment. This personalized approach not only augments the learning journey but also contributes to the creation of a culture of lifelong learning, essential for thriving in the dynamic landscape of Industry 5.0. The incorporation of personalized learning methods has brought contextually relevant and immersive learning experiences to the forefront. Through the use of simulations, virtual reality, and real-world case studies, learners are immersed in practical scenarios that mirror the complexities of Industry 5.0 environments. This not only hones technical skills but also nurtures problem-solving abilities, critical thinking, and adaptability.

However, while the potential benefits of adaptive and personalized learning are promising, several challenges demand careful consideration. Issues related to data privacy, learner autonomy, and the requisite technological infrastructure must be effectively addressed to ensure the successful implementation of these approaches. As we navigate this transformative educational landscape, collaboration among educators, policymakers, and industry leaders is crucial to overcome these challenges and unleash the full potential of adaptive and personalized learning in Industry 5.0 education.

In short, the convergence of adaptive and personalized learning with the demands of Industry 5.0 offers a strategic pathway to cultivating a workforce that not only meets the evolving needs of industries but also drives innovation and sustainable growth. By fostering a harmonious relationship between technology, education, and human potential, adaptive and personalized learning emerges as a linchpin in shaping individuals capable of thriving within the dynamic and ever-evolving landscape of Industry 5.0. As we stand at the nexus of education and industry, embracing these transformative approaches is not just an option, but a necessity, ensuring a future where individuals are prepared, empowered, and eager to shape the world of tomorrow.

REFERENCES

Akundi, A., Euresti, D., Luna, S., Ankobiah, W., Lopes, A., & Edinbarough, I. (2022). State of Industry 5.0 – Analysis and Identification of Current Research Trends. *Applied System Innovation*, *5*(1), 1–14. doi:10.3390/asi5010027

Al-Emran, M., & Al-Sharafi, M. A. (2022). Revolutionizing Education with Industry 5.0: Challenges and Future Research Agendas. *International Journal of Information Technology and Language Studies*, *6*(3), 1–5.

Alvarez-Aros, E. L., & Bernal-Torres, C. A. (2021). Technological competitiveness and emerging technologies in Industry 4.0 and Industry 5.0. *Anais da Academia Brasileira de Ciências*, *93*(1), 1–20. doi:10.1590/0001-3765202120191290 PMID:33886700

Andres, B., Sempere-Ripoll, F., Esteso, A., & Alemany, M. (2022). Mapping between Industry 5.0 and Education 5.0. *EDULEARN22 Proceedings*, 2921–2926. 10.21125/edulearn.2022.0739

Bakir, A., & Dahlan, M. (2023). Higher education leadership and curricular design in industry 5.0 environment: A cursory glance. *Development and Learning in Organizations*, *37*(3), 15–17. doi:10.1108/DLO-08-2022-0166

Bigan, C. (2022). *Trends in Teaching Artificial Intelligence for Industry 5.0*. Advances in Sustainability Science and Technology. doi:10.1007/978-981-16-7365-8_10

Coronado, E., Kiyokawa, T., Ricardez, G. A. G., Ramirez-Alpizar, I. G., Venture, G., & Yamanobe, N. (2022). Evaluating quality in human-robot interaction: A systematic search and classification of performance and human-centered factors, measures and metrics towards an industry 5.0. *Journal of Manufacturing Systems*, *63*(9), 392–410. doi:10.1016/j.jmsy.2022.04.007

Duggal, A. S., Malik, P. K., Gehlot, A., Singh, R., Gaba, G. S., Masud, M., & Al-Amri, J. F. (2021). A sequential roadmap to Industry 6.0: Exploring future manufacturing trends. *IET Communications*, *16*(5), 521–531. doi:10.1049/cmu2.12284

Ghobakhloo, M., Iranmanesh, M., Mubarak, M., Mubarik, M., Rejeb, A., & Nilashi, M. (2022). Identifying industry 5.0 contributions to sustainable development: A strategy roadmap for delivering sustainability values. *Sustainable Production and Consumption*, *33*, 716–737. doi:10.1016/j.spc.2022.08.003

Goyal, D. T. (2019). A Look at Top 35 Problems in the Computer Science Field for the Next Decade. *Proceedings of 4th international conference on information and communication technology for competitive strategies(ICTCS)*, 379 – 396.

Grabowska, S., Saniuk, S., & Gajdzik, B. (2022). Industry 5.0: Improving humanization and sustainability of Industry 4.0. *Scientometrics*, *127*(6), 3117–3144. doi:10.100711192-022-04370-1 PMID:35502439

Gürdür Broo, D., Kaynak, O., & Sait, S. M. (2021). Rethinking engineering education at the age of industry 5.0. *Journal of Industrial Information Integration*, *25*(8).

Javed, A. R., Shahzad, F., & Rehman, S. (2022). *Future smart cities requirements, emerging technologies, applications, challenges, and future aspects* (Vol. 129). Cities.

Nahavandi, S. (2019). Industry 5.0—A Human-Centric Solution. *Sustainability (Basel)*, *11*(16), 4371. doi:10.3390u11164371

Nair, M. M., Tyagi, A. K., & Sreenath, N. (2021). The Future with Industry 4.0 at the Core of Society 5.0: Open Issues, Future Opportunities and Challenges. *International Conference on Computer Communication and Informatics (ICCCI)*, 1–7. 10.1109/ICCCI50826.2021.9402498

Pramod, Naicker, & Tyagi. (2022). Emerging Innovations shortly Using Deep Learning Techniques. *Advanced Analytics and Deep Learning Models*.

Praveen, K. R. M., Pham, Q.-V., & Prabadevi, B. (2022). Industry 5.0: A survey on enabling technologies and potential applications. *Journal of Industrial Information*, *26*, 1–8.

Praveen, P., Khan, A., Verma, A. R., Kumar, M., & Peoples, C. (2023). *Smart Village Infrastructure and Rural Communities*. IGI Global. doi:10.4018/978-1-6684-6418-2.ch001

Ruppert, Darányi, Medvegy, & Csereklei. (2022). Demonstration Laboratory of Industry 4.0 Retrofitting and Operator 4.0 Solutions: Education towards Industry 5.0. *Sensors, 23*(1), 283.

Saxena, A., Pant, D., Saxena, A., & Patel, C. (2020). Emergence of Educators for Industry 5.0 - An Indological Perspective. *International Journal of Innovative Technology and Exploring Engineering*, *9*(12), 359–363. doi:10.35940/ijitee.L7883.1091220

Tyagi, A. K., Dananjayan, S., Agarwal, D., & Thariq Ahmed, H. F. (2023). Blockchain—Internet of Things Applications: Opportunities and Challenges for Industry 4.0 and Society 5.0. *Sensors (Basel)*, *23*(2), 947. doi:10.339023020947 PMID:36679743

Usmaedi, U. (2021). Education curriculum for society 5.0 in the next decade. *Jurnal Pendidikan Dasar Setiabudhi*, *4*(2), 63–79.

Varsha, R., Nair, S. M., Tyagi, A. K., Aswathy, S. U., & RadhaKrishnan, R. (2020). The Future with Advanced Analytics: A Sequential Analysis of the Disruptive Technology's Scope. *Hybrid Intelligent Systems*, *1375*, 565–579. doi:10.1007/978-3-030-73050-5_56

Chapter 2
Advancing Agriculture With Industry 5.0-Enabled Crop-Type Prediction

Abhikalp Mishra
G.B. Pant Institute of Engineering and Technology, India

Pushkar Praveen
G.B. Pant Institute of Engineering and Technology, India

Ananya Subudhi
G.B. Pant Institute of Engineering and Technology, India

Somil Majila
G.B. Pant Institute of Engineering and Technology, India

Sachin Negi
https://orcid.org/0000-0003-2564-6070
G.B. Pant Institute of Engineering and Technology, India

ABSTRACT

Agriculture is the global foundation of food security and economic stability, crucial for sustenance, employment, and environmental sustainability. Its significance transcends borders. Type of crop produced plays a key role in the agricultural yield. A key problem faced by the farmers is lack of knowledge of the type of crop to be produced as well as the amount of fertilizer required in their particular soil. Farmers think that the higher the fertilizer used, the greater the productivity. But it is not correct. The soil uses the exact amount it needs and leaves the rest. Over utilization leads to leaching and decrease in the natural soil fertility and many such problems. Also, always farming the same crop in the farmland makes the cropland barren; hence, the produce does not yield much profit for the farmers.

DOI: 10.4018/979-8-3693-0782-3.ch002

I. INTRODUCTION

Agriculture, one of the most ancient and fundamental traditions in human history has profoundly shaped societies, economies, and the course of human civilization. This chapter explores the history of agriculture, tracing its evolution from humble beginnings to its contemporary significance in the modern world. The origins of agriculture date back over 10,000 years to the Neolithic Revolution. Before this period, our ancestors primarily relied on hunting and gathering for sustenance. However, the need for a more sustainable food source became increasingly apparent with growing populations and changing climates.

The Fertile Crescent, situated in the modern-day Middle East, is widely acknowledged as the birthplace of agriculture. It was here that early humans initially domesticated crops such as wheat, barley, and lentils, along with animals like goats, sheep, and cattle. The shift from hunting and gathering to agriculture marked a significant turning point in human history, facilitating the establishment of settled communities, surplus food production, and the emergence of more complex societies. The significance of agriculture in the rise of ancient civilizations cannot be overstated. In Egypt, the annual flooding of the Nile River deposited nutrient-rich silt on its banks, making it an ideal location for agriculture. The ancient Egyptians cultivated crops like wheat and barley and raised livestock such as cattle and sheep. The surplus food generated by agriculture allowed for the construction of monumental structures like the pyramids and the growth of advanced societies H. Pathak et al. (2020).

Similarly, in Mesopotamia, located between the Tigris and Euphrates Rivers, agriculture thrived due to the development of irrigation systems. This fertile region was the birthplace of the Sumerian, Akkadian, and Babylonian civilizations, where agriculture served as a means of sustenance and as the foundation of its political power. Exploring how agriculture played a critical role in the growth and development of ancient civilizations throughout history is truly fascinating. The Nile River and Mesopotamia show how natural resources and innovative techniques were crucial in developing agriculture and civilization. Sophisticated farming techniques also emerged in other parts of the world, such as the Indus Valley, China, and the Americas. The Harappan civilization in modern-day Pakistan and India utilized advanced drainage and water management systems, while China made significant contributions to agricultural technology, including the invention of the iron plow and the use of fertilizers. In the Americas, the Maya, Aztec, and Inca civilizations cultivated crops such as maize, beans, and squash. They relied on innovative farming techniques like terracing and crop rotation to sustain their populations.

The yield of a crop depends on various factors. In a study by K. Krishna Kumar et al. (2004), the intricate relationship between climate variables and their effects on Indian agriculture is examined. The researchers analyze data from various meteorological sources to understand the impact of climate change and variability on crop yields, cropping patterns, and overall agricultural productivity in India. This paper discusses how changes in temperature, precipitation, and monsoon patterns can significantly affect crop growth and food production. It highlights the vulnerability of Indian agriculture to climate fluctuations and the potential consequences for food security and livelihoods. The authors also explore the importance of adaptive strategies and policy interventions to mitigate the adverse effects of climate change on agriculture in India, emphasizing the need for timely and targeted interventions to ensure the resilience and sustainability of the agricultural sector in the face of a changing climate. Another research study by X. Zhang and E. A. Davidson (2018) addresses the complex challenge of optimizing nitrogen and water usage to enhance crop productivity while minimizing environmental impacts. The authors analyze the current state of nitrogen and water management in U.S. agriculture and propose strategies for improvement. The paper underscores the significance of efficient nutrient management and water conservation in agriculture, considering their direct impacts on crop yield, food security, and environmental sustainability. It explores innovative techniques, technologies, and policies that can be implemented nationally to achieve these goals. Furthermore, the authors discuss the potential benefits of precision agriculture, data-driven decision-making, and integrating remote sensing and modeling techniques to enhance nitrogen and water management practices. They also highlight the importance of interdisciplinary collaboration among scientists, policymakers, and farmers to address these challenges effectively. In a recent publication, P. Praveen et al. (2023) discussed a comprehensive exploration of smart farming, smart irrigation, smart dairy, and intelligent agricultural waste management. Furthermore, insights were shared on how rural populations can elevate both their social life and economic prospects.

Recent years have witnessed increased attention to technological processes in agriculture. These methods include (a) irrigation, (b) consumption of fertilizers and manure, (c) improved seed, and (d) agricultural implements. This chapter focuses on irrigation, the consumption of fertilizers and manure, and the efficient prediction of suitable crops for specific soil conditions. Agriculture remains an essential driver of global food production and economic stability. Precision agriculture has emerged thanks to modern technological advancements, utilizing data, sensors, and automation to improve farming techniques. Furthermore, sustainable agricultural practices have become increasingly significant in addressing environmental challenges and ensuring food security for future generations.

II. LITERATURE REVIEW

Smart agriculture, also known as precision agriculture or digital farming, reveals a rapidly evolving field at the intersection of agriculture, technology, and data science. Researchers have extensively investigated various aspects of smart agriculture, ranging from integrating Internet of Things (IoT) devices and sensors to using big data analytics and artificial intelligence (AI) algorithms in farm management. Studies have shown that smart agriculture has the potential to revolutionize traditional farming practices by enabling real-time monitoring of crop conditions, soil health, and weather patterns. Furthermore, the literature highlights the benefits of data-driven decision-making in optimizing resource allocation, such as precise irrigation and fertilization, ultimately leading to improved crop yields, resource efficiency, and sustainability M.S. Farooq et al. (2019). However, challenges related to data privacy, infrastructure limitations, and the digital divide in rural areas have also been identified as important considerations in adopting smart agriculture technologies. As the field continues to advance, it is essential to explore the technical aspects and the socioeconomic and environmental implications of smart agriculture to ensure its successful implementation and broader societal benefits.

B. Kashyap and R. Kumar (2021) explained the critical role of soil moisture and nutrient content in agricultural productivity and sustainability. It outlines the significance of accurate and timely data on these parameters for optimizing crop growth and resource management. The authors emphasize that modern sensing technologies are pivotal in achieving these objectives. The authors extensively review various sensing methodologies, including ground-based sensors, remote sensing, and wireless sensor networks, highlighting their principles of operation, advantages, and limitations. It also discusses the integration of advanced technologies, such as the Internet of Things (IoT) and data analytics, to enhance the accuracy and efficiency of soil monitoring. Furthermore, the authors explore the potential applications of these sensing methodologies in precision agriculture, irrigation management, and nutrient optimization. They discuss real-world examples and case studies to illustrate the practical implementation of these technologies. A comprehensive analysis has been carried out by D. Ghosh et al. (2022) on utilizing IoT technologies in agriculture, focusing on their effect on real-time soil fertility evaluation. They delve into the various aspects of IoT systems, including sensors and communication protocols, and also shed light on their advantages and limitations. This thorough review is an indispensable guide for practitioners and researchers in the agriculture sector, providing valuable perspectives on how IoT could revolutionize soil management and enhance crop productivity via data-driven decision-making.

R. A. Kalpana et al. (2020) focus on the significance of efficient agricultural irrigation and soil nutrient management to enhance crop yield and resource utilization. The authors present a novel approach to this challenge by proposing an automated irrigation system regulating crops' water supply and analysing soil nutrient content in real-time. The paper discusses the technical aspects of the automated irrigation system, including the use of sensors and data processing techniques for soil nutrient detection. It highlights the importance of precision agriculture in ensuring that crops receive the right water and nutrients at the right time, optimizing resource use, and minimizing wastage. Furthermore, the authors present the results of their system's performance and discuss its potential applications in modern farming practices. They emphasize the role of such technology in sustainable agriculture, as it can contribute to increased crop productivity while conserving water and reducing the environmental impact of excessive nutrient application.

I. Kumar et al. (2022) introduced a system that leverages IoT technology to manage motor pumps in agriculture remotely. The primary objective of their design is to create a more efficient and automated method for managing motor pumps, particularly in agricultural and industrial settings where pump operation is crucial for various processes. The authors detail the intricacies of the technology's design and implementation, underscoring its capacity to enhance irrigation efficiency and minimize the need for manual labor. By granting users the ability to control pumps through IoT-enabled devices, this system affords unparalleled convenience and real-time monitoring capabilities. Ultimately, this research represents a significant stride in advancing smart agriculture, as it tackles water resource management challenges and paves the way for sustainable farming practices.

In the research of A. Jain, A. Saify, and V. Kate (2020), the color variations in soil samples were utilized by the researchers to predict the levels of Nitrogen (N), Phosphorus (P), and Potassium (K) using a color sensor identified as TCS3200. The focus of this work was on the development and implementation of an innovative sensor-based system to enable rapid and cost-effective nutrient assessment. This research showcases the feasibility of predicting nutrient content using this system, making it a promising tool for precision agriculture. The results of this study are significant as they contribute to the advancement of efficient and accessible soil analysis techniques, which are crucial for optimizing crop management practices. H. Pallevada et al. (2021) developed a real-time soil nutrient detection and analysis system, highlighting its significance in modern agriculture. This technology likely involves sensors and data analysis techniques to provide instant nutrient content assessments, aiding farmers in making informed decisions about fertilization. Such innovations are essential for sustainable farming, as they enable precise and timely management of soil nutrients, ultimately leading to improved crop yields and reduced environmental impact. This research underscores the growing importance of technology in optimizing agricultural practices.

M.Pyingkodi et al. (2022) describe an innovative system that leverages the Internet of Things (IoT) for soil nutrient analysis and monitoring. This technology likely utilizes sensors and data analytics to provide real-time data on soil nutrient levels. This data can empower farmers to make informed decisions about nutrient management, leading to more efficient and sustainable agricultural practices. The authors

J. Parmar et al. (2018) discussed the development of a weather monitoring system based on the Internet of Things (IoT). This innovative system likely employs various sensors and data collection techniques to gather real-time weather data. By leveraging IoT technology, it provides accurate and timely weather information that can be crucial for various applications, from agriculture to disaster management. This research underscores the importance of IoT in enhancing our understanding and response to weather conditions.

A real-time weather monitoring system based on the Internet of Things (IoT) was developed by Vanmathi C. et al. (2020). This system likely incorporates IoT sensors and data collection techniques to provide up-to-date and accurate weather information. Such technology can be valuable for various sectors, including agriculture, transportation, and disaster management, as it enables timely decision-making based on current weather conditions. This research highlights the growing significance of IoT in enhancing weather monitoring and forecasting capabilities. G. N. Reddy et al. (2018) discussed an innovative approach that combines automatic irrigation with soil quality testing. This system likely utilizes sensors to monitor soil conditions, including moisture levels and nutrient content, enabling precise irrigation based on real-time data. Such a technology can significantly enhance agricultural efficiency, conserving water resources while ensuring optimal soil conditions for crop growth. This research represents a noteworthy advancement in the field of smart agriculture. K. Pernapati (2018) introduced a cost-effective smart irrigation system leveraging the Internet of Things (IoT). This system likely incorporates IoT sensors and real-time data analysis to monitor soil moisture levels and weather conditions. Doing so optimizes irrigation practices, conserves water, and reduces operational costs for farmers. The system incorporates real-time soil nutrient Nitrogen(N), Phosphorus(P), and Potassium(K) detection along with a smart irrigation system that monitors and regulates the soil moisture in real time.

M. Ayaz et al. (2019) and A. D. Boursianis et al. (2020) provide valuable insights into the transformative potential of Internet of Things (IoT) technologies in the field of smart agriculture. Their studies underscore the critical role that IoT-based solutions play in modernizing and optimizing agricultural practices. They illustrate the growing importance of IoT in agriculture, offering a promising avenue for addressing the challenges of food security, resource optimization, and environmental sustainability. The work by Benyezza et al. (2021) shows the development of a zoning irrigation smart system that utilizes fuzzy control technology and IoT. This system represents a notable step forward in optimizing agriculture's water usage and energy efficiency. By enabling precise and adaptive irrigation strategies based on real-time data, it has the potential to significantly reduce water wastage and energy consumption, promoting sustainability in agriculture. Zahid et al. (2019) introduce a novel approach to characterizing living plant leaves using terahertz waves. This technology opens up exciting possibilities for non-invasive plant health assessment. By utilizing terahertz waves, researchers can obtain detailed information about the water content of plant leaves, which is a crucial indicator of their health and hydration status; on the other hand, Uzal et al. (2018) focus on the application of deep learning for seed-per-pod estimation in plant breeding. This research demonstrates how advanced machine learning algorithms can streamline breeding by accurately estimating seed-per-pod, a critical factor in plant breeding programs. Such automation accelerates breeding efforts and ensures the production of healthier and more productive crop varieties. The research of Ray P.P. et al. (2017), Partel V. (2019), and A. Villa-Henriksen et al. (2020), shows the transformative potential of IoT and advanced technologies in agriculture. They highlight the importance of data-driven decision-making, precision agriculture, and sustainable farming practices in addressing food security and resource conservation challenges. As the agricultural sector evolves, these studies provide valuable insights and serve as guiding lights for researchers, policymakers, and practitioners seeking to leverage IoT to create a more efficient, productive, and environmentally responsible future for agriculture. The integration of IoT in agriculture represents a promising path forward, driving innovation and progress in an industry vital to human well-being and the planet's sustainability.

METHODOLOGY

This work mainly arose with the motivation to promote the sustainability of human beings. Thus, the work that required development and improvement was agriculture, without which the food we take would not have been possible. Now, with this being our main idea, we moved forward with further systematic steps to create a change in the agricultural sector with adequate crop-type prediction using the analysis of the soil nutrients and the constant moisture monitoring of the soil A. Al-Naji et al. (2021), D. Brunelli et al. (2019). The methodology followed is simple: first, analyzing the nutrients present in the soil and categorizing them as nitrogen, phosphorous, and potassium as detected. Also, the soil moisture is constantly monitored and replenished as per the requirement. Singh A.K. et al. (2023) developed a

Figure 1. Flowchart depiction of the project

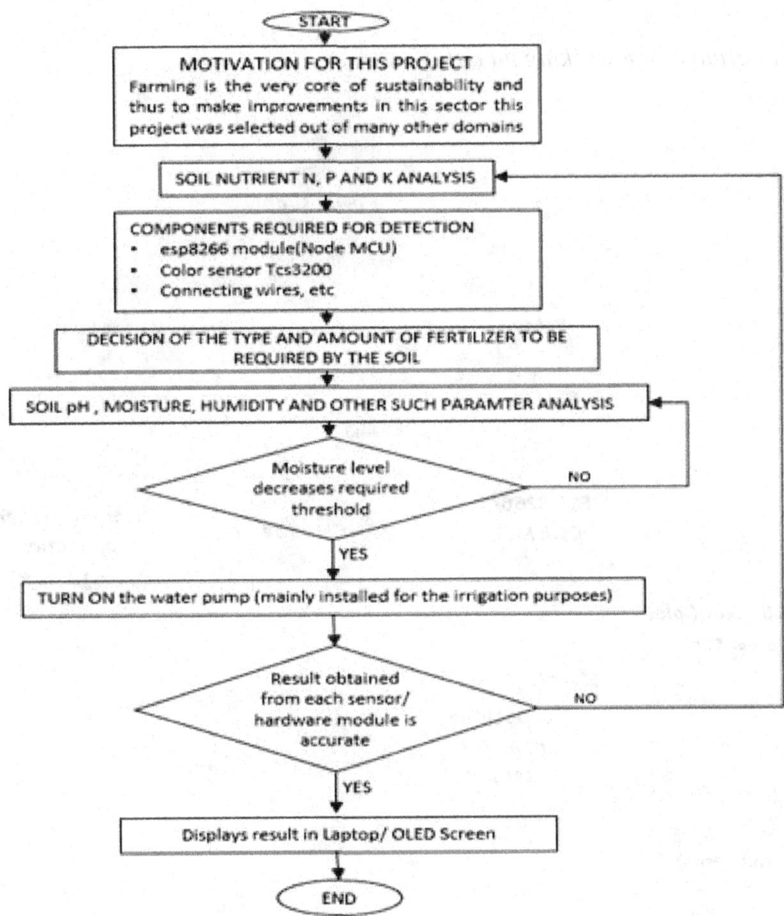

smart irrigation technology used in the agriculture sector for proper monitoring of the humidity of soil as per the requirement. In the present work, we are taking the concept of smart irrigation technology along with the content of nutrients available inside the soil. For this purpose, the hardware equipment is used to achieve the set goals. Figure 1 shows the operation of the different parts of the designed system, starting from motivation to display results to the farmer. They can manage their farm accordingly using the instructions or data available on the screen.

(A) Working Model

The block diagram in Figure 2 shows the overall working of the model. Initially, the soil undergoes monitoring using the TCS 3200 color sensor and Capacitive moisture sensor. Data from these sensors is collected by the device and transmitted to the ESP 8266 module. Subsequently, this data is sent to the cloud, where it can be accessed by the client. The information is then displayed in an application on the end user's mobile phone. Additionally, the data from the capacitive moisture sensor is utilized to control the water pump. The end user, typically a farmer, can issue commands through the application. These commands are transmitted to the cloud, and from there, the data is relayed to the NODE MCU/ ESP 8266 via the built-in Wi-Fi module. This enables the user to remotely turn the water pump on or off, as the pump is connected to the NODE MCU via a relay.

Figure 2. Block diagram of the working model

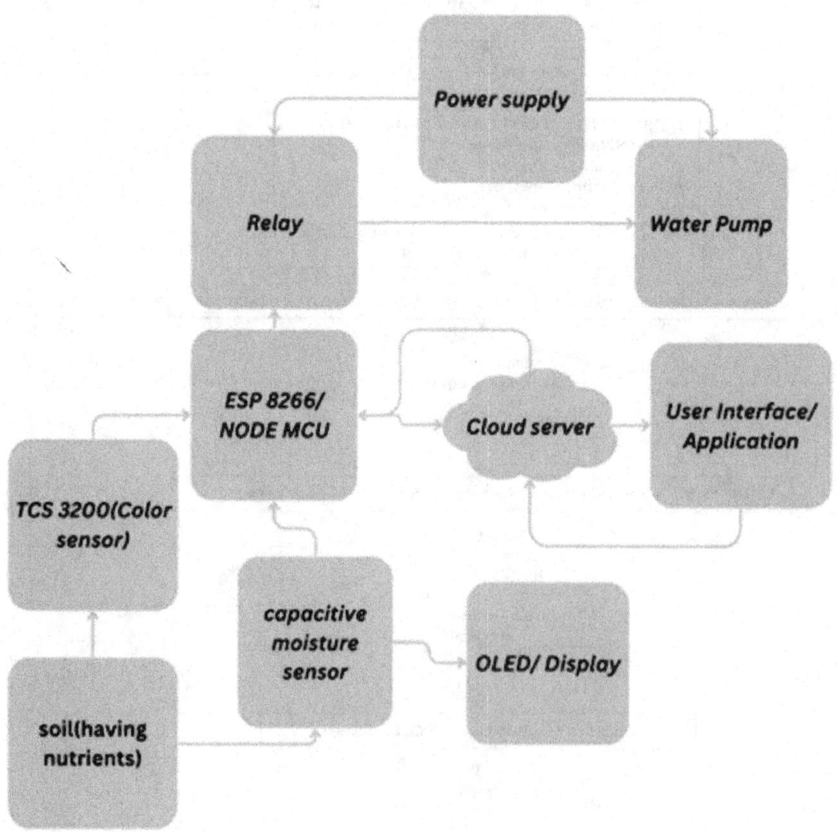

(B) Description Of Different Technical Components Used

This venture into smart agriculture is a harmonious blend of software and hardware ingenuity, which will be meticulously explored in the forthcoming sections dedicated to their intricate details.

(1) SOFTWARE

(a) ***Arduino Software (IDE):*** The Arduino IDE (Integrate Development Environment) is software that is used to program microcontrollers. It consists of a code editor, a compiler, a debugger, a serial plotter, and a serial monitor. It is an open-source software. With the help of the Arduino IDE, we can program different microcontrollers like the Arduino Uno, Node MCU, and even Raspberry Pi boards. The latest version also supports micro-python for programming different boards. It has a large library of support that can help users program different boards and sensors easily.

(b) ***BLYNK Software:*** IoT is the main goal of Blynk Cloud's design. The Blynk platform allows for the online control of Arduino, Raspberry Pi, and other devices through IOS and Android applications. It can do many things, such as save and visualize data and display sensor data, and remotely control devices.

(2) HARDWARE

(a) ***ESP8266 Module:*** In an era of the Internet of Things (IoT) and the seamless interconnectivity of devices, the NodeMCU/ESP8266 WiFi module stands as a beacon of innovation and versatility. This diminutive yet powerful piece of hardware has revolutionized how we interface with the digital world, enabling enthusiasts, makers, and professionals to infuse Wi-Fi connectivity into their projects effortlessly. The NodeMCU/ESP8266 WiFi module, colloquially called the "NodeMCU" or simply "ESP8266," is an affordable and open-source microcontroller that has garnered immense popularity in the electronics and IoT communities. Developed by the Chinese company Espressif Systems, this remarkable piece of technology has catapulted the concept of IoT development into the mainstream. At its core, the NodeMCU/ESP8266 is a compact microcontroller with built-in Wi-Fi capabilities. Its small size and low power consumption make it ideal for various applications, from home automation and environmental monitoring to remote sensing and data logging. This module offers a cost-effective and accessible solution for anyone looking to bring their creative ideas to life in the digital realm.

(b) ***Color Sensor Module:*** In modern agriculture and environmental monitoring, the TCS3200 Color Sensor has emerged as a remarkable innovation, revolutionizing how we assess soil nutrient content. As our world grapples with the critical challenges of sustainable agriculture and efficient resource management, the TCS3200 Color Sensor is an indispensable tool, enabling precise, real-time analysis of soil composition and nutrient levels. This chapter delves into the capabilities and applications of the TCS3200 Color Sensor in soil nutrient assessment, shedding light on its significance in promoting healthier, more productive crops while conserving vital resources. The TCS3200 Color Sensor represents a convergence of cutting-edge technology and the age-old need to optimize agricultural practices. This sensor has become an invaluable asset to farmers, researchers, and environmentalists by leveraging advanced optical sensing principles. Its primary function lies in capturing and quantifying the spectral characteristics of soil, allowing for the determination of essential nutrient levels, pH, and other vital parameters. The journey ahead promises to unravel the profound impact of the TCS3200 Color Sensor on modern agriculture, highlighting its role as a key driver of precision agriculture and sustainable farming. By harnessing the power of this technology, we can unlock new avenues for maximizing crop yields, minimizing resource wastage, and ultimately contributing to a more food-secure and environmentally conscious world where science and agriculture converge for a brighter, more sustainable future.

(c) ***OLED Display:*** In modern technology, one innovation has been illuminating our lives – literally and metaphorically – with its breathtaking display capabilities: Organic Light Emitting Diode, or OLED. This groundbreaking technology has redefined how we perceive and interact with screens, offering a visual experience that transcends the limitations of traditional displays. As we embark on this journey to explore the enchanting realm of OLED, we will uncover the science, the magic, and the myriad applications that have made it an indispensable component of contemporary electronic devices. At their core, OLEDs are composed of organic materials that emit light when an electric current is applied. This fundamental principle sets them apart from conventional displays, such as Liquid Crystal Displays (LCDs) or Light Emitting Diodes (LEDs), which require backlighting for illumination. Intriguingly, the beauty of OLED lies not only in its vibrant colors and high contrast but also in its design's simplicity. At the heart of an OLED display are individual OLED pixels, each capable of emitting its own light independently. These pixels are organized in rows and columns, forming the display matrix. OLED displays feature a series of pins or connectors that enable precise manipulation of the electrical current flowing through them to control the light emitted by each pixel.

Figure 3. (a) ESP8266 hardware, (b) color sensor TCS3200, (c) OLED display, (d) Arduino UNO, (e) water pump, (f) capacitive soil moisture sensor v2.0, (g) 12V 8-channel relay module, (h) Breadboard

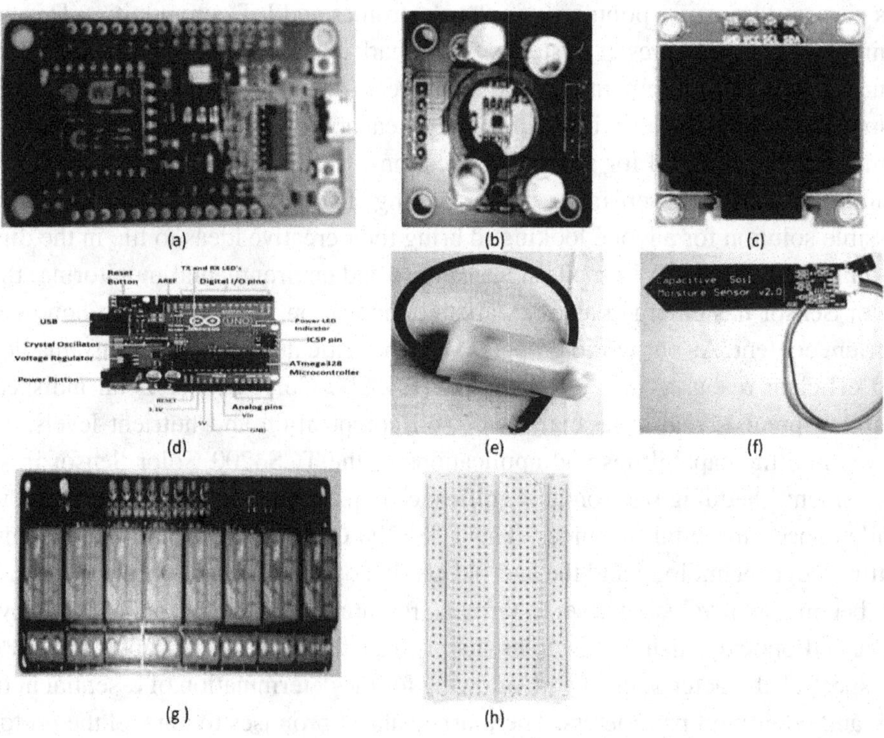

(d) ***Arduino Uno:*** The Arduino Uno is a popular and versatile microcontroller board widely used in electronics and DIY (Do-It-Yourself) projects. It's an excellent choice for beginners and experienced electronics enthusiasts alike. The Arduino Uno is a microcontroller board based on the ATmega328P microcontroller chip. It was developed by Arduino, an open-source electronics platform, to provide an easy and affordable way for hobbyists, students, and professionals to create interactive electronic projects. The Arduino Uno is part of a family of boards that includes various models catering to different needs, but the Uno is one of the most popular and widely used.

(e) ***Pump:*** In engineering and fluid dynamics, pumps are pivotal in moving liquids, gases, or slurries from one place to another. These remarkable machines have been instrumental in countless industrial processes, water distribution systems, and our daily lives. Whether you're filling up your car's gas tank, circulating water in a heating system, or ensuring the smooth operation of a chemical plant, chances are you're relying on a pump to get the job done.
Pumps are a class of mechanical devices designed to transfer fluids by creating pressure and directing the flow of the substance in a controlled manner. They come in various shapes, sizes, and configurations, each tailored to specific applications and requirements. From the simple hand-operated water pump used in rural communities to the sophisticated centrifugal pumps employed in large-scale industrial processes, the diversity of pumps is truly astounding.

(f) ***Capacitive Moisture Sensor:*** A capacitive soil moisture sensor is a device designed to measure soil or other porous materials' moisture content by utilizing the capacitance principle. These sensors are commonly used in agriculture, horticulture, and environmental monitoring to help optimize irrigation, ensure proper plant hydration, and gather data for research purposes.
Principle of Operation: Capacitive soil moisture sensors work on the principle that the dielectric constant of a material (in this case, soil) changes with its moisture content. As soil moisture increases, its dielectric constant also increases, leading to a change in the capacitance of the sensor.*Measurement*: When the sensor is inserted into the soil, it measures the capacitance between its electrodes. This capacitance is directly related to the moisture content of the soil. Higher capacitance indicates higher moisture levels, while lower capacitance suggests drier soil.

(g) ***Relay Module:*** A relay is an electromechanical device that controls electrical circuits by using a low-power signal to switch electrical loads. It is a switch that can be remotely controlled, allowing you to control a circuit or device from a distance. Relays are widely used in various applications and industries, from home automation to industrial machinery, automotive systems, and more.
Basic Operation: A relay consists of several key components, including an electromagnet, an armature, and electrical contacts. When a small electrical current (control signal) is applied to the relay's coil (the electromagnet), it generates a magnetic field that pulls the armature, closing or opening a set of electrical contacts. This action allows or interrupts the flow of a larger electrical current (load current) through the relay's contacts.

(h) ***Breadboard:*** A breadboard is a rectangular, plastic board with a grid of small holes into which electronic components like resistors, capacitors, integrated circuits (ICs), and wires can be inserted. These holes are typically arranged in rows and columns and connected internally in a specific pattern. Breadboards are used to create temporary electrical connections between components without soldering. This makes them an invaluable tool for prototyping, experimenting, and testing electronic circuits. The term "breadboard" originates from the early days of electronics when wooden breadboards were often used for this purpose.

III. IMPLEMENTATION

(A) Circuit Diagram

By providing a visual representation through a comprehensive circuit diagram, we will elucidate the core hardware components and their interconnections, shedding light on the system's operational intricacies and the synergy between its elements.

***Automation for Irrigation and Fertilization*:** This system's core lies in automation, streamlining irrigation and fertilizer management. The system continuously assesses real-time soil composition data and soil moisture levels to determine each crop's optimal irrigation and fertilizer application patterns.

***Remote Accessibility*:** A standout feature is the cloud-based control system, enabling users to access remotely and manage the entire agricultural system via a user-friendly interface. Users gain convenient and adaptable control over their agricultural operations through smartphones, tablets, or computers.

***Real-Time Soil Composition Data*:** To make data-driven decisions, the system integrates real-time soil composition data, including nutrient levels, pH, and other vital soil parameters. This ensures that crops receive the ideal mix of nutrients, fostering healthy growth and maximizing yields.

***Water Pump Control*:** The system autonomously manages the water pump. The system activates the water pump when soil moisture levels drop below the optimal range. Once the desired soil conditions are achieved, the system automatically ceases these operations, optimizing resource utilization.

Figure 4. Circuit diagram of the complete set-up

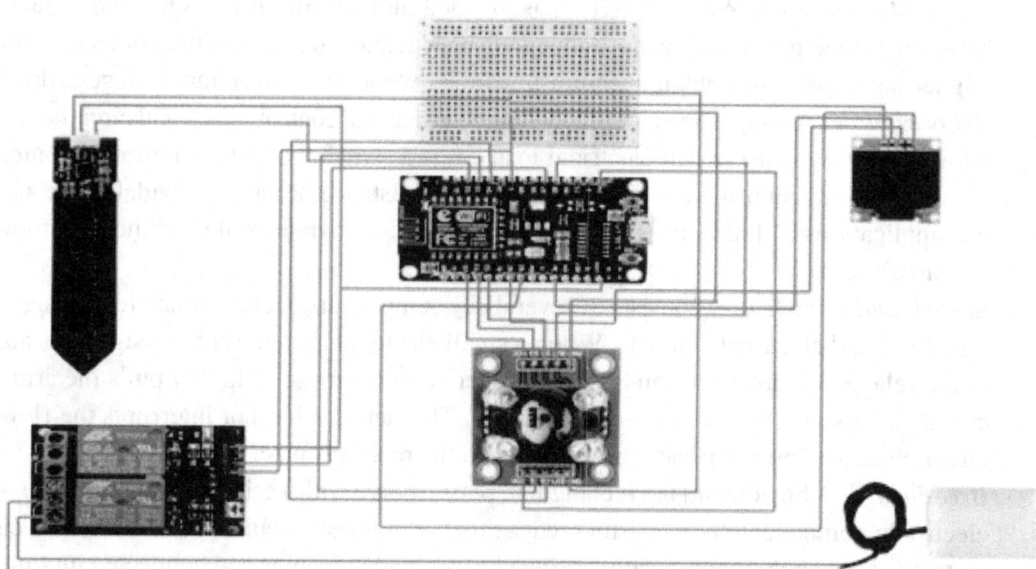

(B) Hardware Design

The experiment involves assessing soil samples using a color sensor placed at a distance of 3 cm from the soil. To determine the optimal distance, we vary the length between the sample and reflector, selecting the most accurate distance for incidence and reflection angles. In the sensor, the LED and photodiode are aligned in parallel, both facing the soil sample. Four LEDs illuminate the soil, and the photodiode collects reflected light after it's absorbed by the soil. The schematic diagram of the experimental setup is presented in Figure 4.

Figure 5. (a) Testing of moisture sensor, (b) testing of TCS 3200(color)sensor, (c) testing of humidity sensor, (d) testing of temperature sensor, (e) testing of whole model, (f) working of the whole model

Three types of soil samples from different areas are tested to evaluate soil quality based on nutrient composition. By comparing the standard wavelength range absorbed by specific nutrients with various colors, we identify the most accurate color wavelength, which approximates the light absorbed by nutrients. Activating the photodiode filter ensures precise measurement of nutrient levels and path length in the soil samples. The distance between the sample and reflector is adjusted to achieve the most accurate readings for specific color wavelengths, optimizing the incidence and reflection angles for the sensors.

Maintaining a consistent 3 cm path length between the soil sample and the sensor yields reliable data that can enhance crop management and improve resource allocation in modern agriculture. Figures 5a-f illustrate the testing and operation of various sensors and the overall model.

(C) Result Discussion

The project is a comprehensive soil parameter monitoring and automated irrigation system with the NodeMCU ESP8266 microcontroller at its core. This microcontroller serves as the central control unit for the entire system, facilitating communication with various sensors, connecting to Wi-Fi networks, sending data to the cloud, and controlling the water pump through a relay. Figure 6 shows the result of the TCS3200 Sensor for the soil sample taken and the BLYNK App showing the performance of the Smart-irrigation system.

The Color Sensor plays a pivotal role in evaluating soil composition. By measuring the color properties of the soil, it indicates variations in nutrient levels. This data helps users make informed decisions

Figure 6. (a) Result of TCS3200 sensor for the taken sample, (b) BLYNK App showcasing the smart-irrigation system

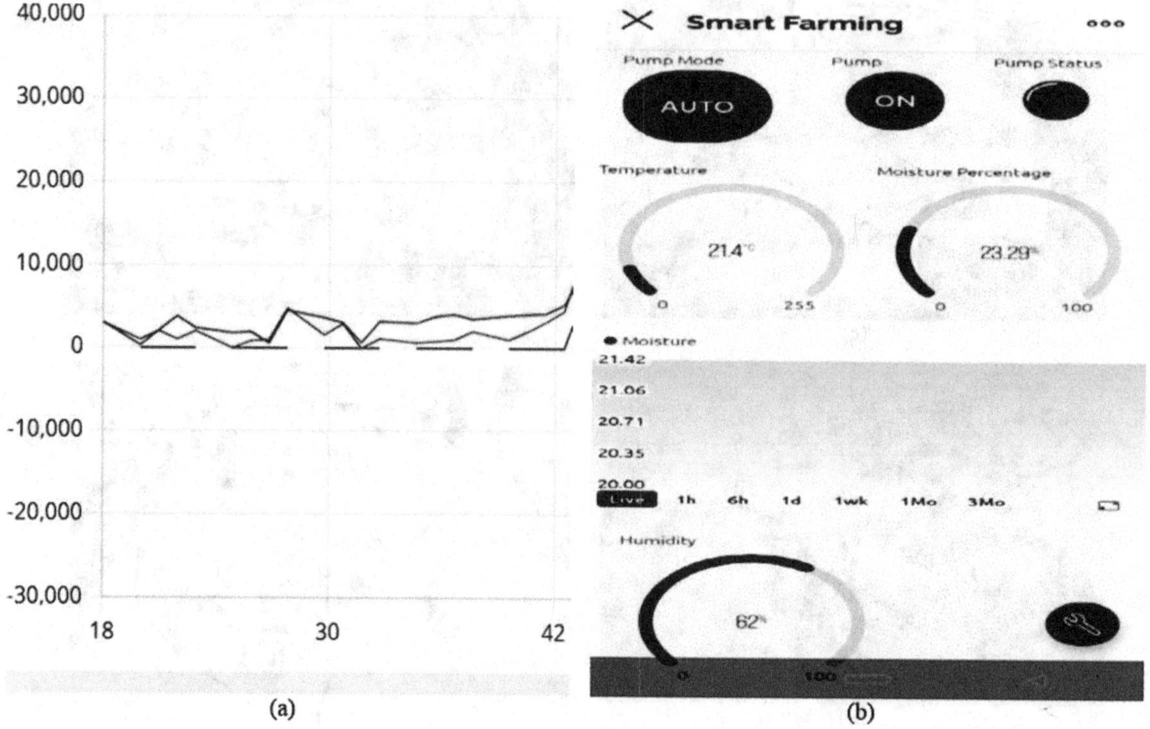

about soil amendments and treatments, ensuring optimal conditions for plant growth. On the other hand, the Capacitive Moisture Sensor tracks soil moisture levels, enabling informed decisions regarding irrigation. By continuously monitoring moisture content, the system ensures that plants receive the right amount of water, promoting healthy growth while conserving water resources.

The NodeMCU connects to Wi-Fi networks, enabling data transmission to the cloud; data collected from the sensors is forwarded to the cloud for further analysis. In the cloud, advanced analytics are performed to determine optimal irrigation needs and the best suitable crop or fertilizers a farmer can use. Processed data, including soil composition insights from the color sensor, is returned to the NodeMCU for display on the OLED screen. Moreover, the NodeMCU, in conjunction with a relay, controls the water pump based on sensor data and cloud-based analysis; it automatically manages the water pump, ensuring efficient irrigation. Users also have the option to manually control the water pump through a smartphone application, providing flexibility and control over the system.

IV. CONCLUSION AND FUTURE SCOPE

The findings of this study have unveiled an innovative solution using the TCS3200 color sensor paired with Node MCU to ascertain the nutrient levels of Nitrogen (N), Phosphorus (P), and Potassium (K) within soil samples collected from the local vicinity. This research not only addresses the longstanding predicaments faced by farmers in nutrient assessment but does so at a fraction of the cost compared to conventional methods. Moreover, it champions sustainability by curbing excessive fertilizer usage, achieved by precisely gauging nutrient absorption through the color sensor. The resultant NPK value informs farmers about their soil's health and guides crop selection based on soil nutrients, substantially diminishing the need for unnecessary fertilizers. Furthermore, incorporating a moisture sensor augments the agricultural landscape by conserving water resources and providing crops with the right amount of hydration, effectively streamlining farming operations.

As we look ahead, our future endeavors are geared towards enhancing user-friendliness within our application. We aspire to integrate valuable additional data and information, such as the latest government policies on soil testing and fertilizer usage. Additionally, we aim to provide real-time seed prices sourced from authoritative government platforms. We envision incorporating weather forecasting capabilities to elevate the device's utility further, empowering farmers with even more comprehensive and actionable insights for efficient and sustainable agriculture.

REFERENCES

A. P. G. V., Sree, Meera, & Kalpana. (2020). Automated Irrigation System and Detection of Nutrient Content in the Soil. *International Conference on Power, Energy, Control and Transmission Systems (ICPECTS)*, 1-3. 10.1109/ICPECTS49113.2020.9336990

Al-Naji, A., Fakhri, A. B., Gharghan, S. K., & Chahl, J. (2021). Soil color analysis based on a RGB camera and an artificial neural network towards smart irrigation: A pilot study. *Heliyon*, *7*(1), e06078. Advance online publication. doi:10.1016/j.heliyon.2021.e06078 PMID:33537493

Ayaz, M., Ammad-Uddin, M., Sharif, Z., Mansour, A., & Aggoune, E. M. (2019). Internet-of-Things (IoT)-based smart agriculture: Toward making the fields talk. *IEEE Access : Practical Innovations, Open Solutions*, *7*, 129551–129583. doi:10.1109/ACCESS.2019.2932609

Benyezza, H., Bouhedda, M., & Rebouh, S. (2021). Zoning irrigation smart system based on fuzzy control technology and IoT for water and energy saving. *Journal of Cleaner Production*, *302*, 127001. Advance online publication. doi:10.1016/j.jclepro.2021.127001

Boursianis, A. D., Papadopoulou, M. S., Diamantoulakis, P., Liopa-Tsakalidi, A., Barouchas, P., Salahas, G., Karagiannidis, G., Wan, S., & Goudos, S. K. (2020). Internet of Things (IoT) and agricultural unmanned aerial vehicles (UAVs) in smart farming: A comprehensive review. *Internet of Things : Engineering Cyber Physical Human Systems*, (100187). Advance online publication. doi:10.1016/j.iot.2020.100187

Brunelli, D., Albanese, A., Acunto, D., & Nardello, M. (2019). Energy neutral machine learning based IoT device for pest detection in precision agriculture. *IEEE Internet Things Mag.*, *2*(4), 10–13. doi:10.1109/IOTM.0001.1900037

Farooq, M. S., Riaz, S., Abid, A., Abid, K., & Naeem, M. A. (2019). A survey on the role of IoT in agriculture for the implementation of smart farming. *IEEE Access : Practical Innovations, Open Solutions*, *7*, 156237–156271. doi:10.1109/ACCESS.2019.2949703

Ghosh, D., Anand, A., Gautam, S. S., & Vidyarthi, A. (2022). Soil Fertility Monitoring with Internet of Underground Things: A Survey. *IEEE Micro*, *42*(1), 8–16. doi:10.1109/MM.2021.3121496

Jain, A., Saify, A., & Kate, V. (2020). *Prediction of Nutrients (N, P, K) in soil using Colour Sensor (TCS3200). International Journal of Innovative Technology and Exploring Engineering.*

Kashyap, B., & Kumar, R. (2021). Sensing Methodologies in Agriculture for Soil Moisture and Nutrient Monitoring. *IEEE Access : Practical Innovations, Open Solutions*, *9*, 14095–14121. doi:10.1109/ACCESS.2021.3052478

Krishna Kumar, K., Rupa Kumar, K., Ashrit, R. G., Deshpande, N. R., & Hansen, J. W. (2004). Climate impacts on Indian agriculture. *International Journal of Climatology*, *24*(11), 1375–1393. doi:10.1002/joc.1081

Kumar, I., Mishra, Z., Rajput, A. S., & Parmar, O. (2022). IoT based Motor Pump Control System. *2022 IEEE International Conference on Current Development in Engineering and Technology (CCET)*, 1-5. 10.1109/CCET56606.2022.10080813

Nithin Reddy, Danish, Babu, & Koperundevi. (2018). Automatic Irrigation and Soil Quality Testing. *International Conference on Recent Innovations in Electrical, Electronics & Communication Engineering*.

Pallevada, Potu, Munnangi, Rayapudi, Gadde, & Chinta. (2021). Real-time Soil Nutrient detection and Analysis. *International Conference on Advance Computing and Innovative Technologies in Engineering (ICACITE).*

Parmar, J., Palav, P., Nagda, T., & Lopes, H. (2018). IOT Based Weather Intelligence. *International Conference on Smart City and Emerging Technology (ICSCET).*

Partel, V., Kakarla, S. C., & Ampatzidis, Y. (2019). Development and evaluation of a low-cost and smart technology for precision weed management utilizing artificial intelligence. *Computers and Electronics in Agriculture*, *157*, 339–350. doi:10.1016/j.compag.2018.12.048

Pathak, Pal, & Mohapatra. (2020). Mahatma Gandhi's Vision of Agriculture: Achievements of ICAR. Indian Council of Agricultural Research.

Pernapati, K. (2018). IoT based low cost smart irrigation system. *Proc. 2nd Int. Conf. Inventive Commun. Comput. Technol. (ICICCT)*, 1312–1315.

Praveen, P., Khan, A., Verma, A. R., Kumar, M., & Peoples, C. (2023). *Smart Village Infrastructure and Sustainable Rural Communities*. IGI Global. doi:10.4018/978-1-6684-6418-2.ch001

Pyingkodi, M., Thenmozhi, K., Karthikeyan, M., Kalpana, T., & Suresh Palarimath, G. (2022). IoT based Soil Nutrients Analysis and Monitoring System for Smart Agriculture. *Proceedings of the Third International Conference on Electronics and Sustainable Communication Systems*.

Ray, P. P. (2017). Internet of Things for smart agriculture: Technologies, practices and future direction. *Journal of Ambient Intelligence and Smart Environments*, *9*(4), 395–420. doi:10.3233/AIS-170440

Singh, A. K., Praveen, P., Tripathi, D., Pandey, V. P., & Verma, P. (2023). *Revolutionizing Agriculture with Cloud-Connected Irrigation Technology. In Smart Village Infrastructure and Sustainable Rural Communities*. IGI Global. doi:10.4018/978-1-6684-6418-2.ch010

Uzal, L. C., Grinblat, G. L., Namías, R., Larese, M. G., Bianchi, J. S., Morandi, E. N., & Granitto, P. M. (2018). Seed-per-pod estimation for plant breeding using deep learning. *Computers and Electronics in Agriculture*, *150*, 196–204. doi:10.1016/j.compag.2018.04.024

Vanmathi, C., Mangayarkarasi, R., & Jaya Subalakshmi, R. (2020). Real Time Weather Monitoring using Internet of Things. *2020 International Conference on Emerging Trends in Information Technology and Engineering*.

Villa-Henriksen, A., Edwards, G. T. C., Pesonen, L. A., Green, O., & Sørensen, C. A. G. (2020). Internet of Things in arable farming: Implementation, applications, challenges and potential. *Biosystems Engineering*, *191*, 60–84. doi:10.1016/j.biosystemseng.2019.12.013

Zahid, A., Abbas, H. T., Imran, M. A., Qaraqe, K. A., Alomainy, A., Cumming, D. R. S., & Abbasi, Q. H. (2019). Characterization and water content estimation method of living plant leaves using terahertz waves. *Applied Sciences (Basel, Switzerland)*, *9*(14), 2781. doi:10.3390/app9142781

Zhang, X., & Davidson, E. A. (2018). *Improving Nitrogen and Water Management in Crop Production on a National Scale*. American Geophysical Union.

Chapter 3
AI and Machine Learning in Smart Education:
Enhancing Learning Experiences Through Intelligent Technologies

Pawan Kumar Goel
https://orcid.org/0000-0003-3601-102X
Raj Kumar Goel Institute of Technology, Ghaziabad, India

Amit Singhal
Raj Kumar Goel Institute of Technology, Ghaziabad, India

Shailendra Singh Bhadoria
Inderprastha Engineering College, Ghaziabad, India

Birendra Kumar Saraswat
Raj Kumar Goel Institute of Technology, Ghaziabad, India

Arvind Patel
Raj Kumar Goel Institute of Technology, Ghaziabad, India

ABSTRACT

This chapter explores the applications, benefits, and challenges of integrating AI and ML in smart education, focusing on how these technologies can enhance learning experiences, personalize education, and improve learning outcomes. It also addresses ethical and privacy concerns, highlighting the need for robust policies and guidelines to mitigate them and protect students' rights. AI and ML can enable personalized learning experiences, tailor content, delivery, and assessment to individual needs, and support competency-based education. However, the chapter acknowledges the challenges of privacy, security, algorithmic biases, teacher training, and ethical implications. By embracing these recommendations, educators and policymakers can harness the full potential of AI and ML technologies in creating a smarter and more effective educational environment.

DOI: 10.4018/979-8-3693-0782-3.ch003

I. INTRODUCTION

1.1 Background and Significance of AI and ML in Education

The field of education has always been focused on imparting knowledge and nurturing the growth of learners. With the rapid advancements in technology, particularly in the realm of AI and ML, there has been a significant shift in the way education is approached. AI refers to the development of intelligent machines that can perform tasks that typically require human intelligence, while ML involves the ability of machines to learn from data and improve their performance over time without explicit programming (Moghaddam et al., 2022).

The integration of AI and ML in education has opened up new possibilities for personalized and adaptive learning experiences. These technologies have the potential to cater to the individual needs of learners, provide targeted support, and enable educators to make data-driven decisions. Furthermore, AI and ML can automate administrative tasks, enhance content creation, and facilitate efficient assessment processes (Embarak, 2018). As such, understanding the applications, benefits, and challenges of AI and ML in education is of paramount importance in shaping the future of smart education.

1.2 Research Objectives and Methodology

The primary objective of this research paper is to explore the various aspects of integrating AI and ML in smart education. The specific research objectives include:

a) Investigating the applications of AI and ML in smart education, including personalized learning, learning analytics, automated grading, and intelligent content creation.
b) Examining the benefits and impacts of AI and ML in smart education, such as enhanced learning outcomes, personalized learning experiences, and data-driven decision-making.
c) Identifying the challenges and considerations associated with the implementation of AI and ML in education, such as privacy concerns, ethical considerations, and the need for teacher-student redefinition.
d) Providing real-world case studies and examples of successful AI and ML implementations in smart education.
e) Discussing future directions and recommendations for effectively leveraging AI and ML in the field of education.

To achieve these objectives, a comprehensive literature review will be conducted to gather relevant information and insights from existing research and studies in the field of AI, ML, and education. Additionally, case studies and examples of successful AI and ML implementations in education will be analyzed to provide practical insights and outcomes (Kim et al., 2020).

The research will also consider potential ethical and privacy concerns associated with AI and ML in education. The findings will be synthesized, and future directions and recommendations will be provided based on the analysis of the research.

II. AI AND MACHINE LEARNING: FOUNDATIONS AND CONCEPTS

2.1 Definitions and Basic Concepts

2.1.1 Artificial Intelligence (AI): AI refers to the development of computer systems that can perform tasks that typically require human intelligence. It involves the creation of intelligent machines capable of reasoning, learning, problem-solving, and decision-making.

2.1.2 Machine Learning (ML): ML is a subset of AI that focuses on the development of algorithms and statistical models that enable computers to learn from data and improve their performance over time without explicit programming. ML algorithms extract patterns and insights from data to make predictions and decisions.

2.1.3 Data: Data is the raw information used by AI and ML systems to learn and make predictions. It can be structured (organized and labeled) or unstructured (lacks a predefined format), such as text, images, or videos.

2.1.4 Training Data: Training data is a labeled dataset used to train ML models. It consists of input data (features) and corresponding output data (labels) that the model learns from.

2.1.5 Supervised Learning: Supervised learning is a type of ML where the algorithm learns from labeled training data to make predictions or classify new, unseen data. The model maps input features to corresponding output labels.

2.1.6 Unsupervised Learning: Unsupervised learning is a type of ML where the algorithm learns from unlabeled data to discover patterns, relationships, or structures within the data.

2.1.7 Reinforcement Learning: Reinforcement learning is a type of ML where an agent learns through interactions with an environment. The agent receives feedback in the form of rewards or penalties based on its actions, allowing it to learn optimal strategies.

2.2 Machine Learning Algorithms and Techniques

2.2.1 Decision Trees: Decision trees are tree-like structures that make sequential decisions based on input features. They split the data based on attribute values to reach a prediction or classification.

2.2.2 Naive Bayes: Naive Bayes is a probabilistic algorithm that applies Bayes' theorem to calculate the probability of a class given input features. It assumes independence among features.

2.2.3 Support Vector Machines (SVM): SVM is a supervised learning algorithm that separates data points into different classes by finding the optimal hyperplane with maximum margin between the classes.

2.2.4 Neural Networks: Neural networks are computational models inspired by the human brain. They consist of interconnected nodes (neurons) organized in layers. Each neuron performs a mathematical operation on input data, and the network learns to recognize patterns through adjusting the connection weights.

2.2.5 Clustering Algorithms: Clustering algorithms group similar data points together based on their inherent similarities or distances. Examples include k-means clustering and hierarchical clustering.

2.3 Deep Learning and Neural Networks

2.3.1 Deep Learning: Deep learning is a subfield of ML that focuses on training neural networks with multiple hidden layers. Deep learning algorithms can automatically learn hierarchical representations of data, enabling them to extract complex patterns and features (Zhou et al., 2020).

2.3.2 Convolutional Neural Networks (CNN): CNNs are specialized neural networks for processing grid-like data, such as images or videos. They use convolutional layers to extract spatial features and pooling layers to reduce dimensionality.

2.3.3 Recurrent Neural Networks (RNN): RNNs are neural networks that process sequential data by incorporating feedback connections. They can capture temporal dependencies and are suitable for tasks like natural language processing and speech recognition.

2.3.4 Generative Adversarial Networks (GAN): GANs consist of two neural networks, a generator and a discriminator, which compete against each other. GANs are used for generating synthetic data that resembles real data distributions.

2.3.5 Transfer Learning: Transfer learning is a technique where a pre-trained neural network model is used as a starting point for a new task. The knowledge gained from the previous task is transferred to the new task, improving efficiency and performance.

Understanding these foundational concepts and algorithms provides the basis for applying AI and ML techniques in the context of smart education.

III. SMART EDUCATION: DEFINITION AND SCOPE

3.1 The Evolution of Educational Technology

The field of education has witnessed a remarkable transformation with the advancement of technology. Traditional educational practices have been supplemented, and in some cases replaced, by innovative technological tools and platforms (Kim et al., 2020). This evolution can be categorized into several stages:

a) Computer-Assisted Instruction (CAI): In the early stages, computers were introduced in education to provide computer-assisted instruction. This involved using software and interactive programs to enhance teaching and learning processes.
b) Multimedia-Based Learning: The emergence of multimedia technologies brought about new opportunities for incorporating visuals, audio, and interactive elements into educational materials. Multimedia-based learning facilitated a more engaging and dynamic learning experience.
c) Online Learning and E-Learning: With the advent of the internet, online learning and e-learning gained popularity. These approaches allowed learners to access educational resources, participate in virtual classrooms, and engage in remote learning.
d) Mobile Learning: The proliferation of smartphones and tablets gave rise to mobile learning. Mobile devices enabled learners to access educational content anytime, anywhere, and facilitated personalized and self-paced learning experiences.
e) Smart Education: Smart education represents the current phase in the evolution of educational technology. It encompasses the integration of advanced technologies like AI, ML, Internet of

Things (IoT), and data analytics into the learning environment to create intelligent and adaptive educational systems.

3.2 Characteristics and Goals of Smart Education

Smart education is characterized by the following key features:

a) Personalization: Smart education aims to provide personalized learning experiences tailored to individual needs, learning styles, and preferences. AI and ML technologies enable adaptive learning pathways, customized content delivery, and personalized feedback.
b) Interactivity and Engagement: Smart education leverages interactive and immersive technologies to enhance learner engagement. Virtual reality (VR), augmented reality (AR), and gamification techniques are used to create interactive and stimulating learning environments.
c) Data-Driven Decision Making: Smart education relies on data analytics and learning analytics to gather insights about learners, their progress, and learning patterns. This data-driven approach allows educators to make informed decisions about instructional strategies and interventions.
d) Lifelong Learning: Smart education recognizes the importance of continuous learning beyond traditional educational institutions. It promotes lifelong learning by providing accessible and flexible learning opportunities through online platforms and microlearning modules.
e) Collaboration and Social Learning: Smart education emphasizes collaborative learning experiences, fostering communication, and interaction among learners. Social learning platforms and online communities facilitate peer-to-peer collaboration and knowledge sharing.

The goals of smart education include:

- Enhancing learning outcomes and academic performance
- Promoting learner engagement and motivation
- Facilitating self-directed and lifelong learning
- Enabling personalized and adaptive learning experiences
- Supporting effective assessment and feedback mechanisms

3.3 The Role of AI and ML in Smart Education

AI and ML technologies play a pivotal role in enabling the vision of smart education. They offer intelligent solutions and capabilities that enhance various aspects of the educational process, such as:

a) Personalized Learning: AI and ML algorithms analyze learner data and behavior to provide personalized recommendations, adaptive content, and tailored learning paths. Intelligent tutoring systems and virtual assistants use ML to deliver individualized support and guidance to learners.
b) Learning Analytics: AI and ML enable the analysis of vast amounts of educational data to derive insights and patterns. Learning analytics helps educators monitor learner progress, identify areas of improvement, and make data-driven decisions to optimize teaching strategies.

Figure 1. Role of AI and ML in smart education

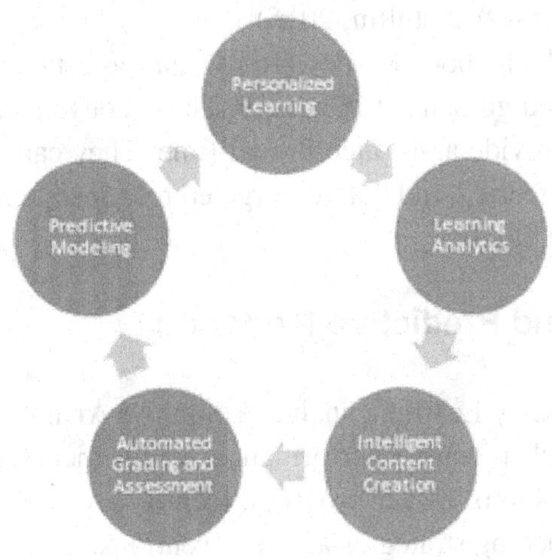

c) Intelligent Content Creation: AI and ML technologies can automate the generation of educational content, including textbooks, quizzes, and exercises. They can also curate and recommend relevant learning resources based on individual learner preferences and performance.
d) Automated Grading and Assessment: AI and ML algorithms can automate the grading and assessment process, reducing the burden on educators and providing immediate feedback to learners. Automated essay scoring and multiple-choice question analysis are examples of such applications.
e) Predictive Modeling: ML algorithms can analyze learner data and historical patterns to make predictions about future performance and identify students who may require additional support. Early warning systems can alert educators to intervene and provide timely interventions.

Overall, AI and ML technologies empower smart education by enabling personalized, adaptive, and data-driven learning experiences. They facilitate efficient and effective teaching and learning processes, improving educational outcomes for learners.

IV. APPLICATIONS OF AI AND ML IN SMART EDUCATION

4.1 Personalized Learning

4.1.1 Adaptive Learning Platforms: Adaptive learning platforms utilize AI and ML algorithms to deliver personalized learning experiences. These platforms collect and analyze data on learners' performance, preferences, and learning styles. Based on this analysis, the system adapts the content, pace, and difficulty level to meet each learner's specific needs. It provides customized learning paths, individualized feedback, and targeted interventions (Embarak, 2019).

4.1.2 Intelligent Tutoring Systems: Intelligent tutoring systems (ITS) leverage AI and ML to provide individualized instruction and support. These systems incorporate pedagogical expertise and domain knowledge to offer tailored guidance, monitor learner progress, and provide immediate feedback. ITS can adapt to learners' knowledge gaps, provide additional resources, and scaffold learning to facilitate mastery (Kim et al., 2020; Merzon & Ibatullin, 2016).

4.1.3 Virtual Assistants and Chatbots: AI-powered virtual assistants and chatbots are designed to provide personalized support and guidance to learners. These conversational agents can answer questions, offer explanations, and provide assistance in real-time. They can engage in interactive dialogue with learners, understand their queries, and deliver targeted responses, creating a personalized learning experience (Merzon & Ibatullin, 2016).

4.2 Learning Analytics and Predictive Modeling

4.2.1 Data-driven Decision-Making: Learning analytics employs AI and ML techniques to analyze large volumes of educational data, such as learner performance, engagement, and behavior. By analyzing this data, educators can gain insights into learner progress, identify areas of improvement, and make data-driven decisions to optimize teaching strategies and interventions.

4.2.2 Learning Management Systems and Analytics Tools: Learning management systems (LMS) integrated with AI and ML capabilities provide comprehensive analytics dashboards and tools for educators. These systems track learner activities, generate real-time reports, and visualize data for educators to assess the effectiveness of instructional materials, identify struggling students, and adapt instructional strategies accordingly.

4.2.3 Early Warning Systems: AI and ML algorithms can be utilized to develop early warning systems that identify students who may be at risk of academic difficulties or disengagement. By analyzing data on attendance, grades, and learner behavior, these systems can provide timely alerts to educators, enabling them to intervene and provide targeted support to struggling students (Merzon & Ibatullin, 2016).

4.3 Automated Grading and Assessment

4.3.1 Automated Essay Scoring: AI and ML algorithms can automatically assess and score essays based on predefined criteria. These algorithms analyze the text for grammar, coherence, and content relevance. Automated essay scoring systems provide timely and consistent feedback to learners, saving educators significant time in grading.

4.3.2 Multiple-Choice Question Analysis: AI and ML techniques can analyze responses to multiple-choice questions to assess learner understanding, misconceptions, and areas of weakness. These algorithms can identify patterns in learner responses and provide targeted feedback or adaptive follow-up questions to address specific learning needs.

4.3.3 Formative and Summative Assessments: AI and ML can enhance formative and summative assessments by automating the grading process and providing immediate feedback to learners. These technologies can also help identify patterns and trends in assessment data, supporting educators in evaluating instructional effectiveness and making data-informed decisions.

4.4 Intelligent Content Creation

4.4.1 AI-Generated Content and Textbooks: AI and ML can be utilized to automatically generate educational content, including textbooks, exercises, and practice questions. These algorithms can analyze existing content, extract key concepts, and generate new content that aligns with curriculum standards and learning objectives.

4.4.2 Adaptive Learning Materials: AI-powered adaptive learning materials can dynamically adjust their content and difficulty level based on learners' progress and performance. These materials provide customized learning experiences, ensuring that learners are appropriately challenged and supported throughout their learning journey.

4.4.3 Content Curation and Recommendation Systems: AI and ML algorithms can curate and recommend relevant learning resources based on learners' preferences, interests, and learning history. These systems analyze learner data to provide personalized recommendations for articles, videos, or interactive materials, enhancing learners' access to diverse and engaging content.

The applications described above demonstrate how AI and ML technologies are revolutionizing smart education by personalizing learning experiences, providing data-driven insights, automating assessments, and creating adaptive learning environments (Lotfy Abdrabou & Salem, 2010; Siemens & Baker, 2019).

V. BENEFITS AND IMPACTS OF AI AND ML IN SMART EDUCATION

5.1 Personalization and Adaptivity

- AI and ML technologies enable personalized and adaptive learning experiences in smart education. By analyzing learner data and behavior, these technologies can tailor instruction to individual needs, preferences, and learning styles. The benefits include:
- Customized Learning Paths: AI and ML algorithms can create personalized learning paths for each learner, presenting content and activities that match their current knowledge level and progress. This adaptivity ensures that learners are appropriately challenged and engaged, maximizing their learning potential (Sawyer et al., 2006).
- Individualized Feedback and Support: Intelligent tutoring systems and virtual assistants powered by AI can provide immediate and targeted feedback to learners. This personalized feedback helps learners identify their strengths and weaknesses, enabling them to focus on areas that require improvement.
- Accommodation of Learning Styles: AI and ML systems can identify learners' preferred learning styles and adapt instructional approaches accordingly. Visual learners may receive more visual aids and diagrams, while auditory learners may receive audio explanations or podcasts. This accommodation enhances learners' understanding and retention of information (Luckin, 2018).

5.2 Enhanced Learning Outcomes

The integration of AI and ML in smart education has demonstrated significant impacts on learning outcomes. The benefits include (Larusson et al., 2014; Knight et al., 2017):

- Mastery-based Learning: AI and ML technologies facilitate mastery-based learning, where learners progress at their own pace and only move forward after mastering specific concepts or skills. This approach ensures a deep understanding of the subject matter and fosters a sense of achievement.

- Improved Retention and Engagement: Personalized and adaptive learning experiences increase learner engagement and motivation, resulting in improved information retention. AI-powered systems can leverage gamification techniques, interactive content, and real-time feedback to make learning more enjoyable and effective.
- Targeted Remediation: AI and ML algorithms can identify knowledge gaps or areas of weakness for individual learners. This enables educators to provide targeted remediation, offering additional resources or interventions to address specific learning needs. As a result, learners can overcome challenges and progress more effectively.

5.3 Efficient Resource Allocation

AI and ML technologies optimize resource allocation in smart education, benefiting both learners and educators. The advantages include:

- Adaptive Resource Provision: AI-powered systems can dynamically allocate learning resources based on learners' needs and progress. This ensures that learners receive the most relevant and appropriate content, reducing time spent on irrelevant or redundant materials.
- Personalized Support and Intervention: AI and ML algorithms can identify students who require additional support or interventions, enabling educators to allocate resources effectively. By focusing on learners who need assistance the most, educators can maximize the impact of their support.
- Automation of Administrative Tasks: AI and ML can automate routine administrative tasks, such as grading assessments and generating progress reports. This automation saves educators time, allowing them to dedicate more attention to personalized instruction and support.

5.4 Continuous Assessment and Feedback

AI and ML technologies support continuous assessment and feedback processes, contributing to ongoing improvement and growth. The benefits include:

- Real-time Feedback: AI-powered systems can provide immediate feedback to learners, enabling them to correct mistakes and reinforce their understanding promptly. This real-time feedback promotes active learning and self-correction.
- Formative Assessment: AI and ML algorithms can analyze learner responses in real-time, providing educators with insights into learners' comprehension and misconceptions. Educators can then adjust their teaching strategies and provide targeted interventions to address areas of weakness.
- Progress Tracking: AI-powered analytics tools can track and monitor learner progress over time. This data enables educators to identify patterns and trends, assess the effectiveness of instructional methods, and make informed decisions to optimize learning experiences.

5.5 Data-Driven Decision Making

AI and ML technologies empower educators with data-driven insights, enhancing decision-making processes. The benefits include:

- Educational Insights: AI-powered analytics can analyze large datasets to reveal patterns, correlations, and trends in learner performance and behavior. Educators can gain deep insights into individual and group progress, identify effective teaching strategies, and customize instruction accordingly.
- Predictive Analytics: ML algorithms can predict future performance and learning outcomes based on historical data. Early warning systems can identify at-risk students, allowing educators to intervene proactively and provide timely support.
- Evidence-Based Instructional Design: AI and ML technologies enable evidence-based instructional design by analyzing data on learner interactions, preferences, and outcomes. Educators can use this information to develop effective instructional materials and strategies.

The integration of AI and ML in smart education brings numerous benefits, including personalized and adaptive learning experiences, improved learning outcomes, efficient resource allocation, continuous assessment and feedback, and data-driven decision making. These advantages contribute to a more effective and engaging educational experience for learners and educators alike (Baker & Yacef, 2009; Siemens, 2013).

VI. TRANSFORMATION OF THE EDUCATIONAL SYSTEM THROUGH SMART DIGITALIZATION

6.1 The transformation of the educational system through smart digitalization is a key driver in preparing students for the challenges and opportunities of the digital age. Smart digitalization refers to the integration of technology, data, and innovative pedagogical approaches to enhance the learning experience, improve outcomes, and develop essential digital skills. This transformation encompasses various aspects (Kim et al., 2020; Siemens & Baker, 2019):

- Technology Integration: Smart digitalization involves the integration of technology tools and resources into the learning process. This includes providing students with access to digital devices, interactive learning platforms, educational apps, and online resources. By leveraging technology, educators can create engaging and personalized learning experiences, fostering student creativity, critical thinking, and collaboration.
- Blended Learning and Online Education: Smart digitalization promotes a blended learning approach that combines traditional face-to-face instruction with online and digital resources. Online learning platforms, video lectures, and virtual classrooms enable anytime, anywhere access to educational content. This flexibility accommodates diverse learning styles, enhances accessibility, and allows for self-paced learning (Sawyer et al., 2006).
- Personalized Learning: Smart digitalization enables personalized learning experiences tailored to the individual needs and abilities of each student. Adaptive learning platforms and intelligent tutoring systems use data and algorithms to deliver customized content, track progress, and provide targeted feedback. This approach promotes self-directed learning, identifies areas for improvement, and maximizes student engagement.
- Data-Driven Decision Making: Smart digitalization leverages data analytics and learning analytics to inform decision-making processes. Educational institutions can gather and analyze data on

student performance, learning patterns, and engagement to identify areas of improvement, adapt instructional strategies, and provide targeted support. Data-driven insights enable evidence-based decision making to enhance educational outcomes.

- Collaboration and Communication: Smart digitalization facilitates collaboration and communication among students, teachers, and stakeholders. Online collaboration tools, virtual classrooms, and social learning platforms enable students to collaborate on projects, share ideas, and engage in peer learning. Additionally, digital communication platforms foster effective communication between educators, students, and parents, enhancing engagement and involvement in the learning process.
- Digital Skills Development: Smart digitalization aims to develop essential digital skills that are vital for success in the digital era. This includes digital literacy, information literacy, critical thinking, problem-solving, coding, and cybersecurity awareness. By integrating these skills into the curriculum, educational institutions prepare students to navigate and thrive in a technology-driven world.
- Lifelong Learning and Professional Development: Smart digitalization promotes lifelong learning and continuous professional development for educators. Online courses, webinars, and virtual communities enable educators to stay updated with emerging pedagogies, technologies, and best practices. Continuous learning empowers educators to effectively leverage smart digitalization in the classroom and adapt to evolving educational trends.

The transformation of the educational system through smart digitalization enhances access, engagement, and learning outcomes. It equips students with the digital skills and competencies necessary for the workforce of the future. Additionally, it empowers educators to innovate and personalize instruction, fostering a culture of lifelong learning. By embracing smart digitalization, educational institutions can create inclusive, dynamic, and future-ready learning environments that prepare students to thrive in the digital age (Koller & Ng, 2016; Siemens, 2013).

6.2 To successfully develop the digital economy, the transformation of the educational system is crucial. The establishment of smart education using various smart educational technologies is necessary across three interrelated areas:

- Strengthening Engineering Training: Universities need to focus on strengthening and expanding engineering programs to increase the number of specialists capable of creating new smart solutions for the digital economy. This involves providing comprehensive training in emerging technologies such as artificial intelligence, data analytics, cybersecurity, Internet of Things (IoT), and cloud computing. By producing a skilled workforce, universities can meet the demands of the digital economy.
- Mass Introduction of Smart Educational Technologies: Smart educational technologies should be widely implemented across all areas of education. These technologies enhance the quality of the educational process and improve the competency levels of graduates. By incorporating smart ICT tools, such as virtual reality, augmented reality, adaptive learning platforms, and online collaboration tools, students can acquire the skills necessary to thrive in the digital age. Smart educational technologies enable personalized learning experiences, interactive content delivery, and real-world application of knowledge.

- Establishment of Educational Institutions of the New Formation: Smart technologies enable the transformation of both the content of education and the management of educational institutions. New-generation educational institutions should leverage smart technologies to adapt their curricula to the needs of the digital economy. This involves incorporating emerging technologies into courses, fostering interdisciplinary collaboration, and providing experiential learning opportunities. Additionally, the management of educational institutions can be optimized through smart systems for student administration, learning analytics, and resource allocation.

By addressing these three areas, the educational system can effectively prepare students for the challenges and opportunities of the digital economy. Strengthening engineering training ensures a skilled workforce capable of innovation. Mass introduction of smart educational technologies enhances the learning experience and equips students with essential digital skills. The establishment of educational institutions of the new formation leverages smart technologies to adapt to the changing educational landscape and meet the demands of the digital economy (Baker & Yacef, 2009).

This transformation in education is necessary to bridge the skills gap, foster innovation, and drive economic growth in the digital era. It requires collaboration between academia, industry, and policymakers to align educational initiatives with the needs of the digital economy and ensure a seamless transition into a smart and digitally-driven educational ecosystem.

VII. CHALLENGES AND CONSIDERATIONS

7.1 Privacy and Ethical Concerns

The integration of AI and ML in smart education raises privacy and ethical concerns regarding the collection, storage, and usage of learner data. Key considerations include:

- Data Security: AI and ML systems collect vast amounts of learner data, including personal information and learning progress. Ensuring secure storage, data encryption, and protection against unauthorized access is crucial to safeguard learner privacy.
- Informed Consent: Clear and transparent communication is essential to obtain informed consent from learners and their guardians regarding the collection and usage of their data. Learners should be aware of how their data will be used and have the option to opt out if desired.
- Responsible Data Usage: Educators and institutions must handle learner data responsibly and ethically. Data should only be used for educational purposes and not shared or sold to third parties without explicit consent. Anonymization and aggregation techniques can be employed to protect individual privacy.
- Bias Mitigation: AI and ML algorithms can inadvertently perpetuate biases present in the training data, leading to discriminatory outcomes. Efforts must be made to mitigate algorithmic biases and ensure fairness and equal opportunities for all learners.

7.2 Algorithmic Bias and Fairness

AI and ML systems are susceptible to algorithmic bias, which can perpetuate and amplify existing social, cultural, and gender biases. Considerations include (Li et al., 2020):

- Bias in Training Data: Biases present in training data, such as gender, race, or socioeconomic biases, can result in biased outcomes. Careful attention must be given to the quality and representativeness of the training data to mitigate such biases.
- Regular Auditing and Evaluation: Ongoing auditing and evaluation of AI and ML systems are crucial to identify and rectify algorithmic biases. This process should involve diverse stakeholders to ensure fairness and inclusivity in the educational context.
- Explainability and Transparency: AI and ML algorithms should be designed to provide explanations for their decisions and predictions. Transparent and interpretable models enable educators to understand and address biases effectively.

7.3 Teacher-Student Relationship

Integrating AI and ML technologies in education raises questions about the impact on the teacher-student relationship. Considerations include:

- Pedagogical Role: Educators need to adapt their roles in the learning process when AI and ML are involved. The focus shifts from traditional instruction to curating and contextualizing educational content, providing guidance, and facilitating meaningful interactions with learners.
- Emotional and Social Support: AI-powered systems may lack the ability to provide the emotional and social support that human teachers offer. Maintaining a balance between technology-mediated instruction and nurturing human connections is important for learner well-being.
- Ethical Responsibility: Educators must ensure that the use of AI and ML technologies respects the dignity and autonomy of learners. They should guide learners in using technology responsibly and critically, addressing ethical considerations and potential biases (Blikstein, 2013).

7.4 Technical Infrastructure and Accessibility

The successful implementation of AI and ML in smart education requires robust technical infrastructure and considerations for accessibility. Key points to address include:

- Reliable Connectivity: Access to high-speed internet and reliable connectivity is essential to ensure seamless access to digital learning resources and platforms. Adequate infrastructure is necessary to support online learning and the use of AI-powered tools.
- Digital Divide: Addressing the digital divide is crucial to ensure equitable access to smart education. Efforts must be made to bridge the gap in technology access and provide equal opportunities for learners from diverse backgrounds.
- Assistive Technologies: Considerations should be given to learners with disabilities or special needs. AI and ML technologies can be leveraged to develop assistive tools, such as speech recognition or text-to-speech applications, to support inclusive education.

7.5 Professional Development and Training

The successful integration of AI and ML in education requires ongoing professional development and training for educators. Considerations include:

- Technology Competence: Educators need training to develop proficiency in using AI and ML technologies effectively. This includes understanding how to interpret and utilize data, leveraging educational software and tools, and adapting pedagogical approaches to incorporate technology.
- Ethical Use of Technology: Educators must be aware of the ethical considerations and potential biases associated with AI and ML. Training should emphasize responsible data handling, algorithmic fairness, and privacy protection.
- Pedagogical Integration: Educators need support and training to effectively integrate AI and ML technologies into their teaching practices. This involves aligning technology use with instructional goals, designing meaningful learning experiences, and leveraging technology to enhance pedagogy.

Addressing these challenges and considerations is essential to ensure the ethical, inclusive, and effective implementation of AI and ML in smart education. By doing so, the potential benefits of these technologies can be maximized while minimizing risks and ensuring equity and student well-being (Zhang et al., 2017).

VIII. CASE STUDIES AND EXAMPLES

8.1 Success Stories of AI and ML in Smart Education

8.1.1 Duolingo: Duolingo is a language learning platform that utilizes AI and ML algorithms to personalize and adapt language instruction for individual learners. The platform analyzes learners' responses, progress, and behavior to tailor the content and exercises to their specific needs. Duolingo's intelligent system tracks performance and adjusts difficulty levels, resulting in improved language acquisition and learner engagement.

8.1.2 Carnegie Learning: Carnegie Learning is an adaptive learning platform that combines AI and ML to provide personalized mathematics education. The platform delivers customized learning paths, adaptive assessments, and real-time feedback to learners. Through data analytics, Carnegie Learning identifies knowledge gaps and provides targeted interventions, resulting in improved learning outcomes and increased student achievement.

8.1.3 DreamBox Learning: DreamBox Learning is an AI-powered online math learning platform designed for K-8 students. The platform employs ML algorithms to assess learners' mathematical proficiency and adaptively deliver personalized lessons. DreamBox Learning's intelligent system provides individualized feedback, tracks progress, and adjusts the learning path based on learner performance, leading to improved mathematical understanding and engagement (Broussard et al., 2021; Redmon et al., 2016).

8.2 Real-World Implementations and Outcomes

8.2.1 Georgia State University: Georgia State University implemented an AI-driven system called "Pounce" to enhance student success and retention. The system analyzes data on student performance, engagement, and behaviors to identify at-risk students. Through early warning alerts and targeted interventions, the university has achieved a significant increase in student retention rates, especially for historically underrepresented student populations.

8.2.2 New York City Department of Education: The New York City Department of Education has implemented an AI-based recommendation system called "WeTeach NYC Recommends" to support teacher professional development. The system analyzes teachers' instructional practices, preferences, and learner outcomes to recommend relevant resources, strategies, and professional learning opportunities. This implementation has helped teachers access personalized resources and improve their instructional practices.

8.2.3 Bridge International Academies: Bridge International Academies, operating in several African countries, uses AI and ML technologies to provide quality education at scale. Through a combination of digital content, AI-powered assessments, and real-time data analysis, Bridge International Academies personalizes instruction, identifies areas for improvement, and supports teacher professional development. This approach has led to improved learning outcomes and increased access to quality education for underserved communities.

These case studies and real-world implementations demonstrate the positive impact of AI and ML in smart education. They showcase the potential for personalized learning experiences, improved student outcomes, targeted interventions, and scalable educational solutions. By leveraging AI and ML technologies effectively, educational institutions and platforms can enhance the quality and accessibility of education for learners worldwide (Lee et al., 2001; Tsai et al., 2006).

IX. FUTURE DIRECTIONS AND RECOMMENDATIONS

9.1 Lifelong Learning and AI-powered Education

Future directions in smart education involve a shift towards lifelong learning facilitated by AI-powered education systems. Recommendations include (Li et al., 2015; Wojke et al., 2017):

- Continuous Skill Development: AI and ML can support personalized learning pathways that adapt to individual needs and career goals. By providing targeted resources, microlearning modules, and adaptive assessments, AI-powered systems can enable individuals to develop and update their skills throughout their lives continuously.
- Personal Learning Assistants: AI-powered personal learning assistants can accompany learners throughout their educational journey, offering guidance, feedback, and recommendations. These assistants can provide tailored learning plans, suggest relevant resources, and facilitate self-directed learning, empowering learners to take ownership of their lifelong learning.

9.2 Integration of Augmented Reality (AR) and Virtual Reality (VR)

The integration of AR and VR technologies in smart education offers immersive and interactive learning experiences. Future directions and recommendations include:

- Virtual Laboratories and Field Trips: AR and VR can simulate real-world laboratory experiments and field trips, providing learners with hands-on experiences regardless of physical constraints. This integration can enhance understanding, engagement, and exploration in scientific and geographical contexts.
- Virtual Classrooms: AR and VR technologies can create virtual classroom environments where learners can interact with teachers and peers in a shared virtual space. This integration can foster collaboration, cultural exchange, and inclusive education by transcending physical boundaries and enabling global connections.

9.3 Blockchain and Decentralized Learning Platforms

Blockchain technology offers opportunities for secure and decentralized learning platforms. Future directions and recommendations include:

- Learner Credentials and Certifications: Blockchain can provide secure and tamper-proof storage of learner credentials, certifications, and achievements. This enables learners to have ownership and control over their digital records, facilitating lifelong learning and employability.
- Trust and Verification: Blockchain can ensure the integrity and authenticity of educational content, certificates, and qualifications. By providing transparent and verifiable records, blockchain-based platforms enhance trust among learners, employers, and educational institutions.

9.4 Human-AI Collaboration in the Learning Process

Future directions in smart education involve effective collaboration between humans and AI systems. Recommendations include:

- Hybrid Learning Environments: AI and ML technologies should be integrated into learning environments that foster human-AI collaboration. Educators should guide learners in effectively utilizing AI tools, fostering critical thinking, creativity, and ethical considerations (Broussard et al., 2021).
- Emotional Intelligence and Support: AI systems should be designed to recognize and respond to learners' emotional states. By incorporating emotional intelligence, AI-powered systems can provide empathetic support, encouragement, and mental well-being resources.

9.5 Addressing the Digital Divide in Smart Education

To ensure equitable access to smart education, future directions and recommendations include:

- Infrastructure Development: Efforts should be made to expand access to reliable internet connectivity and digital devices in underserved communities. Governments, educational institutions, and organizations can collaborate to bridge the digital divide and ensure equal opportunities for all learners (Redmon et al., 2016).
- Content Localization and Multilingual Support: Smart education platforms should consider content localization and provide multilingual support to accommodate diverse linguistic backgrounds and promote inclusivity.
- Partnerships and Community Engagement: Collaboration among stakeholders, including governments, NGOs, and private sector organizations, is crucial to address the digital divide. Partnerships can help provide resources, infrastructure, and training to underserved communities and ensure their inclusion in smart education initiatives.

By embracing these future directions and implementing the recommendations, smart education can advance towards lifelong learning, immersive experiences, secure and decentralized platforms, effective human-AI collaboration, and equitable access for all learners. These advancements will shape a transformative educational landscape, empowering individuals to thrive in the digital age.

X. CONCLUSION

10.1 Summary of Key Findings

This research paper has explored the applications, benefits, challenges, and future directions of integrating AI and ML in smart education. Key findings include:

- AI and ML technologies enable personalized and adaptive learning experiences, catering to individual learner needs and preferences.
- Smart education powered by AI and ML improves learning outcomes, enhances engagement, and supports continuous skill development.
- Challenges include privacy and ethical concerns, algorithmic biases, and the need to address the digital divide.
- Successful implementations of AI and ML in smart education, such as Duolingo and Carnegie Learning, have showcased positive impacts on learner achievement and engagement.
- Future directions include lifelong learning, integration of AR and VR, blockchain-based learning platforms, human-AI collaboration, and addressing the digital divide.

10.2 Implications for the Future of Education

The integration of AI and ML in smart education has profound implications for the future of education. It offers opportunities for personalized, adaptive, and inclusive learning experiences. The key implications include:

- Shift towards Lifelong Learning: AI-powered education systems can support individuals in continuously developing their skills throughout their lives, adapting to the evolving demands of the workforce.
- Enhanced Engagement and Learning Outcomes: AI and ML technologies foster learner engagement, improve retention, and provide personalized support, leading to improved learning outcomes and academic achievement.
- Transformation of Teaching Roles: Educators' roles will evolve to encompass curating educational content, guiding learners, and fostering critical thinking, creativity, and ethical considerations in an AI-infused learning environment.
- Equity and Inclusion: AI and ML can address barriers to education by providing personalized resources, bridging the digital divide, and accommodating diverse learning needs, ensuring equitable access and inclusive learning environments.

10.3 Recommendations for Effective Implementation

To effectively implement AI and ML in smart education, the following recommendations are proposed:

- Ethical and Privacy Considerations: Educators and institutions should prioritize data security, informed consent, responsible data usage, and mitigation of algorithmic biases to protect learner privacy and ensure ethical AI implementations.
- Professional Development and Training: Educators need ongoing professional development and training to effectively integrate AI and ML technologies into their instructional practices, promoting pedagogical integration, ethical considerations, and proficiency in using educational technology.
- Collaboration and Partnerships: Collaboration among educational institutions, policymakers, technology providers, and stakeholders is crucial to address challenges, share best practices, and create comprehensive frameworks for AI-powered smart education.
- Learner-Centric Design: AI and ML technologies should prioritize learner needs, preferences, and diverse learning styles, ensuring that personalized learning experiences are accessible, engaging, and supportive of learner agency and autonomy.
- Continuous Evaluation and Improvement: Educational institutions should regularly evaluate the impact of AI and ML implementations, seek learner and educator feedback, and iterate on the design and functionality of smart education systems to continuously enhance their effectiveness.

By following these recommendations, educational systems can harness the full potential of AI and ML in smart education, providing learners with personalized, adaptive, and transformative learning experiences that prepare them for success in the digital age.

REFERENCES

Baker, R. S., & Inventado, P. S. (2014). Educational data mining and learning analytics. In J. Larusson & B. White (Eds.), *Learning Analytics: From Research to Practice* (pp. 61–75). Springer. doi:10.1007/978-1-4614-3305-7_4

Baker, R. S., & Yacef, K. (2009). The State of Educational Data Mining in 2009: A Review and Future Visions. *Journal of Educational Data Mining*, *1*(1), 3–17.

Blikstein, P. (2013). Multimodal Learning Analytics and Education Data Mining: Using Computational Technologies to Measure Complex Learning Tasks. *Journal of Learning Analytics*, *1*(2), 185–189.

Broussard, D. M., Rahman, Y., Kulshreshth, A. K., & Borst, C. W. (2021). An Interface for Enhanced Teacher Awareness of Student Actions and Attention in a VR Classroom. *2021 IEEE Conference on Virtual Reality and 3D User Interfaces Abstracts and Workshops (VRW)*, 284-290. 10.1109/VRW52623.2021.00058

Embarak, O. (2019). *Demolish falsy ratings in recommendation systems. In 2019 Sixth HCT Information Technology Trends*. ITT. doi:10.1109/ITT48889.2019.9075130

Embarak, O. H. (2018). *Three Layered Factors Model for Mining Students Academic Performance. In 2018 Fifth HCT Information Technology Trends*. ITT. doi:10.1109/CTIT.2018.8649491

Kim, S.-H., Lee, C., & Youn, C.-H. (2020). An Accelerated Edge Cloud System for Energy Data Stream Processing Based on Adaptive Incremental Deep Learning Scheme. *IEEE Access : Practical Innovations, Open Solutions*, *8*, 195341–195358. doi:10.1109/ACCESS.2020.3033771

Knight, S., Wise, A. F., & Chen, B. (2017). Seeing through the glass: Using automated feedback to uncover learners' feedback literacy. *Computers in Human Behavior*, *71*, 275–285.

Koedinger, K. R., & Corbett, A. T. (2006). Cognitive tutors: Technology bringing learning science to the classroom. In R. K. Sawyer (Ed.), *The Cambridge Handbook of the Learning Sciences* (pp. 61–78). Cambridge University Press.

Koller, D., & Ng, A. (2016). The Online Revolution: Education for Everyone. *Daedalus*, *145*(1), 62–69.

Lee, Lee, & Chung. (2001). Face recognition using Fisherface algorithm and elastic graph matching. *Proceedings 2001 International Conference on Image Processing*, 998-1001.

Li, H., Liu, D. Y., & Wu, X. (2020). Artificial intelligence in education: A review. *Journal of Computer Assisted Learning*, *36*(1), 6–27.

Li, Q., Li, R., Ji, K., & Dai, W. (2015). Kalman Filter and Its Application. *2015 8th International Conference on Intelligent Networks and Intelligent Systems (ICINIS)*, 74-77. 10.1109/ICINIS.2015.35

Lotfy Abdrabou, E. A. M., & Salem, A.-B. M. (2010). A breast cancer classifier based on a combination of case-based reasoning and ontology approach. *Proceedings of the International Multiconference on Computer Science and Information Technology*, 3-10. 10.1109/IMCSIT.2010.5680045

Luckin, R. (2018). Machine Learning and Human Intelligence: The Future of Education for the 21st Century. *Zeitschrift für Psychologie mit Zeitschrift für Angewandte Psychologie*, *226*(2), 82–93.

Merzon, E. E., & Ibatullin, R. R. (2016). Architecture of smart learning courses in higher education. *2016 IEEE 10th International Conference on Application of Information and Communication Technologies (AICT)*, 1-5. 10.1109/ICAICT.2016.7991809

Moghaddam, M. T., Muccini, H., Dugdale, J., & Kjægaard, M. B. (2022). Designing Internet of Behaviors Systems. *2022 IEEE 19th International Conference on Software Architecture (ICSA),* 124-134. 10.1109/ICSA53651.2022.00020

Redmon, J., Divvala, S., Girshick, R., & Farhadi, A. (2016). You Only Look Once: Unified, Real-Time Object Detection. *2016 IEEE Conference on Computer Vision and Pattern Recognition (CVPR),* 779-788. 10.1109/CVPR.2016.91

Siemens, G. (2013). Learning Analytics: The Emergence of a Discipline. *The American Behavioral Scientist, 57*(10), 1380–1400. doi:10.1177/0002764213498851

Siemens, G., & Baker, R. (Eds.). (2019). *Educational Data Mining: Applications and Trends*. Springer.

Tsai, C. C., Cheng, W. C., Taur, J. S., & Tao, C. W. (2006). Face Detection Using Eigenface and Neural Network. *2006 IEEE International Conference on Systems, Man and Cybernetics,* 4343-4347. 10.1109/ICSMC.2006.384817

Wojke, N., Bewley, A., & Paulus, D. (2017). Simple online and realtime tracking with a deep association metric. *2017 IEEE International Conference on Image Processing (ICIP),* 3645-3649. 10.1109/ICIP.2017.8296962

Zhang, X., Wu, C.-W., Fournier-Viger, P., Van, L.-D., & Tseng, Y.-C. (2017). Analyzing students' attention in class using wearable devices. *2017 IEEE 18th International Symposium on A World of Wireless, Mobile and Multimedia Networks (WoWMoM),* 1-9. 10.1109/WoWMoM.2017.7974306

Zhou, X., Liang, W., Wang, K. I.-K., Wang, H., Yang, L. T., & Jin, Q. (2020, July). Deep-Learning-Enhanced Human Activity Recognition for Internet of Healthcare Things. *IEEE Internet of Things Journal, 7*(7), 6429–6438. doi:10.1109/JIOT.2020.2985082

Chapter 4
AI-Driven Skill Development: Bridging Students With Industry 5.0

Vijay Prakash Gupta
https://orcid.org/0000-0001-6110-5998
Institute of Business Management, GLA University, Mathura, India

ABSTRACT

The rapid advancement of Industry 5.0 demands a skilled workforce, but many students lack the necessary competencies. This chapter explores the transformative potential of AI-driven skill development, aligning education with Industry 5.0 needs. AI analyzes data to personalize skill development. The study aims to bridge the skills gap, emphasizing positive outcomes like enhanced workforce preparedness. AI's potential offers a promising solution for the evolving job market. Investing in AI-driven skill development is imperative for a skilled and adaptable workforce in Industry 5.0.

1.0 BACKGROUND AND CONTEXT

The emergence of Industry 5.0, which is characterised by the incorporation of cutting-edge technologies such as artificial intelligence (AI), robots, the Internet of Things (IoT), and big data (Chander, B., Pal, S., De, D., & Buyya, R. 2022), has resulted in substantial shifts throughout the workforce in every region of the world. The need for workers with advanced levels of expertise has grown in parallel with the increasing acceleration of industry change (Acemoglu, D. 2002). AI-driven skill development initiatives have emerged as a potentially useful approach to close the gap that exists between the education that students get and the needs of Industry 5.0 (Miao, F., Holmes, W., Huang, R., & Zhang, H. 2021). These programmes make use of the potential of AI to give students with individualised and dynamic learning experiences that equip them with the skills they will need to be successful in the workforce of the future (Samavedham, L., & Ragupathi, K. 2012).

Various industries, including education, have undergone a transformation thanks to artificial intelligence (AI), which provides creative ways to improve learning and skill development. There is an increasing need to connect students with industry requirements as Industry 5.0, which is focused on the seamless integration of humans and robots, emerges. (Broo, D. G., Kaynak, O., & Sait, S. M. 2022). By

DOI: 10.4018/979-8-3693-0782-3.ch004

supplying students with the skills and abilities required by the changing labour market, AI-driven skill development is essential in tackling this challenge.

1.1 The Role of AI-Driven Skill Development

Artificial intelligence-driven skill development makes use of artificial intelligence tools and technology to improve learning and prepare students for Industry 5.0. To create personalized and adaptable learning experiences, it utilises AI algorithms, machine learning, natural language processing, and data analytics. The purpose is to prepare students for the difficulties and possibilities that Industry 5.0 will provide. AI-powered skill development programmes provide students with personalized learning pathways, adaptive tests, and real-time feedback, allowing them to study at their own speed (Tapalova, O., & Zhiyenbayeva, N. 2022). Furthermore, AI allows for the analysis of massive volumes of data in order to detect skill gaps and give tailored interventions for improvement. This tailored approach improves the efficacy and efficiency of skill development programmes (Kamalov, F., Santandreu Calonge, D., & Gurrib, I. 2023).

1.2 Benefits of AI-Driven Skill Development

Learning experiences that are personalised are made possible by AI-driven skill development platforms that can identify each student's specific strengths and shortcomings. Artificial intelligence (AI) encourages more successful learning outcomes by tailoring the curriculum and pace of instruction to the student's learning preferences and capabilities.

Real-Time Feedback: Students can get quick feedback on their performance thanks to AI algorithms, which enables them to spot their weak points and make the required corrections. The learning process is improved, and skill development is accelerated, by this real-time feedback loop.

Enhanced Engagement: AI-powered learning platforms include interactive components that enhance learning through gamification, simulations, virtual reality, and augmented reality. Students become more enthusiastic and motivated as a result, which enhances their ability to retain information and develop new skills.

Industry-Relevant Content: AI-driven programmes for skill development can adapt their curricula to meet the changing needs of the labour market. AI algorithms can discover the most in-demand skills by examining market trends and job advertisements, ensuring that students gain the abilities they need to succeed in their chosen industries.

Bridging Students With Industry 5.0: To close the gap between academic and industrial requirements, Industry 5.0 will incorporate AI-driven talent development. It makes it easier for students to move from academic institutions to the industry and gives them the information and skills they need to succeed in a world that is becoming more automated and reliant on technology.

Industry Collaboration: Initiatives for skill development powered by AI encourage cooperation between businesses and educational institutions. Universities can learn about the most recent market trends, job requirements, and skill gaps by collaborating with businesses. Through this partnership, educational programmes are kept current and pertinent to market demands.

Customized Training Programs: Platforms powered by AI make it possible to design specialized training courses that are tailored to the particular demands of different sectors. Institutions can discover the talents that are in demand and create programmes that meet those requirements by utilizing

AI technologies. Students are given the training they need to become useful assets to potential employers because to this personalization.

Lifelong Learning: Industry 5.0 emphasizes the importance of ongoing skill development and lifetime learning. AI-driven skill development systems give professionals the chance to learn new talents or improve their current ones. AI makes sure that people can adjust to the shifting requirements of the sector throughout their careers by providing personalized recommendations and access to pertinent learning resources.

2.0 INDUSTRY 5.0: A PARADIGM SHIFT IN WORKFORCE REQUIREMENTS

Artificial intelligence-driven skill development makes use of artificial intelligence tools and technology to improve learning and prepare students for Industry 5.0. To create personalised and adaptable learning experiences, it utilises AI algorithms, machine learning, natural language processing, and data analytics. The purpose is to prepare students for the difficulties and possibilities that Industry 5.0 will provide. AI-powered skill development programmes provide students with personalised learning pathways, adaptive tests, and real-time feedback, allowing them to study at their own speed. Furthermore, AI allows for the analysis of massive volumes of data in order to detect skill gaps and give tailored interventions for improvement. This tailored approach improves the efficacy and efficiency of skill development programmes.

- **Industry 5.0: Human-Centric Approach:** Industry 5.0 is a shift from Industry 4.0's mostly automated and technologically driven emphasis. It acknowledges that human creativity, problem-solving talents, and emotional intelligence are valuable professional advantages. Rather of replacing people with robots, Industry 5.0 intends to incorporate them into an ecosystem where human-advanced technology cooperation flourishes.
- **Demand for Soft Skills**: Soft skills are becoming more important as technology continues to automate everyday jobs. In Industry 5.0, skills like as critical thinking, creativity, flexibility, emotional intelligence, and effective communication are highly prized. These abilities enable workers to excel in complex problem-solving scenarios, collaborate effectively, and interact meaningfully with customers and colleagues.
- **Cross-Disciplinary Skills**: Work in Industry 5.0 necessitates a more multidisciplinary approach. Because of the convergence of many technologies, such as artificial intelligence, robots, biotechnology, and nanotechnology, a workforce capable of working across several domains is required. Employees with a varied skill set and the capacity to adapt to new technologies will be in great demand, as they will be able to bridge the gap between various areas of knowledge.
- **Lifelong Learning and Upskilling:** Continuous learning and upskilling are important for the Industry 5.0 workforce due to the quick speed of technology innovations. Workers must be prepared to learn new skills and adapt to changing situations when new technologies arise and old ones change. To guarantee that the workforce stays relevant and competitive, employers, educational institutions, and people must emphasise lifelong learning and offer opportunities for reskilling and upskilling.
- **Human-Machine Collaboration**: Industry 5.0 shows how important it is for people and tools to work together. Instead of being seen as replacements for people, tools are seen as partners that add to what people can do. This teamwork lets people focus on more important jobs that require

complex decision-making, imagination, and understanding, while robots do the boring and routine work. In Industry 5.0, it will be important for workers to know how to use and connect with technology well.

- **Ethical and Responsible Work Practices:** As technology becomes more and more important in the workplace, ethics and doing the right thing become even more important. Industry 5.0 focuses on the right way to use technology, keeping data private, and making good decisions. Workers need to know a lot about these things to make sure that technology advances are used for the good of society and that any bad effects are kept to a minimum.

Human intelligence and cooperation drive Industry 5.0. Workforce needs are changing as technology shapes our society. In Industry 5.0, soft skills, cross-disciplinary knowledge, lifelong learning, and machine collaboration are crucial (Broo, D. G., Kaynak, O., & Sait, S. M. 2022). By adopting this new paradigm and training the workforce, we can unleash human genius and build a future where technology and mankind coexist.

3.0 CHANGING SKILL LANDSCAPE IN INDUSTRY 5.0

Introduction: Industry 5.0, also known as the fifth industrial revolution, represents a significant shift in the manufacturing landscape, driven by advancements in automation, artificial intelligence, and digital technologies. This transformative era brings forth new challenges and opportunities, reshaping the skill requirements for the workforce.

Automation and Robotics: With the integration of automation and robotics in Industry 5.0, the demand for traditional manual labor has decreased, while the need for skilled technicians and engineers proficient in operating and maintaining automated systems has risen. According to the International Federation of Robotics, the global sales of industrial robots reached 422,000 units in 2020, showcasing a 20% increase compared to the previous year (Olszewski, M. 2020). Collaborative robots, or cobots, are becoming increasingly prevalent in manufacturing environments. These robots work alongside human workers, enhancing efficiency and productivity. Skilled workers are needed to program and operate cobots safely, ensuring seamless human-robot collaboration.

Artificial Intelligence and Machine Learning: The integration of artificial intelligence (AI) and machine learning (ML) algorithms has opened up new possibilities in Industry 5.0, enabling predictive maintenance, quality control, and data-driven decision-making. Consequently, there is a growing demand for professionals skilled in AI and ML, including data scientists, AI engineers, and algorithm developers.

In the automotive industry, AI-powered computer vision systems are being used for quality inspection on assembly lines. These systems can detect defects and anomalies in real-time, reducing errors and improving overall product quality. Skilled AI engineers are required to develop and deploy such systems.

Data Analytics and Visualization: The abundance of data generated by interconnected systems in Industry 5.0 necessitates effective data analytics and visualization. Organizations seek professionals capable of extracting valuable insights from complex datasets and presenting them in a meaningful manner.

In the energy sector, smart grids equipped with advanced sensors collect vast amounts of data related to energy consumption patterns. Skilled data analysts are essential for interpreting this data, identifying trends, and optimizing energy distribution to improve efficiency and reduce costs.

Cybersecurity and Data Privacy: As Industry 5.0 relies heavily on interconnected systems and data sharing, the importance of cybersecurity and data privacy becomes paramount. The rising frequency of cyber threats and data breaches calls for skilled professionals who can safeguard critical infrastructure, develop secure protocols, and mitigate risks.

In the healthcare industry, the adoption of telemedicine and digital health platforms has increased, necessitating stringent security measures to protect patient data and ensure the integrity of remote consultations. Cybersecurity experts are crucial for implementing robust security frameworks in this context.

Soft Skills and Adaptability: While technical skills are vital in Industry 5.0, soft skills such as adaptability, critical thinking, creativity, and collaboration become equally important. As technology continues to evolve rapidly, professionals need to adapt to changing roles and work collaboratively with intelligent machines.

Cross-functional teamwork is becoming more prevalent in Industry 5.0. For instance, engineers, data scientists, and operations personnel collaborate to develop and implement AI-driven solutions. Effective communication, collaboration, and adaptability are essential for success in such multidisciplinary environments.

Table 1. Top technical skills in Industry 5.0

Skill	Description
Advanced Robotics and Automation	Proficiency in designing and programming advanced robotic systems and automated processes.
Artificial Intelligence and Machine Learning	Understanding and applying AI and ML algorithms for automation, data analysis, and decision-making.
Big Data Analytics	Analyzing and interpreting large volumes of data to extract valuable insights and drive strategic decisions.
Internet of Things (IoT)	Knowledge of IoT architecture, connectivity, and application development for interconnected systems.
Augmented and Virtual Reality	Familiarity with AR/VR technologies and their applications in various industries, including training and visualization.
3D Printing and Additive Manufacturing	Skills in using 3D printers and additive manufacturing techniques for rapid prototyping and production.

Source: Constructed by Authors based on own knowledge.

Table 2. Top soft skills in Industry 5.0

Skill	Description
Creativity and Innovation	Ability to generate novel ideas, think outside the box, and develop innovative solutions to complex problems.
Critical Thinking and Problem-Solving	Capacity to analyze information critically, identify problems, and apply logical reasoning to find effective solutions.
Adaptability and Flexibility	Willingness to embrace change, learn new technologies, and adapt to evolving work environments and challenges.
Collaboration and Communication	Proficiency in working effectively with diverse teams, communicating ideas clearly, and fostering collaboration.
Emotional Intelligence	Ability to understand and manage one's emotions, empathize with others, and navigate social dynamics in the workplace.
Leadership and Management	Skills in guiding and inspiring teams, setting strategic goals, and driving organizational success.

Source: Constructed by Authors based on own knowledge.

Table 3. Top Cross-disciplinary skills in Industry 5.0

Skill	Description
Computational Thinking	Aptitude for approaching problems with a logical and algorithmic mindset, breaking them down into solvable components.
Data Literacy	Ability to understand, interpret, and communicate insights derived

Source: Constructed by Authors based on own knowledge.

4.0 ROLE OF ARTIFICIAL INTELLIGENCE IN SKILL DEVELOPMENT

Artificial intelligence (AI) has had a transformative impact on skill development, providing new chances for people to improve their aptitude and learn new skills (Sofia, M., et.al.2023). AI-powered systems have the potential to revolutionise the way we learn and acquire skills by personalising learning experiences, offering tailored feedback, and providing adaptive information. Let's look at some concrete instances and accompanying information to better grasp how AI is affecting skill development.

Personalized Learning: By analysing a tremendous quantity of data on specific learners, AI offers personalised learning experiences. Depending on the needs and preferences of the learners, it can change the tempo, difficulty levels, and content. For instance, sites like Khan Academy and Coursera make course and learning material recommendations based on each learner's interests and skill level using AI algorithms..

Intelligent Tutoring Systems: By analysing a tremendous quantity of data on specific learners, AI offers personalised learning experiences. Depending on the needs and preferences of the learners, it can change the tempo, difficulty levels, and content. For instance, sites like Khan Academy and Coursera make course and learning material recommendations based on each learner's interests and skill level using AI algorithms.

Adaptive Assessments: By adjusting the content and difficulty level based on student performance, AI can improve assessments. Learners can concentrate on areas that require improvement with the use of adaptive exams, which enable accurate evaluation and assist in identifying knowledge gaps. An AI-based test called CogAT (Cognitive Abilities Test) adjusts its questions based on the test-taker's skills.

Language Learning and Translation: AI algorithms are used by AI-powered language learning services like Duolingo to evaluate learners' competence levels, give individualised exercises, and provide real-time feedback on grammar and pronunciation. Additionally, by offering rapid translations, AI-based translation technologies like Google Translate help students understand foreign languages.

Upskilling and Reskilling: AI has the potential to be extremely useful in reskilling and upskilling the workforce. Platforms like Udacity and LinkedIn Learning make use of AI to pinpoint skill gaps, suggest pertinent courses, and offer personalised learning routes to assist users in developing new abilities or changing careers.

Virtual Reality (VR) and Augmented Reality (AR) Training: Training experiences that are immersive and engaging can be provided by combining AI with VR and AR technology. Medical students, for instance, can practise intricate surgical operations in virtual settings to hone their skills before carrying them out on actual patients.

5.0 THE IMPORTANCE OF EMBRACING MODERN TECHNOLOGY IN EDUCATION

In today's technologically advanced society, educational institutions must constantly adapt to the changing environment by keeping up with new developments. Although schools have made remarkable progress integrating technology into their instructional strategies, there is still a glaring gap in providing students with the skills they need to succeed in an increasingly digital world. A fundamental knowledge of HTML, CSS, and other traditional coding languages is still useful, but it is no longer adequate. This article explores the importance of educational institutions adopting and teaching cutting-edge technology like chatbots and artificial intelligence (AI) to provide students with the skills they need to succeed in the modern world.

5.1 The Importance of AI and Chatbots in Education

From healthcare to finance and beyond, a variety of industries have been transformed by AI and chatbots. These technologies are actively influencing our present and future and are no longer just far-off ideas. Schools unintentionally deprive students of a profound grasp of the technologies that will drive innovation and affect their lives in the years to come by failing to teach them about AI and chatbots.

5.2 Preparing Students for a Job Market Driven by AI

With the employment market undergoing fast transformation, there is an increasing demand for workers with knowledge in AI and related technologies. Introducing AI and chatbot education in schools at an early stage will successfully prepare students for prosperous jobs in this burgeoning industry. By exposing students to AI, educational institutions may pique their interest, develop creative thinking, and cultivate problem-solving skills, all of which are in high demand in today's digital society. This early introduction to AI education provides students with the knowledge and skills they need to thrive in a quickly changing work market.

5.3 Improving Critical Thinking and Problem-Solving Skills

The incorporation of AI and chatbots into education provides students with valuable possibilities to participate in complex problem-solving scenarios and develop their critical thinking skills. Students learn the ability to recognise patterns, analyse data, and make sound judgements by participating in AI projects. These abilities are not just useful for aspiring AI experts, but also necessary for navigating a data-centric society in which critical thinking is more important. Students learn the required abilities to flourish in a future where critical thinking is a highly sought-after asset by immersing themselves in AI activities.

5.4 Promoting Ethical and Responsible AI Application

It is essential to instill in kids a sense of ethical responsibility as AI grows more prevalent. Discussions about AI ethics, bias reduction, privacy issues, and responsible AI use should be included in school curricula. By educating children about the possible dangers and societal repercussions of AI, we may help them develop into responsible digital citizens who can use technology to create a better future.

7.0 HOW AI-DRIVEN TECHNOLOGY PROVES BENEFICIAL IN SKILL DEVELOPMENT AND CONNECTING STUDENTS WITH INDUSTRY 5.0

AI-driven technology has revolutionized various sectors, including education and industry, providing numerous benefits for skill development and bridging the gap between students and Industry 5.0 (Rane, 2023).

Below mentioned points explains, how AI is making a significant impact with supporting facts and data:

1. **Personalized Learning:** AI-powered algorithms analyze individual learning patterns and preferences, enabling personalized learning experiences. According to a study by McKinsey, personalized learning can improve student performance by 30-60% compared to traditional classroom methods.
2. **Adaptive Learning Platforms:** AI-powered adaptive learning platforms adjust the difficulty level of educational content based on a student's performance and knowledge gaps. A research study published in the Journal of Educational Technology & Society found that adaptive learning improved student engagement and academic achievement significantly.
3. **Skills Gap Analysis:** AI algorithms can analyze labor market data and identify skill gaps in specific industries. According to the World Economic Forum's Future of Jobs Report, 54% of employees will require significant reskilling by 2022 to keep up with technological advancements.
4. **Virtual Reality (VR) and Augmented Reality (AR) Simulations:** AI-driven VR and AR simulations provide hands-on training experiences, particularly in fields like engineering, healthcare, and manufacturing. A study by PwC found that VR-trained employees are four times faster in acquiring new skills compared to traditional training methods.

Table 4. Various types of AI-powered learning tools and platforms

Tool/Platform	Features and Benefits
Knewton	Personalized learning paths, adaptive assessments, real-time analytics
Carnegie Learning	AI-driven math and language learning, personalized recommendations, data-driven insights
Eklavvya AI Proctoring	Automated online exam proctoring, AI-based cheating detection, secure and scalable platform
Smart Sparrow	Adaptive and interactive courseware, virtual labs, data-driven feedback
DreamBox	AI-powered math learning, adaptive lessons, real-time progress tracking
Coursera	Online courses from top universities, AI-based personalized recommendations, skills development
Querium	AI-powered math tutoring, step-by-step problem-solving guidance, targeted practice
Woot Math	Adaptive math learning, personalized content, actionable insights
GradeSlam	24/7 on-demand tutoring, AI-based learning analytics, personalized support
TutorMe	Online tutoring in various subjects, AI-powered matching with tutors, 24/7 availability
Pearson AI	AI-based educational solutions, personalized learning experiences, data-driven insights
Kaltura	Video-based learning and collaboration platform, interactive video quizzes, analytics
Duolingo	AI-powered language learning, gamified lessons, personalized feedback
Edmentum	Personalized online education, adaptive assessments, competency-based learning
DocuExprt	AI-powered document analysis, automated feedback, plagiarism detection
Cognii Virtual TA	AI-powered virtual teaching assistant, automated grading, personalized support
Turnitin	Plagiarism detection software, feedback and grading tools, academic integrity promotion
Google Bard	AI-generated poetry, language generation, creative writing support
ChatGPT	AI-powered chatbot, natural language understanding, conversational interaction

Source: Constructed by Authors based on own knowledge.

5. **AI-Enhanced Feedback Mechanisms:** AI algorithms can assess and provide real-time feedback on student performance. Research conducted by the University of Michigan showed that students who received AI-generated feedback outperformed those who received traditional instructor feedback.
6. **Industry 5.0 Integration:** AI helps bridge the gap between educational institutions and Industry 5.0 by aligning educational programs with industry requirements. According to a survey by Deloitte, 47% of companies believe that AI technologies play a vital role in closing the skills gap between academia and the industry.
7. **AI-Powered Career Counseling:** AI-driven career counseling platforms analyze student data, interests, and industry trends to provide tailored career advice. A study by the International Journal of Advanced Research in Computer Science and Software Engineering showed that AI-based career guidance improved career decision-making for students.
8. **AI in Industry 5.0:** The implementation of AI in Industry 5.0 results in increased efficiency, productivity, and innovation. A report by Accenture found that AI could potentially double the annual economic growth rates in major economies by 2035.

In conclusion, AI-driven technology has proven to be highly beneficial in skill development and connecting students with Industry 5.0. Through personalized learning, adaptive platforms, skills gap analysis, immersive simulations, and enhanced feedback mechanisms, AI plays a crucial role in preparing students for the demands of the rapidly evolving job market. Additionally, its integration with Industry 5.0 fosters a symbiotic relationship between education and the industry, ensuring a future-ready workforce capable of driving innovation and economic growth.

8.0 INTEGRATING AI IN EDUCATIONAL INSTITUTIONS FOR TRANSFORMING EDUCATION

AI integration in educational institutions has the potential to revolutionize the way we teach and learn, making education more personalized, effective, and accessible. Let's delve into the details, to understand the transformative impact of AI in education:

1. **Personalized Learning:** AI algorithms analyze vast amounts of student data, including learning patterns, preferences, strengths, and weaknesses. This allows educators to create personalized learning pathways for each student, catering to their individual needs and pace of learning. A study conducted by the Rand Corporation found that students who received personalized instruction showed significant learning gains, with a percentile increase of 27 points in mathematics (Callaway, D. 2021).
2. **AI Tutors and Virtual Assistants**: AI-powered tutoring systems and virtual assistants provide real-time support to students, answering their questions, providing explanations, and offering additional resources. A study published in the International Journal of Artificial Intelligence in Education revealed that students who used AI tutors demonstrated improved learning outcomes, with a 30% reduction in time required to master a concept (Chen, L., Chen, P., & Lin, Z. 2020).
3. **Intelligent Content Recommendations:** AI-driven content recommendation systems suggest relevant educational materials, videos, and resources based on a student's learning progress and interests. This helps students discover supplementary materials aligned with their learning objectives. Research by Docebo indicates that AI-driven content recommendations can increase engagement rates by up to 67% (Ghosh, S., Majumder, S., & Peng, S. L. 2023).
4. **Grading Automation:** AI can automate the grading process for multiple-choice questions, quizzes, and assignments. This saves teachers' time and allows them to focus on providing more personalized feedback to students
5. **Predictive Analytics for Student Success:** AI algorithms can predict students at risk of falling behind or dropping out based on their academic performance and engagement data. Early identification of struggling students allows for timely interventions and support. Georgia State University implemented an AI-powered student success program and increased graduation rates by 23 percentage points over a decade (Hannan, E., & Liu, S. 2023).
6. **Enhancing Collaboration and Communication:** AI-driven communication tools and chatbots facilitate seamless communication between students, teachers, and parents. This improves parent-teacher engagement and strengthens the support system for students.
7. **AI-Enhanced Learning Analytics:** AI-driven learning analytics provide valuable insights into student progress and instructional effectiveness. Educational institutions can use this data to identify areas of improvement and optimize teaching strategies.
8. **Virtual Reality (VR) and Augmented Reality (AR) in Education:** AI-powered VR and AR applications create immersive learning experiences, making complex concepts easier to grasp.

According to a report by Technavio, the global education VR market is expected to grow at a CAGR of over 42% between 2021 and 2025 (Begum, S. 2021).

9. **Bridging the Education Gap**: AI integration can help bridge the education gap in underserved areas and remote regions by providing access to quality educational resources.
10. **Future Job Market Preparedness:** By incorporating AI into the curriculum, educational institutions prepare students for the future job market, which increasingly requires AI literacy and skills.A study by the World Economic Forum listed critical thinking, problem-solving, and technology use as the top skills for the workforce of the future (Sousa, M. J., & Wilks, D. 2018)

The integration of AI in educational institutions holds immense potential for transforming education. From personalized learning and AI tutors to grading automation and predictive analytics, AI-driven solutions enhance the learning experience and contribute to better student outcomes. As AI continues to advance, its integration in education will play a pivotal role in creating a more inclusive, adaptive, and future-ready education system.

9.0 BENEFITS OF INCORPORATING AI-DRIVEN CURRICULUM FOR TRANSFORMING EDUCATION

The global market for AI in education is expected to reach $3.68 billion by 2023, with a CAGR of 47.77% from 2018 to 2023 (*AI in Education Market Size, Share, Trends and Industry Analysis 2018-2030*) The widespread adoption of AI-powered educational tools, such as intelligent tutoring systems, learning management platforms, and personalized learning applications, demonstrates the growing interest in AI's potential to transform education.

Personalized Learning: AI-driven curricula can analyze individual student data, learning styles, and progress to create personalized learning paths. A study conducted by SRI International found that students using AI-powered tutoring systems outperformed their peers by 26 percentile points.

Early Identification of Learning Difficulties: AI algorithms can quickly identify students who are struggling and provide timely interventions.

Enhanced Teacher Productivity: AI can automate administrative tasks, grading, and data analysis, allowing teachers to focus more on personalized instruction and support. AI adoption in education could save teachers' time (Bryant, J., Heitz, C., Sanghvi, S., & Wagle, D. 2020)

Lifelong Learning and Upskilling: AI-powered educational platforms can support continuous learning, helping professionals acquire new skills to stay relevant in a dynamic job market.

10.0 ARTIFICIAL INTELLIGENCE FOR ASSESSMENT AND FEEDBACK TO ENHANCE STUDENT SUCCESS IN EDUCATION

Importance of Immediate Feedback: Immediate feedback plays a pivotal role in fostering a deeper understanding of subject matter and facilitating continuous improvement in student performance. AI-powered assessment tools can provide real-time feedback, identifying errors and misconceptions instantly. This allows students to correct their mistakes promptly, ensuring they grasp concepts correctly and preventing the reinforcement of incorrect knowledge.

Valid Feedback and Learning Outcomes: AI-driven assessment platforms can analyze student responses and provide valid feedback tailored to individual learning needs. By understanding the strengths and weaknesses of each student, educators can personalize learning pathways, optimize curriculum design, and address specific areas where students may require additional support. As a result, student learning outcomes are significantly improved, and academic progress is enhanced.

Qualitative Assessment and Holistic Development: Traditional assessment methods often focus on quantitative metrics, such as test scores, which may not adequately capture students' overall abilities and skills. AI offers the potential to assess students qualitatively, incorporating a range of indicators such as critical thinking, problem-solving, creativity, and communication skills. This holistic approach to assessment promotes well-rounded development and better prepares students for real-world challenges.

Adaptability in Online Education: With the surge in online education due to the COVID-19 pandemic, AI-powered assessment and feedback mechanisms have proven to be invaluable. These tools can efficiently manage large-scale assessments, automatic grading, and data analysis. Moreover, AI systems can identify patterns in student performance, enabling instructors to adjust their teaching strategies in real-time to cater to the evolving needs of their students.

Addressing Learning Gaps and Intervention: AI can track students' progress over time, allowing educators to identify persistent learning gaps and intervene proactively. By continuously analyzing data from various assessments, AI systems can generate early warning signals and recommend targeted interventions to help struggling students. This personalized approach not only boosts academic performance but also fosters a sense of support and motivation among learners.

Ethical Considerations and Data Privacy: While AI-based assessment and feedback offer numerous advantages, it is crucial to address ethical concerns and data privacy issues. Educational institutions must ensure that data collected through AI systems are handled securely and with utmost confidentiality. Transparent communication with students and clear consent procedures are essential to maintain trust and uphold ethical standards.

Conclusion: The integration of Artificial Intelligence in assessment and feedback processes has brought about transformative changes in the education sector, particularly in the context of online learning during the COVID-19 pandemic. The immediacy and validity of feedback, coupled with qualitative assessment practices, promote a conducive learning environment and enhance student success. However, the responsible and ethical implementation of AI in education remains paramount to ensure students' data privacy and to harness the full potential of AI in fostering holistic development and academic growth.

REFERENCES

Acemoglu, D. (2002). Technical change, inequality, and the labor market. *Journal of Economic Literature*, *40*(1), 7–72. doi:10.1257/jel.40.1.7

AI in Education Market Size, Share, Trends and Industry Analysis 2018 - 2030. (n.d.). Markets and Markets. https://www.marketsandmarkets.com/Market-Reports/ai-in-education-market-200371366.html

Begum, S. (2021). A Study on growth in Technology and Innovation across the globe in the Field of Education and Business. *International Research Journal on Advanced Science Hub, 3*(6S), 148-156.

Broo, D. G., Kaynak, O., & Sait, S. M. (2022). Rethinking engineering education at the age of industry 5.0. *Journal of Industrial Information Integration, 25*, 100311. doi:10.1016/j.jii.2021.100311

Bryant, J., Heitz, C., Sanghvi, S., & Wagle, D. (2020). How artificial intelligence will impact K-12 teachers. *Retrieved*, (May), 12.

Callaway, D. (2021). *The Impact of Personalized Learning on Achievement in an Elementary School Mathematics Classroom*. Wilkes University.

Chander, B., Pal, S., De, D., & Buyya, R. (2022). Artificial intelligence-based internet of things for industry 5.0. *Artificial intelligence-based internet of things systems*, 3-45.

Chen, L., Chen, P., & Lin, Z. (2020). Artificial intelligence in education: A review. *IEEE Access : Practical Innovations, Open Solutions, 8*, 75264–75278. doi:10.1109/ACCESS.2020.2988510

Ghosh, S., Majumder, S., & Peng, S. L. (2023). An Empirical Study on Adoption of Artificial Intelligence in Human Resource Management. In *Artificial Intelligence Techniques in Human Resource Management* (pp. 29–85). Apple Academic Press. doi:10.1201/9781003328346-3

Hannan, E., & Liu, S. (2023). AI: New source of competitiveness in higher education. *Competitiveness Review, 33*(2), 265–279. doi:10.1108/CR-03-2021-0045

Kamalov, F., Santandreu Calonge, D., & Gurrib, I. (2023). New Era of Artificial Intelligence in Education: Towards a Sustainable Multifaceted Revolution. *Sustainability (Basel), 15*(16), 12451. doi:10.3390u151612451

Miao, F., Holmes, W., Huang, R., & Zhang, H. (2021). *AI and education: A guidance for policymakers*. UNESCO Publishing.

Olszewski, M. (2020). Modern industrial robotics. *Pomiary Automatyka Robotyka, 24*(1), 5–20. doi:10.14313/PAR_235/5

Rane, N. (2023). ChatGPT and Similar Generative Artificial Intelligence (AI) for Smart Industry: Role, Challenges and Opportunities for Industry 4.0, Industry 5.0 and Society 5.0. *Challenges and Opportunities for Industry, 4*.

Samavedham, L., & Ragupathi, K. (2012). Facilitating 21st century skills in engineering students. *Journal of Engineering Education, 26*(1), 38–49.

Sofia, M., Fraboni, F., De Angelis, M., Puzzo, G., Giusino, D., & Pietrantoni, L. (2023). The impact of artificial intelligence on workers' skills: Upskilling and reskilling in organisations. *Informing Science, 26*, 39–68. doi:10.28945/5078

Sousa, M. J., & Wilks, D. (2018). Sustainable skills for the world of work in the digital age. *Systems Research and Behavioral Science, 35*(4), 399–405. doi:10.1002res.2540

Tapalova, O., & Zhiyenbayeva, N. (2022). Artificial Intelligence in Education: AIEd for Personalised Learning Pathways. *Electronic Journal of e-Learning, 20*(5), 639–653. doi:10.34190/ejel.20.5.2597

Chapter 5
Applications of Deep Learning in Healthcare in the Framework of Industry 5.0

Padmesh Tripathi
https://orcid.org/0000-0001-9455-1652
Delhi Technical Campus, India

Nitendra Kumar
https://orcid.org/0000-0001-7834-7926
Amity Business School, Amity University, Noida, India

Krishna Kumar Paroha
Gyan Ganga College of Technology, India

Mritunjay Rai
https://orcid.org/0000-0002-8911-4826
Noida Institute of Engineering and Technology, Greater Noida, India

Manoj Kumar Panda
Women Institute of Technology, Dehradun, India

ABSTRACT

Emergence of deep learning (DL) and its applicability motivated researchers and scientists to explore its applications in their fields of expertise. In medical technology, a huge amount of data is required, and dealing with huge data is a challenging task for researchers. The emergence of neural networks and its modifications like convolutional neural networks (CNN), generative adversarial network (AGN), recurrent neural networks (RNN), and their subcategories has provided a stage to flourish deep learning. DL has been a successful tool in the fields of pattern recognition, natural language processing (NLP), image processing, speech recognition, computer vision, etc. All these techniques have been employed in healthcare. Image processing has been proven to be a fruitful technique for physicians to properly diagnose patients through CT scan, MRI, PET, radiography, nuclear medicine, ultrasound, etc. In this chapter, some applications of DL in healthcare have been envisaged, and it has been concluded that this technique is very successful in healthcare.

DOI: 10.4018/979-8-3693-0782-3.ch005

1. INTRODUCTION

Deep learning (DL), a powerful technology for dealing with large datasets, is a branch of a popular technology known as machine learning. The relation between deep learning, machine learning (ML) and artificial intelligence (AI) is depicted in figure1. Artificial neural networks (ANN) or simply christened as neural networks (NN) are the backbone of DL. The basic difference between the neural network and deep learning is that neural networks consist of three layers: first one the input layer, second one the hidden layer and third one the output layer while deep learning has more than three layers. In deep learning, there is more than one hidden layers (Bengio 2009, Schmidhuber 2015, Kumar, et al., 2016, 2017; Tripathi, et al. 2022; Dhar, et al., 2023).

Depth is the factor which distinguishes the DL networks from the neural network which has only one hidden layer. In DL networks, data passes through more than one hidden layer in pattern recognition in multistep process. This enables DL networks to have higher learning capabilities with more precision. DL networks are very efficient in dealing with the huge amount of data. Hence, they can solve the complex problems with pace and efficiency (Pan and Yang, 2010). The drawback of DL networks is that they are not applicable on small datasets.

With the emergence of computers and medical technology, the last four decades have witnessed great deals of biomedical data which are available in abundance. But it is a difficult task to deal effectively with the biomedical data as these are noisy, sparse, chaotic and with high dimensions. Therefore, to acquire efficient information from biomedical data, researchers were in high need of a technique which could overcome this problem. The emergence of deep learning has shown a ray of hope for researchers. Deep Learning is an efficient technique for speech recognition (Abdel-Hamid et al., 2014), Computer vision (He et al., 2016), NLP (Lan et al., 2020), etc. DL has been extensively applied in several fields of healthcare like drug development, clinical image processing, disease detection, electronic health record, DNA sequencing, etc. The beauty of Deep learning is that it automatically extracts the features from the data and plays a vital role in sparing time and resources.

Figure 1. Comparison between ANN and deep architecture

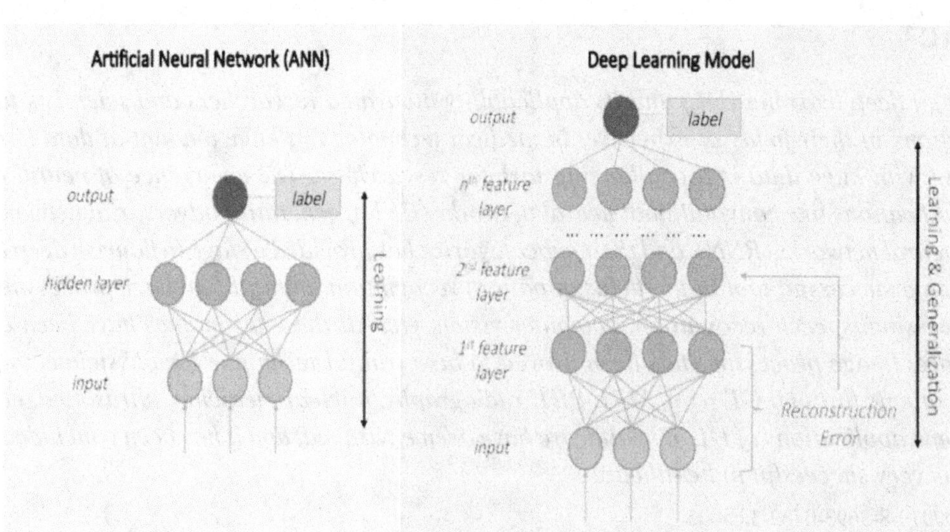

2. COMMON DEEP LEARNING ALGORITHMS

Numerous DL algorithms have achieved popularity in different fields. But the algorithms popular in healthcare are autoencoders, CNN, RNN and belief networks. We will discuss these here in brief.

2.1 Convolutional Neural Networks (CNN)

One of the prominent DL algorithms, CNN has extensively been employed in several fields like financial time series, natural language processing, image classification, medical image analysis, video and image recognition, image classification, recommender systems, image segmentation, brain–computer interfaces, face recognition (Rai et, al. 2022), speech recognition, etc. CNNs are a specific type of ANNs in which the mathematical operation convolution is used rather than matrix multiplication in one of their layers. CNN structure involves three layers: first one is termed as the convolutional layer; second one is termed as the pooling layer and the third one is termed as the fully connected layer. Convolutional layers perform the role of feature extractor, pooling layers perform the role of reducer of dimensions of data whereas fully connected layers perform the role of classifiers (Tripathi, et al. 2022).

Figure 2. Architecture of CNN

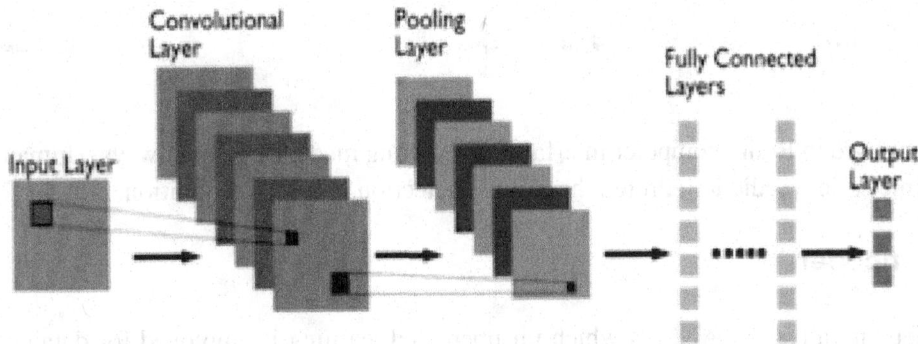

2.2 Recurrent Neural Networks (RNN)

A DL algorithm extensively employed for sequential data like audio data, speech data, video data, time series, financial data, text data, etc. is a recurrent neural network. The architecture of RNN is not different from other ANNs. A RNN structure is made of three layers namely input, hidden and outer layer and they follow a standard sequence. The input layer performs the role of data fetcher performing preprocessing and then filtered data is sent into hidden layer. The hidden layer performs the role of an important information retriever and is composed of NN, algorithms and activation functions. The output layer performs the role of expected outcome provider once the information is received here. RNNs have an internal memory and because of that they remember the inputs. LSTMs are used for creating the units for different layers of RNN.

Figure 3. Architecture of recurrent neural networks

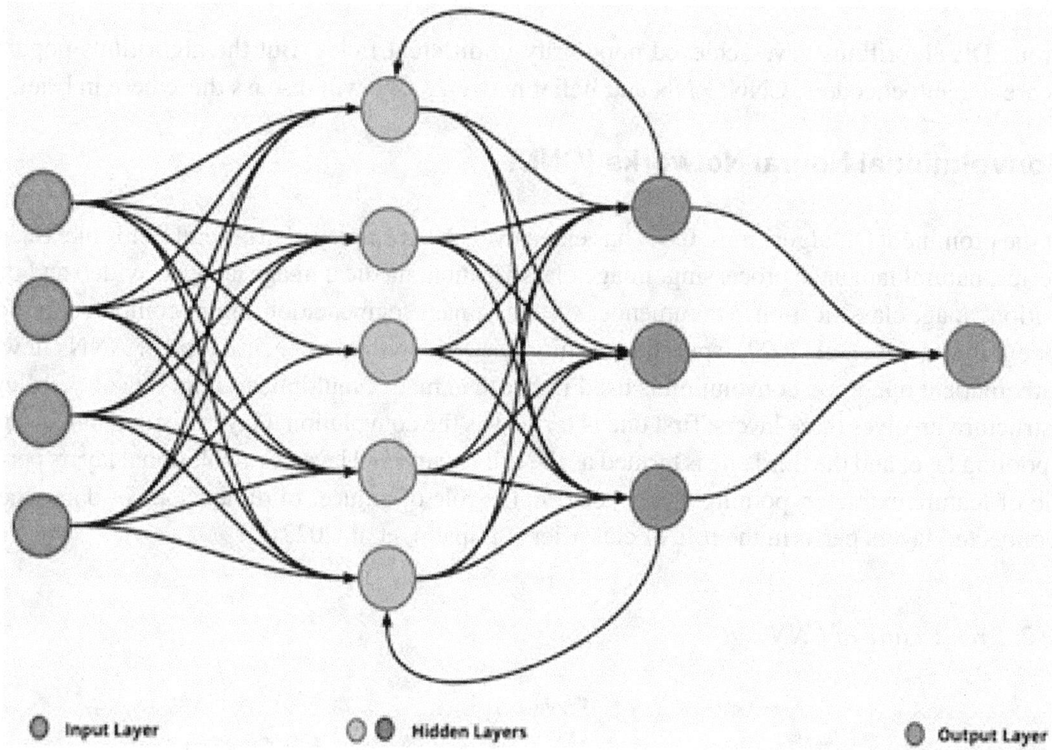

RNNs are used in Brain–computer interfaces, Predicting medical care pathways, Human action recognition, time series prediction, Protein homology detection, Speech recognition, etc.

2.3 Autoencoder

Autoencoders are a type of ANN in which unsupervised learning is employed for data encodings. In Autoencoders, the dimension of input layer is same as of output. Simply saying, output layer and input layer have same number of units. An autoencoder has two components, first one the encoder and second one the decoder which are simply neural networks. The hidden layer has less units in comparison to input and output layers, which is beneficial in data compression and feature learning from data. The encoder plays the role of dimension reducer (data compression) of original data for acquiring a new representation. Next, the decoder plays the role of reconstructing the input data using the acquired new representation.

Different variants of autoencoders available are Regularized autoencoders, Sparse autoencoder, Denoising autoencoder, Contractive autoencoder, Concrete autoencoder, Variational autoencoder, etc. Autoencoders have been employed in many fields like feature detection, attaining the meaning of words, anomaly detection, facial recognition, drug discovery (Zhavoronkov, 2019), disease detection, popularity prediction etc. Martinez-Murcia, et al. (2020) employed autoencoders in Alzemiers detection.

Figure 4. Structure of autoencoders

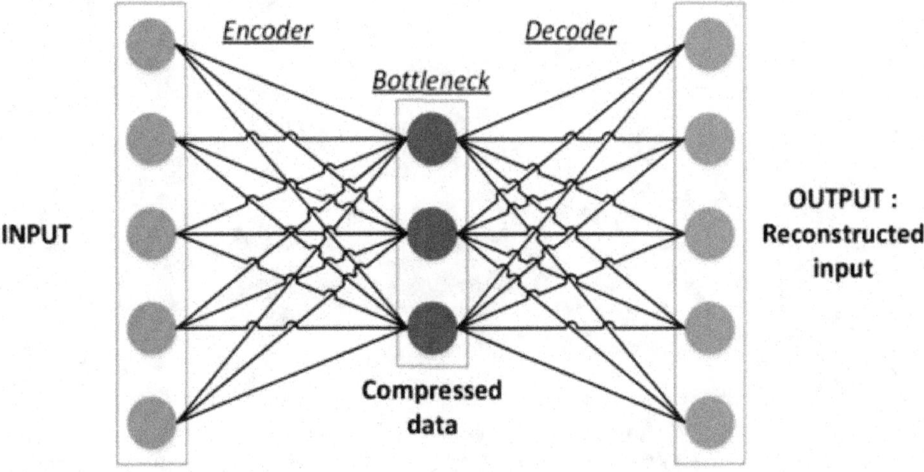

2.4 Deep Belief Network (DBN)

DBN is a probability architecture constructed on restricted Boltzmann machine (RBM). For a set of inputs, RBM is a generative stochastic ANN having the capability to learn a probability distribution. RBM had emerged in 1986, but RBMs got prominence after the development of a robust learning technique for RBMs by Hinton, et al. (2006). In RBM, there are only two layers namely visible layer and hidden layer. Obviously, Visible layer contains of visible units while hidden later contains of hidden units. A visible layer is employed for the input of training data while hidden layer for detecting the features.

A deep belief network is constructed by connecting RBMs in which the hidden layer of the earlier RBM is the visible layer of the subsequent RBM. Meaning by, the output of the earlier RBM is the input of the subsequent RBM. In the training process, RBM in upper layer must be trained fully before the training of RBM in current layer. The process continues until the training of last layer.

Figure 5. Restricted Boltzmann machine

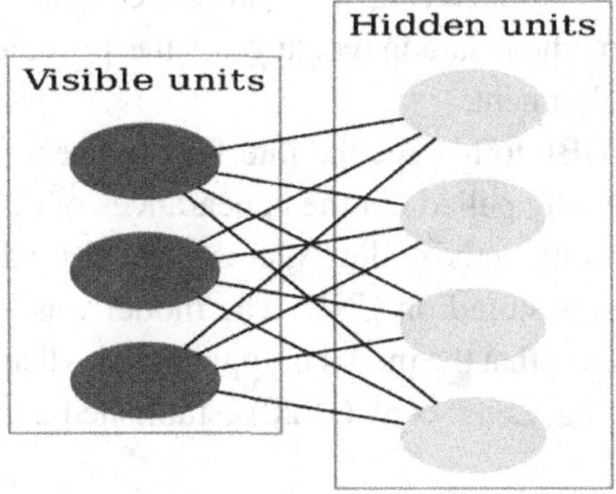

Figure 6. Deep belief network

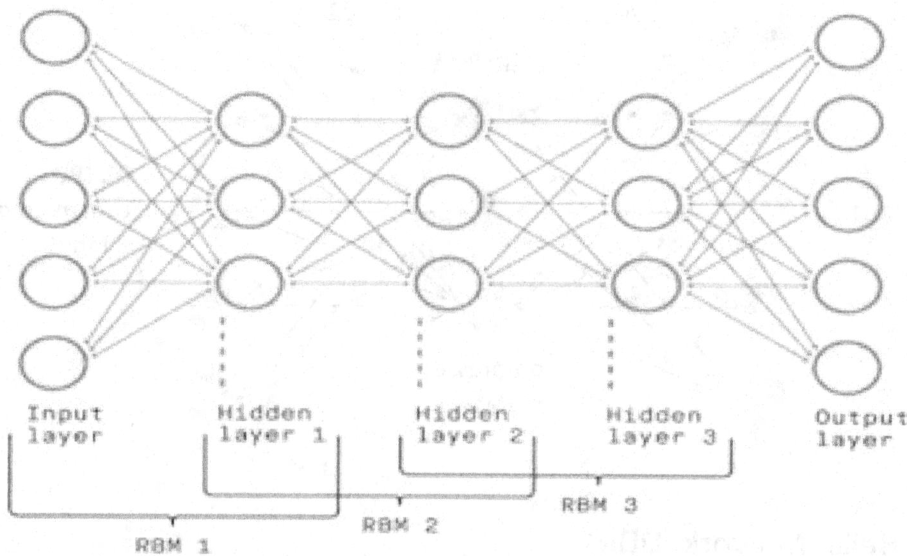

3. APPLICATIONS OF DL IN HEALTHCARE

Numerous DL algorithms have been widely employed in different fields of medical and healthcare. Below are listed some applications:

3.1 Drug Development

With the hasty progression of biomedical data, DL algorithms have evolved a new technique in drug development. With the help of DL algorithms, researchers can carry out research effectively on disease treatment and drug development which are very helpful in precision medicine development.

Drug is developed by two methods: one is the computational method and other is experimental method. The computational method has the benefit of saving time and reducing the loss while experimental method has the drawback of consuming time and efficiency.

One of the supreme steps in drug design and development is the identification of the interface between drug and target. This is a resource saving step and reduces the duration from drug development to market. The abundance of biomedical data in recent years has provided an opportunity to researchers to use deep learning in drug development.

Wen, et al. (2017) employed DBF to foresee the interface between drugs and targets. Their model, in an unsupervised way, automatically pulled out the appearances of drugs and targets. Hu et al. (2019) signified each drug target pair by linking the coding vector of the drug descriptor and the target descriptor. After that, the signification was stored on CNN. The model was tested on DrugBank dataset and accuracy was 0.88. The result reflects that the model is applicable in distinguishing the interface between drug and target. Using deep learning, Zeng, et al. (2020) established a model and used it for recognition of target and reuse of drug. Zhao, et al. (2020) employed two GAN in an unsupervised way to compute binding affinity between drug and target. A CNN was applied for prediction.

Deep Learning has also been used for designing the molecules from scratch, foreseeing the pharmacological characteristics and synergistic impacts of drugs, etc. Deep Learning has been used by Gupta, et al. (2018) for designing molecular weight. . Zeng, et al. (2019) used multimode autoencoders in different networks to learn the advanced features of drugs.

Another technique used in drug development is representation learning which is a combination of unsupervised representation learning and DL techniques. Wan, et al. (2019) employed the combination of DL and unsupervised representation learning to foresee the relations between proteins and unstructured compounds. Karimi, et al. (2020) proposed a deep generation architecture for drug combination design.

Tang (2023) explored the utilities of DL in drug discovery. He also discussed various limitations of DL in drug discovery.

3.2 Medical Imaging

Medical imaging has been proved vital in the fields of health and medicine. With the advances in technologies in medical imaging like CT scan, MRI, PET, etc., doctors have got a great tool to treat patients. These advanced technologies have been very useful in understanding the problems of patients and taking proper decisions in the treatment of patients.

With the availability of a huge amount of medical imaging data, deep learning has shown its capabilities in medicine and health. Deep learning performs several tasks like target recognition, image

Figure 7. MRI of brain

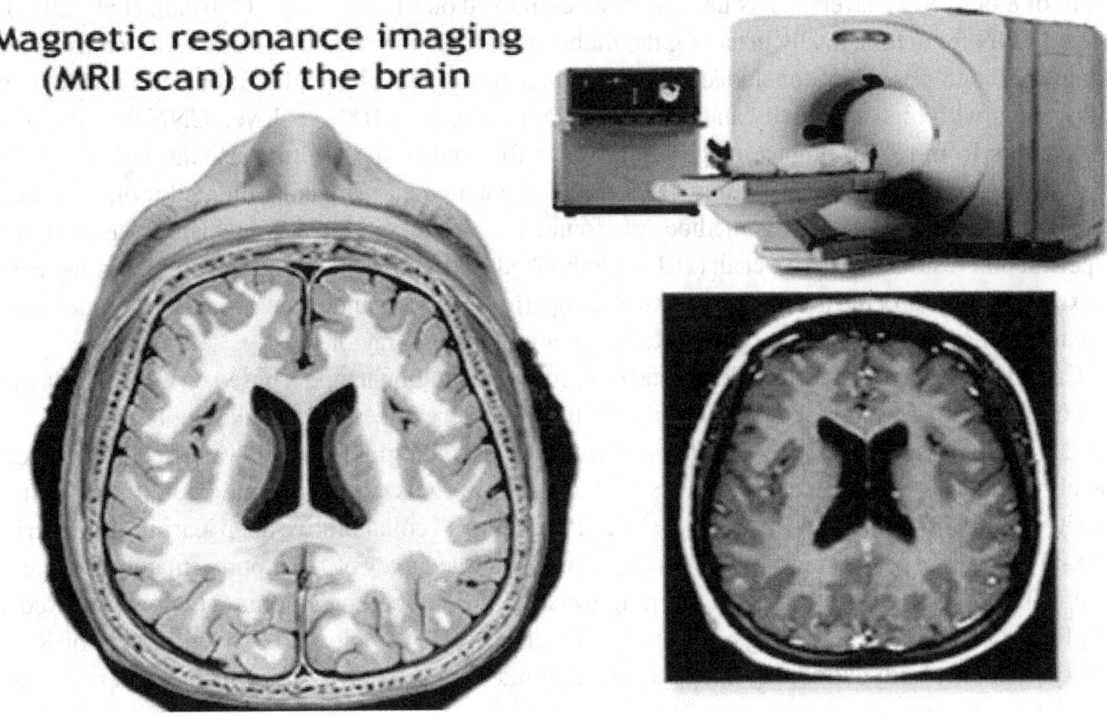

Figure 8. PET of brain

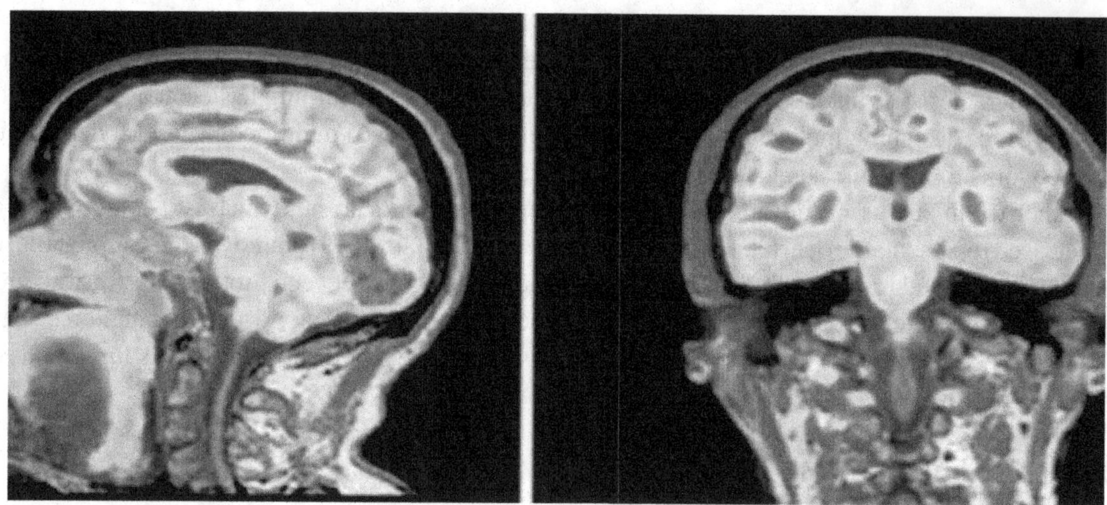

classification, target segmentation performing analysis of images. Now a days, deep learning has been widely employed in analyzing the medical images. DL has been very successful and helpful for doctors to make precise diagnoses and decisions.

Different deep learning methods like CNN, DBF, auto-encoders, RNN, etc. have been employed in medical imaging. But CNN is employed most by researchers and it is very successful in medical imaging.

CNN is appropriate for extracting the image features automatically from the original image. Numerous researchers have employed CNN in medical imaging for different purposes. Shin, et al. (2016) used CNN in case of availability of limited data set. They showed the usefulness of CNN architecture having depth of 8 or even 22 layers. CNN has also been employed on fundus image. Gulshan, et al. (2016) employed CNN for automatically detecting the diabetic macular edema and diabetic retinopathy in retinal fundus photographs. They concluded that the approach was suitable for detection of referable diabetic retinopathy with high specificity and sensitivity. Poplin, et al. (2018) employed CNN for extraction of new results from retinal fundus images and used it for prediction of cardiovascular risk factors. Dai, et al. (2018) proposed a method using CNN to detect microaneurysm and validated it on diabetic retinopathy patients' data sets. This method was found very useful, and 99.7% precision was achieved on experimental results. Kermany, et al. (2018) established a DL model based on transfer learning methods to excellently classify the images of diabetic retinopathy and macular degeneration. This model was also able to distinguish between viral and bacterial pneumonia on X-ray of chest.

Cao, et al. (2016) devised an efficient and effective computational model based on CNN for classification of X-ray image of tuberculosis. The model has been very effective in achieving its goal. Cicero, et al. (2017) employed CNN for classification of frontal chest radiographs which was effective in detecting the abnormalities related to chest. Images of 256 x 256 were used as input and overall specificity and sensitivity of 91% was achieved. Yuan, et al. (2019) employed collaborative deep learning in experiment on X-ray images of chest. They observed an improvement in accuracy of 19% approximately. Bungărdean, et al (2021) used CNN with transfer learning to classify basal cell carcinoma which was proved very helpful for pathologists in diagnosis. Liong-Rung, et al. (2022) employed CNN on images of X-ray of chest of patients with pulmonary edema or pneumonia. The accuracy rate of 83.2% and recall rate of

83.2% was achieved. Suganyadevi and Seethalakshmi (2022) proposed two CNN architectures namely COVID-HybridNetwork1 (CVD-HNet1) and COVID-HybridNetwork2 (CVD-HNet2) for classification of images of chest X-ray of patients with COVID19. The model achieved accuracy of 98%. They concluded that the model was very effective in early detecting and diagnosing COVID 19.

Brinker, et al. (2019) employed CNN to classify clinical melanoma image and compared it with dermatologists. Dermatologists achieved the mean sensitivity and specificity with clinical images as 89.4% and 64.4% while CNN at the same sensitivity achieved 68.2% mean specificity. Tiulpin, et al. (2018) used CNN for diagnosis severity of knee osteoarthritis. Tiulpin and Saarakkala, (2020) proposed a model based on DL employing transfer learning from ImageNet on Osteoarthritis Initiative dataset and achieved satisfactory results.

Apart from CNN, other NN methods like auto-encoders, DBN, RNN, etc. have also been employed in medical imaging. Hu, et al. (2016) used auto-encoders on MRIs of patients with Alzheimers Disease. Results showed that the architecture proposed was successful in classification of patients with Alzheimers Disease and it was even successful in detecting it in earlier stage. Also, they compared it with support machine vectors and concluded that their architecture achieved 25% improvement in precision accuracy. Cheng, et al., (2016) exploited SDAE (stacked de-noising auto-encoder) on the two CADx (computer-aided diagnosis) to differentiate lung CT nodules and breast ultrasound lesions and compared it with conventional CADx techniques. Results showed that proposed model outperformed. Ortiz, et al. (2016) employed deep belief networks on neuroimaging initiative for early detection of Alzheimers Disease for classification. Precision accuracy of 90% was achieved in this model.

Denoised images always produce more accurate results. Hence, in order to have good results, denoising is suggested (Tripathi and Siddiqi 2017, Tripathi 2020, Tripathi, et al. 2020).

3.3 Electronic Health Record (EHR)

EHRs have been very useful and successful for physicians in proper diagnosis and treatment of patients. In EHR, treatment information related to diagnosis, drug, structured demographic and results related to experimental and unstructured clinical tests are kept in electronic form (Jenson, et al. 2012). Excavating EHR improves the quality and effectiveness of diagnosis and encourages medical development. In recent years, it has been extensively used and shown its effectiveness in healthcare.

Different types of neural networks like CNN, RNN, GRU, LSTM in DL have extensively been employed in EHR via NLP techniques. Using deep learning techniques, EHR can be converted into patient representation. Skip-gram and Attention mechanisms have been employed by several researchers in EHR. Choi, et al. (2016) used the skip-gram method in learning medical concepts and it was very useful in decision making. Applying RNN and attention mechanism, Zhang J., et al., (2018) proposed an architecture for learning patients representation from the temporal EHR data. The model was used to predict the risk task of future hospitalization. They concluded that this deep learning based model achieved a precise prediction effect. An unsupervised deep learning technique was employed by Miotto, et al. (2016) for learning the representation of patient from EHR. A combination of RNN and CNN has been employed by Ma, et al. (2018) for extraction of patient information patterns. Rajkomar, et al. (2018) employed the combination of attention mechanism, decision tree with single-layer and LSTM on a dataset for learning the information.

Another model employed to EHR is neural language deep model. This model is extensively used in learning the embedded representations of concepts related to medical that are applied for prediction and analysis of diseases, medications, etc. (Choi, et al. 2016). Tran, et al. (2015) employed RBMs for learning the abstractions of ICD-10 codes on a cohort of 7578 mental health patients for predicting suicide risk. Demoncourt, et al (2016) proposed a framework based on RNN and this was very successful in removal of protected health information from clinical notes.

After discharging a patient from hospital, how likely is that the patient may need to be readmitted has been a big issue. Nguyen, et al (2017) proposed an architecture called deepr (Deep record) which was based on CNN. From medical records, this architecture learns how to extract the features and automatically predicts the future risk. Deepr detects and associates clinical motifs in EHR for stratification of medical risks. Deepr performed better in comparison to traditional techniques and meaningful clinical motifs were detected.

Deep learning has also been popular for carrying out prediction tasks for diseases. As EHR haves information of patients, some characteristics and indicators in information are useful in predicting if a patient has disease. Cheng Y., et al., (2016) proposed an architecture based on CNN for extracting the phenotypes and the validity of model was verified on virtual and real EHR. Nickerson, et al. (2016) proposed deep learning techniques like LSTM on postoperative pain management and compared it with conventional machine learning techniques and concluded that their method worked better than the conventional methods.

3.4 Genomics

Genomics, a field of biology, deals with the studies of the function, mapping, editing, structure and evolution of genomes. Genome refers to the complete set of DNA of an organism including complete set of DNA, comprising of all of its genes together with its structural configuration. In Genomics, intragenomic phenomena like pleiotropy, epistasis, heterosis and other interactions between alletes and loci within the genome are also studied.

Deep learning has been extensively used in genomics due to the authoritative ability of data processing and feature extraction automatically. DL techniques are able to extract richer information, higher dimensional features and other complex structures. Therefore, DL is very advantageous in genomics to perform the tasks like: RNA measurement, gene expression, gene slicing, etc. DL has capability to recognize the development process and the cause of diseases from the molecular level and the interface

Figure 9. Genomics relevancy chart

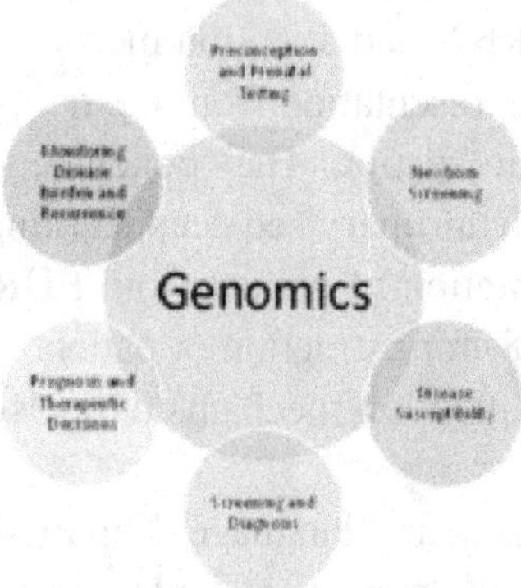

between genes and environment. Applications of DL in Genomics have provided a great tool for physicians for proper and comprehensive diagnosis of diseases in patients. Once, the disease has been diagnosed appropriately, physician can take accurate decisions in treatment of patients. Alharbi and Rashid (2022) have discussed several applications of DL in human genomics. Kanaka, et al. (2023) have envisaged the applications of DL in Genomics.

DL has been used extensively in gene expression. Singh, et al. (2016) developed a model called as DeepChrome, using a deep CNN for classification of gene expression in which histone modification data was used as input. Complex interactions among significant features were extracted automatically in DeepChrome. DeepChrome outperformed SVM and random forests models for classification of gene expression. Sekhon, et al. (2018) developed a DL model called as DeepDiff using LSTM for prediction of differential gene expression using histone modification data. They concluded that DeepDiff outperformed for prediction of differential gene expression. Badsha, et al. (2020) developed two architectures based on autoencoders namely Learning with AuToEncoder (LATE) and TRANSfer learning with LATE (TRANSLATE) for the gene expression levels measurement. Their architecture outperformed the other scRNA-seq imputation techniques. Mostavi, et al. (2020) developed three CNN based models: 2D-Hybrid-CNN, 2D-Vanilla-CNN and 1-D-CNN for detecting cancer types using unstructured gene expression as inputs. Their models achieved outstanding prediction accuracy between 93.9 and 95.0%. Tian, et al. (2021) proposed an autoencoder based on Zero-Inflated Negative Binomial model to impute the discrete scRNA-seq data. This model achieved high imputation accuracy. Also, this model is helpful in increasing the accuracy of differential gene expression. Xu, et al. (2021) developed an imputation method namely AdImpute in which semi-supervised autoencoder was used and comparison of this was done with the existing imputation methods. On the basis of experiments conducted on simulated data and real data, they concluded that their method was more accurate in comparison to existing methods. Zrimec, et al. (2022) designed a deep learning model by using generative adversarial networks and named it as ExpressionGAN for controlling the gene expression. Zhang, et al. (2022) proposed a model for imputation for scRNA-seq count data using autoencoders and named it as NISC. NISC has been very significant imputation approach for dealing sparsity in scRNAseq.

Deep Learning has also been employed in genomic tasks like predicting DNA methylation status, enhancer–promoter interactions, mRNA abundance, RNA-binding proteins. Zhang, et al. (2016) proposed a DL framework for prediction of binding sites of RNA-binding proteins. Angermueller, et al. (2017) proposed a computational technique named DeepCpG based on DNN for prediction of methylation status. Singh et al. (2019) proposed a computational method based on DL and named it as SPEID for prediction of enhancer-promoter interactions. Agarwal and Shendure (2020) proposed a model based on DL and named as Xpresso for prediction of mRNA abundance. Gao, et al. (2022) proposed a deep CNN architecture named as iSEGnet for prediction of mRNA abundance. In this architecture, they used the information on epigenetic modifications and DNA sequences within genes and their cis-regulatory regions. This architecture outstripped other ML models for prediction of mRNA abundance.

4. DISCUSSIONS

Deep learning has emerged as an effective and efficient technique in medical and healthcare. DL has been extremely beneficial in curtailing the time and human resources needed in drug development as well as in acquiring the information which is not easy to have by people. DL has made drug development possible for the diseases that were not treated so far. Variants of DL like CNN, RNN, fully connected NN have been employed in drug development. In medical imaging, for pattern recognition and feature extraction, DL networks like CNN, RNN and autoencoders have been employed. DL offers physicians an automated technology for analysis of videos, pictures, etc. which are useful in diagnosis of diseases and taking the appropriate decisions for treatment of patients. In the field of EHR, RNN and its variants like LSTM and GRU have extensively employed. RNN have shown their applicability in field of EHR as the nature of EHR is sequential and produced more effective results in comparison to customary methods. Researchers have used both supervised and unsupervised learning in digging the information from EHR. The objective of DL methods in EHR is to analyze the patterns and with the help of these learned patterns, several tasks like event prediction, disease prediction, incidence prediction etc., are performed. In field of genomics, DL networks like CNN, RNN, DBN, autoencoders, etc., have extensively been used. 1-D CNN has been employed in proteins and DNA sequencing. Besides, NLP has also been employed in genomics. Thus, DL plays a vital role in extracting the features from genomic data which is beneficial for thoughtful analysis of diseases.

5. CONCLUSION

In this article, various applications of DL in healthcare have been observed. DL and its variants have been very effective in different broad areas like drug development, clinical imaging, electronic health record and genomics in healthcare. DL techniques have extensively employed in early detection of Alzheimer and other brain related diseases and different types of cancers, etc. DL techniques in Clinical imaging have provided a platform for appropriate diagnosis of diseases in patients. The DL techniques in electronic health record provided opportunities to physicians to have medical history of their patients. This has been very helpful for physicians in treatment of the patients.

REFERENCES

Agarwal, V., & Shendure, J. (2020). Predicting mRNA abundance directly from genomic sequence using deep convolutional neural networks. *Cell Reports*, *31*(7), 107663. doi:10.1016/j.celrep.2020.107663 PMID:32433972

Alharbi, W. S., & Rashid, M. (2022). A review of deep learning applications in human genomics using next-generation sequencing data. *Human Genomics*, *16*(1), 26. doi:10.118640246-022-00396-x PMID:35879805

Angermueller, C., Lee, H. J., Reik, W., & Stegle, O. (2017). DeepCpG: Accurate prediction of single-cell DNA methylation states using deep learning. *Genome Biology*, *18*(1), 67–67. doi:10.118613059-017-1189-z PMID:28395661

Badsha, M. B., Li, R., Liu, B., Li, Y. I., Xian, M., Banovich, N. E., & Fu, A. Q. (2020). Imputation of single-cell gene expression with an autoencoder neural network. *Quantitative Biology*, *8*(1), 78–94. doi:10.100740484-019-0192-7 PMID:32274259

Bengio, Y. (2009). Learning Deep Architectures for AI. *Foundations and Trends in Machine Learning*, *2*(1), 1–127. doi:10.1561/2200000006

Brinker, T. J., Hekler, A., Enk, A. H., Klode, J., Hauschild, A., Berking, C., Schilling, B., Haferkamp, S., Schadendorf, D., Fröhling, S., Utikal, J. S., von Kalle, C., Ludwig-Peitsch, W., Sirokay, J., Heinzerling, L., Albrecht, M., Baratella, K., Bischof, L., Chorti, E., ... Schrüfer, P. (2019). A convolutional neural network trained with dermoscopic images performed on par with 145 dermatologists in a clinical melanoma image classification task. *European Journal of Cancer (Oxford, England)*, *111*, 148–154. doi:10.1016/j.ejca.2019.02.005 PMID:30852421

Bungărdean, R. M., Şerbănescu, M. S., Streba, C. T., & Crişan, M. (2021). Deep learning with transfer learning in pathology. Case study: Classification of basal cell carcinoma. *Romanian Journal of Morphology and Embryology*, *62*(4), 1017–1028. doi:10.47162/RJME.62.4.14 PMID:35673821

Cao, Y., Liu, C., Liu, B., Brunette, M. J., Zhang, N., & Sun, T. (2016). Improving tuberculosis diagnostics using deep learning and mobile health technologies among resource-poor and marginalized communities. In *2016 IEEE First International Conference on Connected Health: Applications, Systems and Engineering Technologies (CHASE).* New York, NY: IEEE. 10.1109/CHASE.2016.18

Cheng, J. Z., Ni, D., Chou, Y. H., Qin, J., Tiu, C. M., Chang, Y. C., Huang, C.-S., Shen, D., & Chen, C.-M. (2016). Computer-Aided diagnosis with deep learning architecture: Applications to breast lesions in US images and pulmonary nodules in CT scans. *Scientific Reports*, *6*(1), 24454–24454. doi:10.1038rep24454 PMID:27079888

Cheng, Y., Wang, F., Zhang, P., & Hu, J. (2016). Risk prediction with electronic health records: a deep learning approach. In *Proceedings of the 2016 SIAM International Conference on Data Mining*. SIAM. 10.1137/1.9781611974348.49

Choi, Y., Chiu, C. Y.-I., & Sontag, D. A. (2016). Learning low-dimensional representations of medical concepts. *AMIA Joint Summits on Translational Science Proceedings AMIA Summit on Translational Science*, 41–50. PMID:27570647

Choi, Y., Ms, C. Y.-I. C., & Sontag, D. (2016). Learning low-dimensional representations of medical concepts. *Proceedings of the AMLA Summit on Clinical Research Informatics*.

Cicero, M., Bilbily, A., Colak, E., Dowdell, T., Gray, B., Perampaladas, K., & Barfett, J. (2017). Training and validating a deep convolutional neural network for computer-aided detection and classification of abnormalities on frontal chest radiographs. *Investigative Radiology*, *52*(5), 281–287. doi:10.1097/RLI.0000000000000341 PMID:27922974

Dai, L., Fang, R., Li, H., Hou, X., Sheng, B., Wu, Q., & Jia, W. (2018). Clinical report guided retinal microaneurysm detection with multi-sieving deep learning. *IEEE Transactions on Medical Imaging*, *37*(5), 1149–1161. doi:10.1109/TMI.2018.2794988 PMID:29727278

Demoncourt, F., Lee, J. Y., & Uzuner, O. (2016). De-identification of patient notes with recurrent neural networks. *Journal of the American Medical Informatics Association : JAMIA*, *24*(3), 596–606. doi:10.1093/jamia/ocw156 PMID:26644398

Dhar, K. K., Bhattacharya, P., Kumar, N., & Mitra, A. (2023). Abnormality Detection in Chest Diseases Using a Convolutional Neural Network, Journal of Information and Optimization Sciences. *Taru Publication*, *44*(1), 97–111. doi:10.47974/JIOS-1298

Gao, S., Rehman, J., & Dai, Y. (2022). Assessing comparative importance of DNA sequence and epigenetic modifications on gene expression using a deep convolutional neural network. *Computational and Structural Biotechnology Journal*, *20*, 3814–3823. doi:10.1016/j.csbj.2022.07.014 PMID:35891778

Gao, S., Rehman, J., & Dai, Y. (2022). Assessing comparative importance of DNA sequence and epigenetic modifications on gene expression using a deep convolutional neural network. *Computational and Structural Biotechnology Journal*, *20*, 3814–3823. doi:10.1016/j.csbj.2022.07.014 PMID:35891778

Gulshan, V., Peng, L., Coram, M., Stumpe, M. C., Wu, D., Narayanaswamy, A., Venugopalan, S., Widner, K., Madams, T., Cuadros, J., Kim, R., Raman, R., Nelson, P. C., Mega, J. L., & Webster, D. R. (2016). Development and validation of a deep learning algorithm for detection of diabetic retinopathy in retinal fundus photographs. *Journal of the American Medical Association*, *316*(22), 2402–2410. doi:10.1001/jama.2016.17216 PMID:27898976

Hinton, G. E., Osindero, S., & Teh, Y.-W. (2006). A fast learning algorithm for deep belief nets. *Neural Computation*, *18*(7), 1527–1554. doi:10.1162/neco.2006.18.7.1527 PMID:16764513

Hu, C., Ju, R., Shen, Y., Zhou, P., & Li, Q. (2016). Clinical decision support for Alzheimer's disease based on deep learning and brain network. In *2016 IEEE International Conference on Communications (ICC)*. New York, NY: IEEE. 10.1109/ICC.2016.7510831

Hu, P., Huang, Y., You, Z., Li, S., Chan, K. C. C., & Leung, H. (2019). Learning from deep representations of multiple networks for predicting drug-target interactions. In Lecture Notes in Computer Science: Vol. 11644. *Intelligent Computing Theories, ICIC 2019*. Springer. doi:10.1007/978-3-030-26969-2_14

Kanaka, K. K., Sukhija, N., Sivalingam, J., Goli, R. C., Rathi, P., Jaglan, K., & Raj, C. (2023). Deep Learning in Neural Networks and their Application in Genomics. *Acta Scientific Veterinary Sciences*, *5*(7), 21–26. doi:10.31080/ASVS.2023.05.0683

Kermany, D. S., Goldbaum, M., Cai, W., Valentim, C. C. S., Liang, H., Baxter, S. L., McKeown, A., Yang, G., Wu, X., Yan, F., Dong, J., Prasadha, M. K., Pei, J., Ting, M. Y. L., Zhu, J., Li, C., Hewett, S., Dong, J., Ziyar, I., ... Zhang, K. (2018). Identifying medical diagnoses and treatable diseases by image-based deep learning. *Cell*, *172*(5), 1122–1131. doi:10.1016/j.cell.2018.02.010 PMID:29474911

Kumar, N., Siddiqi, A. H., & Alam, K. (2017). Wavelet Based EEG Signal Classification. *Biomedical and Pharmacology Journal, 10*(4), 2061-2069.

Kumar, N., Tripathi, P., & Alam, K. (2016). Non-Negative Factorization Based EEG Signal Classification. *Indian Journal of Industrial and Applied Mathematics, 7*(2), 2012-219.

Liong-Rung, L., Hung-Wen, C., Ming-Yuan, H., Shu-Tien, H., Ming-Feng, T., Chia-Yu, C., & Kuo-Song, C. (2022). Using Artificial Intelligence to Establish Chest X-Ray Image Recognition Model to Assist Crucial Diagnosis in Elder Patients With Dyspnea. *Frontiers in Medicine*, *9*, 893208. doi:10.3389/fmed.2022.893208 PMID:35721050

Ma, T., Xiao, C., & Wang, F. (2018). Health-ATM: A deep architecture for multifaceted patient health record representation and risk prediction. In *Proceedings of the 2018 SIAM International Conference on Data Mining (SDM).* SIAM. 10.1137/1.9781611975321.30

Martinez-Murcia, F. J., Ortiz, A., Gorriz, J. M., Ramirez, J., & Castillo-Barnes, D. (2020). Studying the Manifold Structure of Alzheimer's Disease: A Deep Learning Approach Using Convolutional Autoencoders. *IEEE Journal of Biomedical and Health Informatics, 24*(1), 17–26. doi:10.1109/JBHI.2019.2914970 PMID:31217131

Miotto, R., Li, L., Kidd, B. A., & Dudley, J. T. (2016). Deep patient: An unsupervised representation to predict the future of patients from the electronic health records. *Scientific Reports, 6*(1), 26094–26094. doi:10.1038rep26094 PMID:27185194

Mostavi, M., Chiu, Y.-C., Huang, Y., & Chen, Y. (2020). Convolutional neural network models for cancer type prediction based on gene expression. *BMC Medical Genomics, 13*(S5), 44. doi:10.118612920-020-0677-2 PMID:32241303

Nguyen, P., Tran, T., Wickramasinghe, N., & Venkatesh, S. (2017). Deepr: A Convolutional Net for Medical Records. *IEEE Journal of Biomedical and Health Informatics, 21*(1), 22–30. doi:10.1109/JBHI.2016.2633963 PMID:27913366

Nickerson, P., Tighe, P., Shickel, B., & Rashidi, P. (2016). Deep neural network architectures for forecasting analgesic response. In *2016 38th Annual International Conference of the IEEE Engineering in Medicine and Biology Society (EMBC).* New York, NY: IEEE. 10.1109/EMBC.2016.7591352

Ortiz, A., Munilla, J., Górriz, J. M., & Ramírez, J. (2016). Ensembles of deep learning architectures for the early diagnosis of the Alzheimer's disease. *International Journal of Neural Systems, 26*(7), 1650025. doi:10.1142/S0129065716500258 PMID:27478060

Pan, S. J., & Yang, Q. (2010). A survey on transfer learning. *IEEE Transactions on Knowledge and Data Engineering, 22*(10), 1345–1359. doi:10.1109/TKDE.2009.191

Poplin, R., Varadarajan, A. V., Blumer, K., Liu, Y., McConnell, M. V., Corrado, G. S., Peng, L., & Webster, D. R. (2018). Prediction of cardiovascular risk factors from retinal fundus photographs via deep learning. *Nature Biomedical Engineering, 2*(3), 158–164. doi:10.103841551-018-0195-0 PMID:31015713

Rajkomar, A., Oren, E., Chen, K., Dai, A. M., Hajaj, N., & Hardt, M. (2018). Scalable and accurate deep learning with electronic health records. *NPJ Digital Medicine, 1*, 18. PMID:31304302

Schmidhuber, J. (2015). Deep Learning in Neural Networks: An Overview. *Neural Networks, 61*, 85–117. doi:10.1016/j.neunet.2014.09.003 PMID:25462637

Sekhon, A., Singh, R., & Qi, Y. (2018). DeepDiff: DEEP-learning for predicting DIFFerential gene expression from histone modifications. *Bioinformatics (Oxford, England), 34*(17), i891–i900. doi:10.1093/bioinformatics/bty612 PMID:30423076

Shin, H.-C., Roth, H. R., Gao, M., Lu, L., Xu, Z., Nogues, I., Yao, J., Mollura, D., & Summers, R. M. (2016). Deep convolutional neural networks for computer-aided detection: Cnn architectures, dataset characteristics and transfer learning. *IEEE Transactions on Medical Imaging*, *35*(5), 1285–1298. doi:10.1109/TMI.2016.2528162 PMID:26886976

Singh, R., Lanchantin, J., Robins, G., & Qi, Y. (2016). Deep chrome: Deep-learning for predicting gene expression from histone modifications. *Bioinformatics (Oxford, England)*, *32*(17), 639–648. doi:10.1093/bioinformatics/btw427 PMID:27587684

Singh, S., Yang, Y., Póczos, B., & Ma, J. (2019). Predicting enhancer-promoter interaction from genomic sequence with deep neural networks. *Quantitative Biology*, *7*(2), 122–137. doi:10.100740484-019-0154-0 PMID:34113473

Suganyadevi, S., & Seethalakshmi, V. (2022). CVD-HNet: Classifying Pneumonia and COVID-19 in Chest X-ray Images Using Deep Network. *Wireless Personal Communications*, *19*(4), 1–25. doi:10.100711277-022-09864-y PMID:35756172

Tang, Y. (2023). Deep learning in drug discovery: Applications and limitations. *Frontiers in Computing and Intelligent Systems*, *3*(2), 118–123. doi:10.54097/fcis.v3i2.7575

Tian, T., Min, M. R., & Wei, Z. (2021). Model-based autoencoders for imputing discrete single-cell RNA-seq data. *Methods (San Diego, Calif.)*, *192*, 112–119. doi:10.1016/j.ymeth.2020.09.010 PMID:32971193

Tiulpin, A., & Saarakkala, S. (2020). Automatic Grading of Individual Knee Osteoarthritis Features in Plain Radiographs Using Deep Convolutional Neural Networks. *Diagnostics (Basel)*, *10*(11), 932. doi:10.3390/diagnostics10110932 PMID:33182830

Tiulpin, A., Thevenot, J., Rahtu, E., Lehenkari, P., & Saarakkala, S. (2018). Automatic knee osteoarthritis diagnosis from plain radiographs: A deep learning-based approach. *Scientific Reports*, *8*(1), 1727. doi:10.103841598-018-20132-7 PMID:29379060

Tran, T., Nguen, T. D., & Phung, D. (2015). Learning vector representation of medical objects via EMR-driven nonnegative restricted Boltzmann machines (eNRBM). *Journal of Biomedical Informatics*, *54*, 96–105. doi:10.1016/j.jbi.2015.01.012 PMID:25661261

Tripathi, P. (2020). Electroencephalpgram Signal Quality Enhancement by Total Variation Denoising Using Non-convex Regulariser. *International Journal of Biomedical Engineering and Technology*, *33*(2), 134–145. doi:10.1504/IJBET.2020.107709

Tripathi, P., Kumar, N., Rai, M., & Khan, A. (2022). Applications of Deep Learning in Agriculture. In *Artificial Intelligence Applications in Agriculture and Food Quality Improvement* (pp. 17–28). IGI Global. doi:10.4018/978-1-6684-5141-0.ch002

Tripathi, P., Kumar, N., & Siddiqi, A. H. (2020). De-noising Raman spectra using total variation de-noising with iterative clipping algorithm. In *Computational Science and its Applications, Taylor and Francis Group* (pp. 225–231). CRC Press. doi:10.1201/9780429288739-14

Tripathi, P., & Siddiqi, A. H. (2017). De-noising EEG signal using iterative clipping algorithm. *Biosciences Biotechnology Research Asia*, *14*(1), 497–502. doi:10.13005/bbra/2470

Tyagi, N., Rai, M., Sahw, P., Tripathi, P., & Kumar, N. (2022). *Methods for the Recognition of Human Emotions Based on Physiological Response: Facial Expressions in Smart Healthcare for Sustainable Urban Development*. IGI Global. doi:10.4018/978-1-6684-2508-4.ch013

Xu, L., Xu, Y., Xue, T., Zhang, X., & Li, J. (2021). AdImpute: An Imputation Method for Single-Cell RNA-Seq Data Based on Semi-Supervised Autoencoders. *Frontiers in Genetics*, *12*, 739677. doi:10.3389/fgene.2021.739677 PMID:34567089

Yang, S., Zhu, F., Ling, X., Liu, Q., & Zhao, P. (2021). Intelligent Health Care: Applications of Deep Learning in Computational Medicine. *Frontiers in Genetics*, *12*, 607471. doi:10.3389/fgene.2021.607471 PMID:33912213

Yuan, D., Zhu, X., Wei, M., & Ma, J. (2019). Collaborative deep learning for medical image analysis with differential privacy. In *2019 IEEE Global Communications Conference (GLOBECOM)*. New York, NY: IEEE. 10.1109/GLOBECOM38437.2019.9014259

Zhang, J., Kowsari, K., Harrison, J. H., Lobo, J. M., & Barnes, L. E. (2018). Patient2Vec: A personalized interpretable deep representation of the longitudinal electronic health record. *IEEE Access : Practical Innovations, Open Solutions*, *6*, 65333–65346. doi:10.1109/ACCESS.2018.2875677

Zhang, S., Zhou, J., Hu, H., Gong, H., Chen, L., Cheng, C., & Zeng, J. (2016). A deep learning framework for modeling structural features of RNA-binding protein targets. *Nucleic Acids Research*, *44*(4), e32. doi:10.1093/nar/gkv1025 PMID:26467480

Zhang, X., Chen, Z., Bhadani, R., Cao, S., Lu, M., Lytal, N., Chen, Y., & An, L. (2022). NISC: Neural Network-Imputation for Single-Cell RNA Sequencing and Cell Type Clustering. *Frontiers in Genetics*, *13*, 847112. doi:10.3389/fgene.2022.847112 PMID:35591853

Zhavoronkov, A., Ivanenkov, Y. A., Aliper, A., Veselov, M. S., Aladinskiy, V. A., Aladinskaya, A. V., Terentiev, V. A., Polykovskiy, D. A., Kuznetsov, M. D., Asadulaev, A., Volkov, Y., Zholus, A., Shayakhmetov, R. R., Zhebrak, A., Minaeva, L. I., Zagribelnyy, B. A., Lee, L. H., Soll, R., Madge, D., ... Aspuru-Guzik, A. (2019). Deep learning enables rapid identification of potent DDR1 kinase inhibitors. *Nature Biotechnology*, *37*(9), 1038–1040. doi:10.103841587-019-0224-x PMID:31477924

Zrimec, J., Fu, X., Muhammad, A. S., Skrekas, C., Jauniskis, V., Speicher, N. K., Börlin, C. S., Verendel, V., Chehreghani, M. H., Dubhashi, D., Siewers, V., David, F., Nielsen, J., & Zelezniak, A. (2022). Controlling gene expression with deep generative design of regulatory DNA. *Nature Communications*, *13*(1), 5099. doi:10.103841467-022-32818-8 PMID:36042233

Chapter 6
Artificial Intelligence and Machine Learning in Smart Education

Sheelesh Kumar Sharma
ABES Institute of Technology, Ghaziabad, India

Ram Jee Dixit
ABES Institute of Technology, Ghaziabad, India

Deepshikha Rai
ABES Institute of Technology, Ghaziabad, India

Shalu Mall
ABES Institute of Technology, Ghaziabad, India

ABSTRACT

Artificial intelligence and machine learning have the potential to transform education by making it more effective, personalized, and engaging. In the context of smarter education, these technologies can be used in many ways to enhance learning for students and teachers: Intelligent Content Creation: AI can create learning content such as quizzes, exercises, and interactive experiments based on the specific needs of the course; Education Analysis: AI can analyze data from a variety of sources, such as student interactions with lessons, attendance records, and assessment results to understand learning patterns and trends. Teachers can use this information to identify struggling students, adjust instruction, and create interventions to improve overall learning; Automatic Grading and Feedback: Machine learning algorithms can grade assignments and tests, providing instant feedback to students. These technologies allow students to explore complex concepts in a practical and engaging way, making learning more effective and memorable.

DOI: 10.4018/979-8-3693-0782-3.ch006

1. INTRODUCTION

By improving its efficiency, individualization, and engagement, artificial intelligence and machine learning have the potential to revolutionize education. These technologies can be applied in a variety of ways to improve learning for both students and teachers in the context of smarter education (Baker et al., 2019). Figure 1 depicts that there are some highlights of how AI and machine learning are changing education for the smarter.

- **Self-Learning**: AI can examine a lot of data regarding students' learning styles, aptitudes, and deficiencies. With this knowledge, the system can provide personalized content, activities, and challenges tailored to each student's needs. This individualized method encourages pupils to learn more by assisting them in independently understanding the material.
- **Intelligent Teaching Tools**: These tools can operate as virtual assistants for students, giving them guidance and help in the present. Based on student achievement, these systems can respond to inquiries, offer explanations, and recommend additional learning. They can also keep an eye on students' progress and modify their training as necessary (Popenici et al., 2017).
- **Intelligent Content Creation**: Depending on the requirements of the course, AI can develop educational materials including quizzes, exercises, and interactive experiments. In addition to saving teachers time, doing this makes ensuring that the content is current with new information and is in line with the learning objectives.
- **Education Analysis**: In order to comprehend learning patterns and trends, AI can analyze data from a range of sources, including student interactions with lectures, attendance records, and assessment outcomes. Teachers can enhance students' overall learning by identifying problematic pupils, modifying their lessons, and developing solutions.
- **Automatic Grading and Feedback**: Algorithms for machine learning can perform exams and grade tasks, giving pupils immediate feedback. This lightens the load on teachers and enables them to concentrate on instructing and providing one-to-one help (Baker et al., 2011).
- **Intelligent Classroom Management**: Teachers can effectively supervise classes with the aid of AI systems. AI-powered cameras, for instance, may track student attentiveness, search for diversions, and warn teachers when they appear distracted. By doing this, the ideal learning environment is maintained.
- **Virtual and Augmented Reality**: By building interactive simulations and experiences, artificial intelligence (AI) can improve learning in virtual reality (VR) and augmented reality (AR). These tools let students explore difficult ideas in a useful and interesting way, which improves learning and retention (Luckin et al., 2016).
- **Predictive Analytics**: AI has the ability to forecast student performance and identify the likelihood of failure by examining historical data and patterns (Sharma et al., 2019). By giving extra support to students who are at risk, this enables instructors to get involved early and boost their chances of success.

While machine learning and artificial intelligence hold enormous promise for smart learning, it's critical to address personal and ethical issues to ensure the responsible and open handling of student data (Sharma et al., 2019). A successful learning environment must also strike a balance between automation and human-computer interaction.

Figure 1. Depicts some highlights of how AI and machine learning are changing education for the smarter

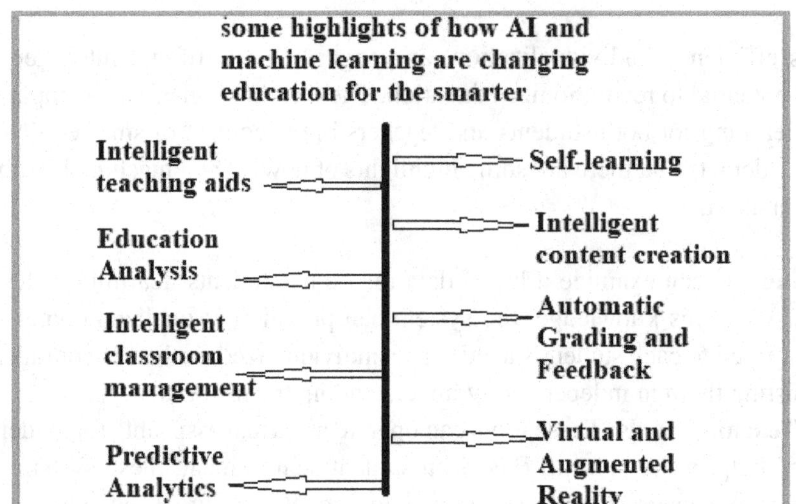

2. SELF-LEARNING

Artificial Intelligence (AI) and Machine Learning (ML) play a crucial role in self-learning systems, which are designed to improve their performance and knowledge without explicit programming or human intervention. Self-learning AI systems have the ability to analyze data, identify patterns, make predictions, and adjust their algorithms based on the data they receive. Here's how AI and machine learning contribute to self-learning:

- **Data Analysis**: Self-learning systems rely on a tremendous amount of data to understand their surroundings and make wise decisions. To glean useful insights from this data, artificial intelligence (AI) tools including natural language processing (NLP), computer vision, and data analytics are applied (Sharma et al., 2019).
- **Machine Learning Algorithms**: Self-learning systems are built on machine learning. These systems process data using a variety of machine learning (ML) techniques, including reinforcement learning, unsupervised learning, and supervised learning.
- **Adaptation and Improvement**: Self-learning By continuously learning from fresh data, AI can adjust to shifting circumstances and enhance its performance. In dynamic contexts where new information is continuously surfacing, this adaptable flexibility is especially helpful.
- **Pattern Recognition**: Algorithms for machine learning are quite good at spotting patterns and trends in data. Self-learning systems make use of this ability to identify abnormalities, categorize things, and forecast outcomes based on past data (Sharma et al., 2019).
- **Reinforcement Learning**: Reinforcement learning is the process by which an AI agent learns to accomplish a task in a complex, uncertain environment by getting feedback in the form of rewards or penalties. This strategy is frequently employed in self-learning systems to enhance decision-making.

- **Deep Learning**: A branch of machine learning called deep learning involves building artificial neural networks with several layers. This method works especially well for self-learning systems' key functions of sequential data analysis, picture recognition, and language comprehension.
- **Neural Networks and Artificial Neural Systems**: Neural networks take their cues from the structure and operation of the human brain. They make it possible for self-learning systems to interpret data and modify their behavior in response to input.
- **Autonomous Systems**: Self-learning In order to create autonomous systems like robots, drones, and self-driving automobiles, AI and machine learning are essential. Without direct human input, these systems employ data and learning algorithms to navigate and make decisions.
- **Personalization**: By learning from user behavior and preferences, self-learning systems can tailor their interactions with users. virtual assistants, recommendation engines, and personalized content distribution all make use of this capacity.
- **Knowledge Base Expansion**: AI-powered self-learning systems can continuously expand their knowledge base by acquiring information from various sources, such as the internet, books, or databases.

Overall, self-learning systems can grow more intelligent, flexible, and capable of carrying out complicated tasks without continual human supervision thanks to the combination of AI and machine learning. These developments have wide-ranging effects on a variety of sectors, including healthcare, finance, education, and transportation. To guarantee the proper deployment of such technologies, it is imperative to carefully evaluate ethical and safety issues (Sharma et al., 2019).

3. AI TEACHING AIDS

AI teaching aids refer to educational tools and technologies that leverage artificial intelligence to enhance the learning experience for students and teachers. These aids can complement traditional teaching methods by providing personalized, interactive, and adaptive learning experiences. Here Figure 2 depicts that there are some common AI teaching aids.

3.1 Intelligent Tutoring Systems (ITS)

Intelligent Tutorial System(ITS) is an educational program that uses artificial intelligence (AI) and machine learning techniques to provide students with personalized and adaptive learning. This method focuses on taking on the role of a human teacher by understanding the individual needs, strengths and weaknesses of the student and adapting the learning process to suit. ITS can be used in many educational settings, including K-12 schools, universities, and workplaces.

Key Features of Intelligent Tutoring Systems:

- Personalization: ITS builds a profile for each student that represents their knowledge, learning preferences, and development using information acquired from their interactions with the system. The system may adjust the pace and substance of training based on this data to match the unique needs of each learner.

Figure 2. Depicts some common AI teaching aids

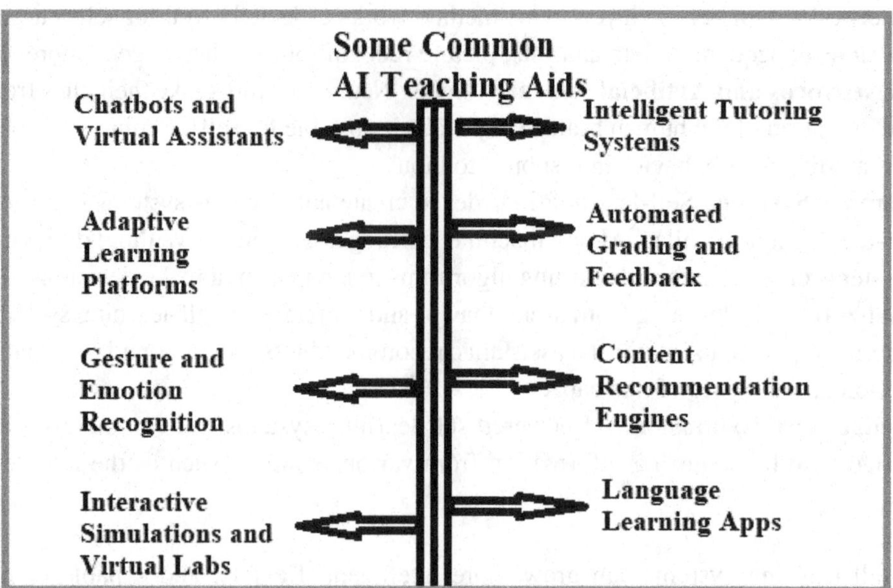

- The feedback is frequently crafted to dispel myths, reaffirm accurate comprehension, and promote critical thought.
- Knowledge Tracing: To monitor a student's progress through various learning ideas and topics, ITS makes use of advanced algorithms. The system can determine the student's knowledge, their regions of difficulty, and the areas that require reinforcing thanks to this knowledge tracing.
- Domain Expertise: ITS includes a wide range of topics, including arithmetic, science, programming, and language development. They are made to have competence in the subject topic they teach, enabling them to deliver precise and thorough education.
- Interactive Learning: By including students in interactive exercises, simulations, and multimedia materials, intelligent tutoring systems improve learning efficiency and enjoyment.
- Real-time Monitoring: As a result of the system's ongoing monitoring of the student's performance, it is possible to spot trends, patterns, and areas that could want improvement.

3.2 Chatbots and Virtual Assistants

Smart learning environments are rapidly incorporating chatbots and virtual assistants to improve the educational process, automate administrative duties, and offer personalized help to students and teachers. These AI-powered technologies have a number of advantages and help traditional education become a more effective and engaging model.

Key Features

- Personalized Learning: Chatbots and virtual assistants can examine student data, such as learning progress, preferences, and performance, to provide recommendations and information that are specifically tailored to each individual student. These tools can enhance learner engagement and information retention by personalizing the educational experience to meet each learner's needs.

- Instant Support: Through chatbots and virtual assistants, teachers and students can get immediate assistance. These AI-based systems are available 24/7, cutting down on reaction times and promoting continuous learning. They can clear up doubts, respond to inquiries, or provide instant information.
- Administrative Efficiency: Chatbots can help with administrative duties including class scheduling, enrollment management, and grading in intelligent educational environments. By automating these procedures, educators gain more time to devote to instruction and mentoring.
- Tutoring and Mentoring: Virtual assistants can serve as personalized tutors or mentors, helping students to understand complex ideas, providing explanations, and providing access to more resources for learning. Students who require additional support outside of the typical classroom context may find this to be very helpful.
- Language Learning: By engaging students in conversation, offering vocabulary drills, and assisting with pronunciation, chatbots can support language acquisition. This interactive method improves fluency and language learning.
- Student Progress Monitoring: Virtual assistants and chatbots can monitor student development and produce reports for teachers and parents. These insights enable early interventions for problematic pupils and help identify areas for improvement.
- Gamification and Interactive Learning: Through the integration of chatbots and virtual assistants into gamified learning platforms, education can become more dynamic and engaging. To boost motivation and engagement, they could take on characters in educational video games.
- Multilingual Support: Multilingual virtual assistants can aid students who are not native English speakers by providing explanations and instructions in their preferred language.
- Campus Navigation: Chatbots can help visitors and students in a university or other major educational institution find their way about the campus and find classrooms, libraries, and other amenities.
- Student Well-being: Virtual assistants can keep an eye on and assess students' wellbeing by identifying stress or emotional discomfort tendencies. They can then offer pertinent resources or notify the proper authorities for more assistance.

The field of smart education can benefit greatly from chatbots and virtual assistants since they give students and teachers individualized attention, quick access to information, and effective administrative support. The potential for chatbots and virtual assistants to improve education will only increase. As AI technology develops, making learning more accessible, interesting, and efficient. To establish a complete and encouraging learning environment, a delicate balance between AI-driven automation and the human touch should be maintained.

3.3 Automated Grading and Feedback

An important use of artificial intelligence in smart education is in automated grading and feedback systems. These systems evaluate student assignments, tests, and exams and offer feedback using machine learning algorithms and natural language processing. Automated grading and feedback implementation comes with a number of benefits, but it also presents some difficulties that must be resolved for effective integration into the educational system.

Benefits of AI-based Automated Grading and Feedback:

- Efficiency: Manual grading assignments and tests takes a lot of time and effort, which automated grading greatly reduces. This enables educators to concentrate more on instructing and giving pupils customized support.
- Immediate Feedback: Students receive feedback on their work right away, allowing them to quickly recognize their strengths and faults. This timely criticism encourages lifelong learning and aids students in developing the necessary skills.
- Uniformity and Objectivity: AI-based systems use uniform grading standards for all entries, removing any potential biases or inconsistent grading that could occur with human grading.
- Scalability: Because automated grading is simple to scale, it is useful in settings like big courses or online learning environments where manual grading would be laborious and difficult to manage.
- Data Insights: The technology generates data insights that assist teachers in locating frequent errors or trouble spots for kids. Decisions about instruction can be made using this information, and tailored interventions can be made.
- Tailored Learning: AI-generated feedback may be customized to take into account each student's unique needs and learning preferences, fostering a more individualized learning environment.
- Lessened Workload: Teachers can better balance their workload by automating grading and feedback, which will boost job satisfaction and overall teaching effectiveness.

Challenges of AI-based Automated Grading and Feedback:

- Subject Complexity: It can be difficult for AI systems to grade and provide feedback for some subjects, especially those that call for creative thinking or subjective judgment.
- Assessment Type Restrictions: AI-based grading systems perform better when multiple-choice questions or structured answers are used. Human evaluation may still be necessary for large projects or free-form responses.
- Language Understanding: Despite advancements in language processing technologies, it may still be difficult to grasp subtleties and context in students' comments, which could result in inaccurate grading.
- Ethical Considerations: Since high-stakes assessments require clear and transparent decision-making, educators and institutions must carefully address the ethical implications of AI-based grading.
- Initial Implementation and Training: The initial setup and training of AI-based grading systems is necessary. This includes supplying a sizable dataset for the AI model to learn from, which can take some time.
- Student Engagement: Relying only on automated feedback may limit possibilities for teachers and students to communicate directly, which could have an impact on motivation and engagement among students.

The evaluation and feedback procedures in smart education could be revolutionized by AI-based automated grading and feedback systems. They can greatly increase productivity, offer quick feedback, and offer insightful data that can help teachers improve their lesson plans. Even though these systems have many advantages, they should be used carefully, taking into account the subject matter, moral issues, and the necessity for individualized human interaction during the learning process. The key to maximizing the promise of AI in smart education will be finding the correct mix between human interaction and AI automation.

3.4 Adaptive Learning Platforms

Adaptive learning platforms are educational technologies that customize each student's learning experience via the use of artificial intelligence (AI) and machine learning algorithms. In order to deliver personalized material and learning pathways, these systems analyze data on students' performance, behaviors, and preferences. By adapting instruction to meet the individual needs of each learner, adaptive learning seeks to maximize learning results.

Key Features of Adaptive Learning Platforms:

- Personalization is a key component of adaptive learning platforms, which identify each student's strengths and weaknesses and design personalized learning routes that take into account their unique skills and learning preferences.
- Continuous Assessment: To adapt and modify the subject matter and degree of difficulty in real-time, these platforms continuously evaluate students' development, learning patterns, and responses to questions.
- Tailored Interventions: Adaptive learning recognizes students' areas of weakness and offers tailored interventions, extra resources, and remedial material to close those gaps in learning.
- Mastery-Based Learning: This approach to learning requires students to show that they have a firm grasp of fundamental ideas before moving on to more complicated subjects.
- Data Analytics: To provide educators and administrators with information about student outcomes and instructional efficacy, these platforms gather and analyze a large quantity of data on student performance, interactions, and development.
- Interactivity and Engagement: To increase student motivation and engagement, adaptive learning platforms frequently include interactive components like gamification and simulations.
- Flexibility and Accessibility: These platforms may frequently be accessed from a variety of devices, enabling students to learn at their own speed and location and encouraging self-directed learning.
- Automated Feedback: Adaptive learning solutions offer automated feedback on student evaluations, assignments, and progress, allowing for quick feedback and shorter learning iterations.

By delivering individualized and data-driven learning experiences, adaptive learning platforms have the potential to completely transform education. These systems may successfully meet each student's specific demands by utilizing AI and machine learning, which encourages engagement, retention, and academic performance.

3.5 Content Recommendation Engines

AI-driven systems known as content recommendation engines analyze user data, behavior, and preferences to recommend pertinent material. To tailor content offers for consumers, these engines are frequently employed in a variety of online platforms, such as streaming services, e-commerce websites, social media, news portals, and educational platforms. By offering individualized and pertinent ideas, content recommendation engines hope to boost user engagement, increase user experience, and increase content consumption.

Key Components and Working of Content Recommendation Engines:

- Data Gathering: Users' interactions, browser histories, search terms, demographic data, and other pertinent data are all collected by the recommendation engine.
- User Profiling: The information gathered is utilized to develop user profiles, which contain information on preferences, interests, previous actions, and any explicit comments made by users.
- Information Analysis: The recommendation engine uses a variety of techniques, such as natural language processing, picture recognition, or collaborative filtering, to analyze the information that is available on the platform, such as articles, videos, products, or services.
- Matching User Profiles with Relevant material: Various recommendation algorithms are used to connect user profiles with pertinent material. Collaborative filtering, content-based filtering, and hybrid algorithms that integrate various strategies are examples of common algorithms.
- Real-time Processing: To create individualized recommendations for each user, the recommendation engine processes user data and content information in real-time.
- Feedback Loop: The engine incorporates user feedback to continuously improve its recommendations. Future recommendations are honed using user interactions with the suggested material, such as clicks, likes, and purchases.

Types of Content Recommendation Engines:

- Collaborative filtering: This strategy suggests content based on the tastes of users who share similar preferences. It recognizes users who behave similarly and proposes goods that people who share your interests have loved or consumed.
- Content-Based Filtering: The focus of content-based recommendations is on the properties of the content itself. On the basis of shared characteristics, keywords, or categories, it proposes products that are comparable to those the user has already interacted with.
- Hybrid Recommendation: To capitalize on the advantages of both methodologies and offer more precise and varied recommendations, hybrid recommendation engines combine collaborative filtering and content-based filtering.
- Context-Aware Recommendation: Context-aware recommendation engines increase accuracy of recommendation by taking into account more contextual aspects, such as time, location, device, and user behavior in various scenarios.

The user experiences on different online platforms are greatly improved by content recommendation systems. These engines provide relevant and personalized content suggestions by utilizing AI and data analytics, which increases user engagement and content consumption. To retain user confidence and assure the ethical usage of recommendation systems, it is crucial to handle privacy issues, biases, and ethical considerations. and user behavior in various circumstances, to enhance suggestion accuracy.

3.6 Gesture and Emotion Recognition

Two areas of research and development in the fields of artificial intelligence and computer vision are gesture and emotion recognition (Raju et al., 2015). These technologies attempt to make it possible for robots to comprehend and interpret human gestures and emotions, opening up a wide range of useful applications in fields including robotics, healthcare, and more.

- **Gesture Recognition**: Gesture recognition refers to a computer system's comprehension and interpretation of human gestures. Usually, cameras and sensors are used to record and examine hand movements, body postures, and facial expressions.
- **Emotion Recognition**: The detection and interpretation of human emotions by AI systems via facial expressions, voice cues, physiological signs, or other behavioural patterns is known as emotion recognition.

Technologies that recognize gestures and emotions have the potential to improve human-computer interaction and make it possible for more intuitive and natural interactions between people and machines. We can anticipate further use cases for these technologies as they develop in a variety of sectors, from gaming and entertainment to healthcare and mental health support.

3.7 Language Learning Apps

Recent years have seen a rise in the popularity of language learning applications, which provide simple and interactive ways for users to learn new languages. These apps make language learning fun, available, and efficient by combining technology, gamification, and personalized learning methods.

Key Features of Language Learning Apps:

- Gamified Learning: Language learning apps often incorporate gamification elements, such as quizzes, challenges, and rewards, to keep users motivated and engaged throughout their language learning journey.
- Personalization: These apps use AI algorithms to create personalized learning paths based on users' proficiency levels, learning preferences, and progress. This ensures that learners receive content that suits their specific needs.
- Interactive Content: Language learning apps provide a variety of interactive content, including audio exercises, videos, flashcards, and speaking exercises, to encourage active participation and improve language skills.
- Real-time Feedback: Users receive instant feedback on their pronunciation, grammar, and vocabulary, allowing them to correct mistakes and improve their language proficiency.
- Multi-platform Access: Language learning apps are accessible across various devices, including smartphones, tablets, and computers, allowing users to learn anytime and anywhere.
- Social Features: Some apps incorporate social elements, such as language exchange forums or peer-to-peer language practice, to facilitate communication and collaboration between learners.

Language learning apps have revolutionized the way people approach learning new languages. With their gamified and interactive approach, personalized learning paths, and accessibility on multiple platforms, these apps offer an engaging and convenient way for learners to acquire language skills. Whether someone is a beginner or looking to advance their proficiency, language learning apps can be a valuable tool in achieving their language learning goals.

3.8 Interactive Simulations and Virtual Labs

Interactive simulations and virtual labs are powerful educational tools that use technology to create virtual environments where students can actively engage in hands-on learning experiences. These tools provide an immersive and interactive learning environment that allows students to explore, experiment, and learn concepts in a safe and controlled setting. Interactive simulations and virtual labs are used in various disciplines, including science, engineering, mathematics, and healthcare, to enhance learning outcomes and improve students' understanding of complex concepts.

Key Features of Interactive Simulations and Virtual Labs:

- Practical Application: Students can interact with virtual items, conduct experiments, and evaluate results in a virtual setting, giving them a hands-on knowledge of theoretical concepts.
- Instantaneous Feedback: Interactive simulations and virtual laboratories frequently give students rapid feedback on their actions and performance, assisting them in correcting their errors and moving forward.
- Flexibility: Virtual labs give students the freedom to carry out experiments and simulations at their own pace, encouraging independent research and experimentation.
- Accessible Anywhere: Virtual laboratories are accessible from any computer or mobile device with an internet connection, allowing for 24/7 accessibility and remote learning.
- Economical: Since virtual labs don't require any real tools or supplies, they are a more cost-effective choice for educational institutions.
- Safe Environment: Unlike real-world labs, which may contain risks, simulations offer students a risk-free setting in which to explore and learn.
- Visualization of Complicated Concepts: Simulations can use visualizations and animations to depict complicated and abstract concepts, making them simpler to comprehend.

Virtual labs and interactive simulations have revolutionized how students' study and interact with challenging ideas. Through immersive and secure environments for experimentation and exploration, these instructional tools promote greater comprehension and active learning. Interactive simulations and virtual labs are anticipated to become more important in education as technology develops, giving students chances to explore the worlds of science, engineering, and other subjects in fresh and meaningful ways.

Incorporating AI teaching aids into educational settings has the potential to revolutionize traditional classrooms, making learning more accessible, engaging, and effective for students of all ages. However, it is crucial to strike a balance between technology and human interactions, ensuring that AI tools complement and support the role of teachers rather than replacing them entirely. Additionally, ethical considerations, data privacy, and security must be taken into account while deploying AI in educational contexts.

4. INTELLIGENT CONTENT CREATION

AI has the ability to produce educational content, which has the potential to transform the industry. Natural language processing (NLP), machine learning, and other AI techniques are used by systems powered by AI to create instructional content, such as text-based articles, tests, exercises, and even interactive multimedia.

There are some ways AI can create learning content:

- **Automated Text Generation**: Coherent and contextually appropriate writing may be produced by AI models like OpenAI's GPT (Generative Pre-trained Transformer) and related language models. On a variety of subjects, they may be used to provide instructional articles, summaries, explanations, and tutorials.
- **Quiz and Assessment Generation**: Quizzes and assessments may be automatically created by AI and customized to meet certain learning objectives. These tests may be adaptive, changing their degree of difficulty in response to how well students do and how much they are learning.
- **Multimedia Content Creation**: AI can be employed to generate multimedia content, such as educational videos, animations, and interactive simulations. For instance, AI can convert text content into visual presentations or create personalized video lessons.
- **Language Learning Support**: AI may be used to create interactive simulations, instructive movies, and other types of multimedia content. For instance, AI may provide individualized video lectures or transform written information into visual presentations.
- **Personalized Learning Paths**: In order to suggest personalized learning routes, AI may examine students' learning preferences and tendencies. These routes may consist of newly created content or a well chosen assortment of educational resources.
- **Content Summarization**: Long instructional materials may be condensed using AI, which makes it simpler for students to efficiently understand important ideas and facts.
- **Automated Translation**: By automatically translating instructional materials into many languages, AI can increase the accessibility of learning resources for a wide range of consumers.
- **Coding and Programming Practice**: AI may produce code problems and workouts, allowing pupils to practice and improve their coding abilities.

It's crucial to remember that while AI-generated content has a lot of potentials, there are certain difficulties as well. It is vital to ensure the accuracy and quality of information produced by AI since it might affect the learning process. AI-generated content can also lack the originality and personalization that expert educators bring to their carefully created teaching materials. To guarantee that the learning materials fulfill educational requirements, are contextually relevant, and are interesting for the learners, a combination of AI-generated content and human curation is frequently advised. Additionally, educators are essential in utilizing AI-generated information well, adapting it to particular learning requirements, and offering students individualized support throughout their learning process.

5. EDUCATION ANALYSIS

AI-powered education analysis can analyze data from a wide range of sources to acquire insightful knowledge about learning trends and patterns (Sharma et al., 2020). AI may give educators and administrators useful information to enhance teaching methods, student support, and overall educational outcomes by using data from student interactions with lessons, attendance records, assessment results, and other educational activities. How AI-driven education analysis functions, process is as follows;

1. **Data Collection**: Learning management systems (LMS), student information systems, internet platforms, and classroom sensors are a few of the sources of data that AI systems acquire. This information may include how students engage with digital course materials, their involvement in class discussions, how they complete assignments, and their attendance statistics.
2. **Data Preprocessing**: The data goes through preparation, where it is cleaned, converted, and organized into a format that is appropriate for AI algorithms, before analysis.
3. **Machine Learning Algorithms**: To analyze the preprocessed data, AI uses machine learning algorithms. Unsupervised learning algorithms may find hidden patterns and categorize students based on their learning behaviors, whereas supervised learning algorithms can forecast student performance based on prior data.
4. **Learning Analytics**: AI-driven learning analytics provide information on the participation, progress, and learning preferences of students. It can spot patterns in students' performance and point out areas that require development.
5. **Personalized Learning**: AI may create personalized learning pathways and offer relevant information for particular students by analyzing data on students' interactions with classes and learning materials.
6. **Early Warning Systems**: AI can spot early indicators of pupils who may be in danger of falling behind or having academic difficulties. This makes it possible for teachers to step in right away and provide specific help.
7. **Course and Curriculum Optimization**: Education analysis may offer insightful input on the success of curricula and courses. It aids teachers in determining areas that require improvement and improving educational techniques.
8. **Feedback and Improvement**: AI may provide comments on exam performance and pinpoint areas where students may need further practice or explanation by analyzing assessment outcomes.
9. **Pedagogical Research**: By delivering data-driven insights into the efficacy of various teaching styles and strategies, AI-powered education analysis may enhance pedagogical research.
10. **Continuous Improvement**: Analysis of education data led by AI is a continuous process. The models may be updated and enhanced to produce more precise and pertinent insights when new data becomes available.

Educators and educational institutions may make data-informed decisions, design more efficient learning experiences, and make sure that every student has the assistance they need to thrive by utilizing AI-driven education analysis (Mann et al., 2022). To guarantee the safety and wellbeing of students and stakeholders, ethical issues related data privacy and appropriate AI usage must constantly be taken into account (Mann et al., 2022).

6. AUTOMATIC GRADING AND FEEDBACK

Machine learning-based automatic grading and feedback is a useful use of AI in education. Without the need for human grading by teachers, machine learning algorithms can analyze and evaluate student responses to assignments, quizzes, and examinations, offering pertinent and unbiased feedback (Baker et al., 2010). This procedure has the following advantages:

1. **Efficiency**: The time and effort needed by teachers to evaluate and grade student work is reduced by automatic grading. It frees up their time so they can concentrate on other facets of education, including giving each student customized help.
2. **Consistency**: Algorithms for machine learning offer uniform and consistent grading for all pupils. By doing this, impartiality is guaranteed and any potential biases brought about by human grading are eliminated.
3. **Real-time Feedback**: Students receive immediate feedback on their performance, allowing them to see problem areas and quickly correct misunderstandings.
4. **Scalability**: Where manual grading would be unfeasible, automatic grading scales readily to meet big class numbers or massive open online courses (MOOCs).

The following processes are commonly included in automated grading and feedback using machine learning:

1. **Data Collection**: As training data, the system requires a collection of graded student replies. The machine learning model is trained using these replies and the accompanying grades.
2. **Feature Extraction**: Relevant characteristics, such as keywords, sentence structure, or semantic representations, are extracted from the replies through preprocessing. The machine learning model uses these features as input.
3. **Model Training**: To discover patterns and correlations between characteristics and grades, machine learning algorithms, such as natural language processing (NLP) methods or neural networks, are trained on the dataset.
4. **Grading and Feedback**: Once the model is trained, it can automatically grade new student responses based on the patterns it has learned. The system can also provide feedback, highlighting areas of improvement or providing explanations for incorrect answers.
5. **Continuous Improvement**: As students turn in their projects and receive feedback, the model may be regularly updated and retrained with fresh data to improve accuracy.

It's important to note that while automatic grading is effective for objective questions (e.g., multiple-choice, fill-in-the-blank), it becomes more challenging for subjective questions (e.g., essays) that require human judgment and creativity. However, recent advances in natural language understanding and advanced AI models, such as transformer-based architectures, have improved the capability of machine learning systems to handle more complex and subjective tasks.

Despite the benefits, it is crucial to keep the ratio of computerized grading to human grading under check. In order to evaluate students' critical thinking, originality, and other qualitative components of their work for some assignments and examinations, human review may still be necessary. The most effective application of automatic grading is in combination with human input and evaluation, fusing the effectiveness of AI with the knowledge of teachers (Perin et al., 2018).

7. INTELLIGENT CLASSROOM MANAGEMENT

Intelligent classroom management is the application of technology, data analytics, and artificial intelligence to improve student engagement and performance in the classroom, improve the overall learning experience, and expedite administrative processes. By utilising intelligent systems and technologies, it seeks to produce a learning environment that is more effective and efficient. Here are some crucial components and illustrations of clever classroom management:

- **Student Performance Monitoring**: Academic performance, behavior, and engagement patterns of students can be tracked and analyzed using AI and data analytics. By assessing learning gaps, identifying challenging pupils, and providing individualized support and interventions, this information aids teachers.
- **Personalized Learning Paths**: Personalized learning routes can be suggested for specific students by intelligent classroom management systems based on their learning preferences, aptitudes, and deficiencies. This guarantees that the material and exercises given to students are adapted to their individual needs and rate of learning.
- **Automated Grading and Feedback**: As was previously said, AI can automatically grade homework, tests, and exams, giving students immediate feedback. Without having to wait for human grading, this enables pupils to comprehend their errors and places for progress.
- **Classroom Resource Management**: Adaptive learning systems modify the content and degree of difficulty of lessons based on the development and performance of each individual student. This method guarantees that students can study at their own pace and are adequately challenged.
- **Behavioral Analytics**: By monitoring student behavior and interactions, intelligent classroom management systems can aid teachers in better understanding the dynamics of the classroom. Utilizing this information can help you spot disruptive behavior, evaluate group dynamics, and enforce better classroom rules.
- **Virtual Assistants and Chatbots**: AI-powered virtual assistants can help teachers and students by answering questions, giving advice on assignments, and promoting dialogue in the classroom.
- **Attendance Management**: Using facial recognition or other technologies, AI can automate attendance tracking, easing the administrative strain on teachers and enhancing accuracy.
- **Early Warning Systems**: AI can detect kids who could be in danger of falling behind or having difficulties by examining historical data. Early intervention and additional support can be given by educators using early warning systems to help avert academic issues.
- **Parent-Teacher Communication**: AI-driven platforms can help instructors and parents communicate effectively by delivering updates on students' academic progress, behaviour, and other crucial data.

While intelligent classroom management has many benefits, it's crucial to remember that privacy protections and ethical data use are paramount. Maintaining trust and creating a successful intelligent classroom environment depend on protecting student data and guaranteeing transparent, ethical usage of AI in education. The personal touch and knowledge of teachers remain crucial to the educational process, and AI should support rather than replace their efforts.

8. VIRTUAL AND AUGMENTED REALITY

Virtual reality (VR) and augmented reality (AR) are immersive technologies that let people interact more naturally or realistically with computer-generated worlds and objects (Perez et al., 2017). While the two technologies are comparable, they also differ in the following ways:

Virtual Reality (VR):

- Users of virtual reality (VR) can engage with a fully computer-generated, immersive world as if they were there in person. It entirely replaces the user's physical surroundings with a digital one that has been simulated.
- VR finds applications in gaming, simulations, training, education, entertainment, architecture, healthcare, and therapy, among other fields. When users wear VR headsets, they are transported to a different reality, blocking out the physical world and engaging their senses (visual, auditory, and sometimes haptic) with the virtual environment.

Augmented Reality (AR):

- AR projects digital content into the physical world, such as virtual information or objects. In contrast to VR, augmented reality (AR) enriches the real world by including virtual components.
- Through gadgets like smartphones, tablets, or smart glasses, AR can be experienced. Users can view both physical and digital content at the same time, fusing the two together effortlessly.
- In addition to entertainment, architecture, healthcare, and therapy, AR is utilized in a variety of applications including games (such as Pokemon Go), navigation, marketing, training and maintenance (such as superimposing instructions on machinery), education, and interior design (Luckin et al., 2016).

Key Differences:

1. **Immersiveness**: Since it completely replaces the real world, virtual reality (VR) offers a more immersive experience than augmented reality (AR), which keeps a connection to the real world.
2. **Interaction**: Utilizing portable controllers or gesture-based devices, users typically interact with the virtual environment in VR. In augmented reality, interactions can take place with actual items or with virtual ones that are superimposed over the real world
3. **Hardware**: Dedicated VR headsets like the Oculus Rift or HTC Vive are typically required, but AR may be enjoyed on a variety of devices, including smartphones and specialized AR glasses.

Combined Reality (Mixed Reality): Another term, "Combined Reality" or "Mixed Reality" (MR), is frequently used to refer to a setting that combines components of both VR and AR. In MR, people can interact with both real and virtual things simultaneously, and virtual items can respond to and interact with the physical surroundings. As a result, the virtual and actual worlds are better integrated.

As VR and AR technologies advance, they present unique potential for a range of sectors, including education, healthcare, entertainment, and business. They do this through increasing user experiences, training and simulations, and real-world problem-solving.

AI can enhance virtual reality (VR) and augmented reality (AR) learning by building engaging and realistic simulations that engage students in learning in a more immersive and individualized way, AI can dramatically improve virtual reality (VR) and augmented reality (AR) learning experiences. The following are some ways AI can improve VR and AR learning issues:

- **Personalized Learning Paths**: AI learner's performance, preferences, and progress can be analyzed by AI to produce personalized learning pathways. Learning experiences can be made more effective by providing students with content that is specifically suited to their requirements and learning preferences.
- **Adaptive Learning Environments**: AI can dynamically change the VR or AR environment's complexity and difficulty based on the learner's performance and interactions. This adaptive method guarantees that students are adequately challenged and inspired to enhance their knowledge and abilities.
- **Natural Language Interaction**: AI-enabled virtual assistants in VR and AR settings can comprehend inquiries and comments from students using natural language processing. This function improves interaction and offers immediate support and direction.
- **Real-time Feedback and Assessment**: During VR and AR simulations, AI may assess students' actions and reactions in real time. This quick feedback assists students in identifying their strengths and limitations, which promotes ongoing development.
- **Intelligent Tutoring Systems (ITS)**: AI-driven ITS can serve as virtual tutors by giving students advice, solutions, and step-by-step explanations when they run into problems in a VR or AR learning environment.
- **Content Generation**: AI is capable of producing interactive learning content, such as 3D models, simulations, and scenarios, automatically for VR and AR scenarios. This expedites the process of creating content and guarantees a comprehensive and varied learning environment.
- **Emotion and Engagement Analysis**: During VR and AR interactions, AI may assess learners' emotions and levels of involvement. This information can be used by educators and developers to improve learning opportunities and pinpoint problem areas.
- **Virtual Collaborative Learning**: By intelligently connecting learners with similar interests and skill levels, AI can support collaborative learning in VR and AR settings. This encourages peer-to-peer learning and builds a sense of community.
- **Performance Analytics and Insights**: During VR and AR learning sessions, AI may produce detailed data on students' performance, development, and behaviour. This information can be used by educators to make data-driven decisions and improve their teaching methods.
- **Natural Environment Generation**: AI can assist in the creation of dynamic, realistic virtual worlds for VR and AR simulations, enhancing the learner experience and making it more engaging.

Educators and developers may create cutting-edge, interactive, and adaptive learning experiences that capture students' attention, increase their understanding, and support knowledge retention by fusing

AI with VR and AR technology. These improved learning opportunities could revolutionise instruction and training in many different professions and sectors.

9. PREDICTIVE ANALYTICS

AI-powered predictive analytics can be a potent tool in education to predict student performance and pinpoint failure risk. AI models may anticipate outcomes for specific students or groups of students by examining previous data and patterns. This enables educators to intervene and offer focused support to children in order to enhance learning outcomes. Using predictive analytics in teaching works like this:

1. **Data Collection**: AI systems require access to a wide range of relevant data in order to make accurate predictions, including past academic achievement, attendance records, engagement levels, assessment scores, and demographic data.
2. **Data Preprocessing**: Preprocessing actions are conducted to clean, transform, and get the data ready for analysis before feeding it to AI models. This guarantees that the data is in a format that the AI algorithms can use.
3. **Machine Learning Algorithms**: The preprocessed data can be used with a variety of machine learning algorithms, including deep learning models, decision trees, random forests, and logistic regression. These algorithms study patterns and connections between variables that affect students' success or failure using historical data.
4. **Feature Selection**: AI algorithms locate pertinent traits or factors that significantly affect student performance. These characteristics could be engagement in extracurricular activities, exam results, homework completion, and attendance.
5. **Training and Validation**: The AI model is first trained on a subset of the data (training data), and then evaluated on a separate subset (validation data), to ensure that it is correct and generalizable to new data.
6. **Predictive Insights**: The AI model can forecast future student performance based on fresh data once it has been trained and validated. For instance, it can spot pupils who are likely to fail a course or are having difficulty with a particular subject.
7. **Early Warning Systems**: Early warning systems, which provide instructors with alerts and notifications about kids who might be at risk of falling behind, can be enhanced with predictive analytics. This makes prompt assistance and interventions possible.
8. **Intervention Strategies**: With the use of predictive insights, instructors can modify intervention plans and offer specialized assistance to the children who need it the most. Additional instruction, mentoring, or counseling may be necessary.
9. **Continuous Improvement**: Iteration is a key component of predictive analytics. AI models can be upgraded and retrained to increase their accuracy and efficiency when new data becomes available.

Predictive analytics can help educational institutions adopt a proactive stance to improve student achievement, lower dropout rates, and enhance learning opportunities. It enables instructors to spot potential issues before they become serious ones, promoting a more favourable learning environment for all kids. To ensure ethical and just deployment, it is crucial to strike a balance between the use of predictive analytics and privacy-protecting measures, openness, and accountability.

10. CONCLUSION AND FUTURE SCOPE

A new era of individualized, interactive, and data-driven learning experiences has been ushered in by artificial intelligence (AI) and machine learning (ML), which have completely changed the landscape of smart education. The way that students learn, professors teach, and educational institutions run has all been revolutionized by these technologies. The use of AI and ML in smart education is widespread and has a significant influence. Examples include adaptive learning platforms, intelligent tutoring systems, and immersive virtual and augmented reality experiences. Improved learning outcomes and lower dropout rates have resulted from the use of AI-driven content development, predictive analytics, and early warning systems by educators.

The potential for further developments and improvements in the field of smart education is enormous. Future prospects that are important include:

- **Enhanced Personalization**: AI and ML will go further, providing even more individualized learning opportunities that take into account each student's learning preferences and areas of strength and weakness.
- **Seamless Integration**: As AI technologies advance, they will easily interface with educational resources, platforms, and hardware, enhancing user-friendliness and accessibility of smart education.
- **Lifelong Learning**: Personalized learning paths and opportunities for upskilling that are specific to each learner's professional goals and interests can be enabled by AI to enhance lifelong learning.
- **Continuous Assessment and Feedback**: AI-powered solutions for continuous assessment and feedback will give educators current information on student development and allow them to modify their teaching methods.
- **Collaborative Learning Environments**: By supporting virtual interactions and group activities, AI can create collaborative learning experiences by encouraging teamwork and problem-solving abilities.
- **Enhanced Teacher Support**: Teachers can concentrate more on individualized instruction and mentoring by using AI to help with administrative duties, grading, and content production.
- **Ethical AI Education**: In order to ensure responsible implementation that respects learners' rights and secures their data, the future of AI in education will emphasize ethical AI usage, data privacy, and transparency.
- **AI-Driven Research in Education**: By analyzing enormous volumes of data, spotting trends, and producing insightful findings that guide the development of practice-based educational strategies, AI and ML will contribute to educational research.
- **Augmented Reality in Skill Training**: The use of AR-powered skill training simulations will increase, allowing students to practice real-world tasks in secure settings.
- **AI for Inclusive Education**: By providing individualized help and accommodations for children with a variety of learning requirements, AI can play a crucial role in advancing inclusive education.

In conclusion, AI and ML integration in smart education have demonstrated tremendous promise and potential. These technologies will transform education in the future as they develop, making it more interesting, efficient, and available to students everywhere. To ensure responsible AI usage and provide an equal and inclusive learning environment for everyone, educators, researchers, policymakers, and technology suppliers must work together to assure effective deployment. Smart education has a promising future, and AI and ML will surely be crucial in determining its course.

REFERENCES

Baker, T., Smith, L., & Anissa, N. (2019). *Educ-AI-tion rebooted? Exploring the future of artificial intelligence in schools and colleges*. Academic Press.

Barker, T. (2010). An automated feedback system based on adaptive testing: Extending the model. *International Journal of Emerging Technologies in Learning*, *5*(2), 11–14. doi:10.3991/ijet.v5i2.1235

Barker, T. (2011). An Automated Individual Feedback and Marking System: An Empirical Study. *Electronic Journal of e-Learning*, *9*(1), 1–14.

Luckin, R., & Holmes, W. (2016). *Intelligence unleashed: An argument for AI in education*. Academic Press.

Mann, S., Pathak, N., Sharma, N., Kumar, R., Porwal, R., Sharma, S. K., & Aung, S. M. Y. (2022). Study of Energy-Efficient Optimization Techniques for High-Level Homogeneous Resource Management. *Wireless Communications and Mobile Computing*. doi:10.1155/2022/1953510

Perez, S., Massey-Allard, J., Butler, D., Ives, J., Bonn, D., Yee, N., & Roll, I. (2017). Identifying productive inquiry in virtual labs using sequence mining. *Artificial Intelligence in Education: 18th International Conference, AIED 2017, Wuhan, China, June 28–July 1, 2017 Proceedings*, *18*, 287–298.

Perin, D., & Lauterbach, M. (2018). Assessing text-based writing of low-skilled college students. *International Journal of Artificial Intelligence in Education*, *28*(1), 56–78. doi:10.100740593-016-0122-z

Popenici, S. A., & Kerr, S. (2017). Exploring the impact of artificial intelligence on teaching and learning in higher education. *Research and Practice in Technology Enhanced Learning*, *12*(1), 1–13. doi:10.118641039-017-0062-8 PMID:30595727

Raju, D., & Schumacker, R. (2015). Exploring student characteristics of retention that lead to graduation in higher education using data mining models. *Journal of College Student Retention*, *16*(4), 563–591. doi:10.2190/CS.16.4.e

Sharma, S. K., & Sharma, N. K. (2019a). Text Document Categorization using Modified K-Means Clustering Algorithm. *International Journal of Recent Technology and Engineering*, *8*(2), 508–511.

Sharma, S. K., & Sharma, N. K. (2019b). Text Classification using Ensemble of Non-Linear Support Vector Machines. *International Journal of Innovative Technology and Exploring Engineering*, *8*(10), 3170–3174. doi:10.35940/ijitee.J9520.0881019

Sharma, S. K., & Sharma, N. K. (2019c). Text Classification using LSTM based Deep Neural Network Architecture. *International Journal on Emerging Technologies.*, *10*(4), 38–42.

Sharma, S. K., Sharma, N. K., & Potter, P. P. (2020, December). Fusion approach for document classification using random forest and svm. In *2020 9th International Conference System Modeling and Advancement in Research Trends (SMART)* (pp. 231-234). IEEE. 10.1109/SMART50582.2020.9337131

Sharma, S. K., Sharma, N. K., & Singh, G. (2019). Unified framework for deep learning based text classification. *International Journal of Scientific Technology Research*, *8*(10), 1479–1483.

Chapter 7
Artificial Intelligence for Monitoring Agricultural Essentials

Sheelesh Kumar Sharma
ABES Institute of Technology, Ghaziabad, India

ABSTRACT

In this chapter, the authors explore the role of AI in monitoring agricultural essentials, such as soil health, water management, and crop diseases. They delve into the various applications and benefits of AI in monitoring agricultural essentials. Through the integration of weather forecasts, soil moisture sensors, and machine learning algorithms, AI systems can determine the precise water requirements of crops in real-time. AI-based monitoring systems equipped with sensors and cameras can detect pest activity and provide real-time alerts to farmers. By harnessing the power of AI-driven monitoring systems, farmers can access real-time data, make informed decisions, and optimize resource allocation for sustainable agriculture. The applications discussed in this chapter, including soil health monitoring, precision water management, crop disease detection, pest monitoring, and yield prediction, highlight the transformative potential of AI in the agricultural sector.

1. INTRODUCTION

The integration of artificial intelligence (AI) into agriculture has changed the way farmers monitor and control the importance of their jobs. In this chapter, we explore the role of AI in monitoring critical agricultural issues such as soil health, water management and crop disease. AI-powered monitoring systems enable farmers to make informed decisions, improve resources and increase overall profitability. Let's look at the many uses and benefits of artificial intelligence in intensive care farming.

Soil health monitoring using artificial intelligence: Monitoring the health of the soil is crucial for sustainable agriculture. Farmers are able to precisely monitor and measure soil using the AI-powered technology. According to (Schmidt et al., 2016), AI algorithms can analyze soil data obtained from various sensors and IoT devices to reveal information on soil moisture, fertility levels, content, and pH

DOI: 10.4018/979-8-3693-0782-3.ch007

balance. Informed judgments on irrigation and crop management by farmers lead to higher agricultural yields and productivity.

Precision water management: Agriculture is difficult due to water constraints. Water efficiency has increased and water purity has been maintained thanks in large part to AI technology. By combining weather predictions, soil moisture sensors, and machine learning algorithms, AI systems can calculate the water requirements of crops at any moment (Hanson et al., 2007). This data-driven strategy decreases the danger of consuming too much or too little water, minimizes water waste, and protects this priceless resource.

Disease Detection and Control: Timely detection and control of crop diseases is important to reduce crop losses. AI-powered surveillance systems can detect early symptoms of diseases, pests, and nutrient deficits using computer vision and machine learning. AI algorithms can detect patterns and symptoms invisible to the naked eye by analyzing the shape of leaves, stems and fruits. This allows farmers to act quickly, such as spraying plans or supplemental food, to prevent disease and protect crops.

Pest Monitoring and Control: Pests pose a serious threat to health and production. AI-based surveillance systems equipped with sensors and cameras can detect pests and provide alerts to farmers. AI algorithms can identify pest species, population characteristics and disease patterns by constantly monitoring pests. This knowledge allows farmers to implement pest management strategies such as integrated pest management (IPM), reducing dependency on pesticides and reducing their environmental impact (Garre and Harish et al., 2018).

Growth and Yield Forecasting: Planning harvest schedules, estimating market demands, and optimizing resource allocation all depend on an accurate assessment of crop growth and output. AI algorithms combined with satellite imagery, weather data and historical data can create growth models and forecasts. By analyzing many variables such as temperature, humidity, precipitation and certain crops, AI can

Figure 1. Role of AI in monitoring critical agricultural issues

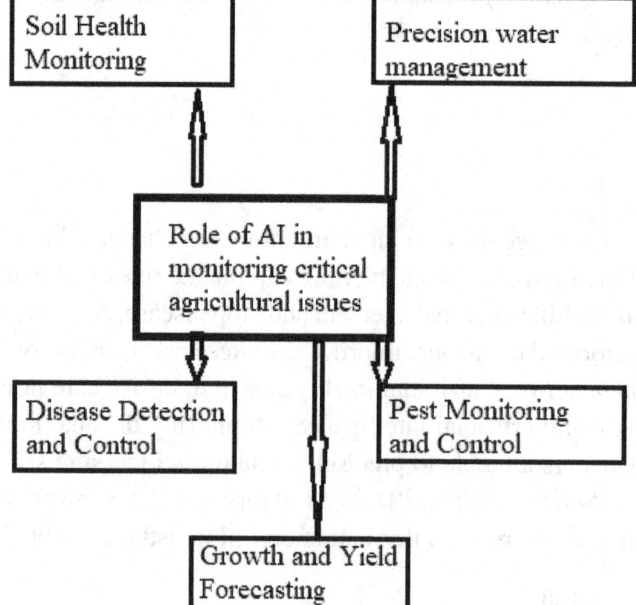

provide farmers with useful information for future crop production (Gandhi et al., 2020). This allows them to make informed decisions about harvesting, storage and marketing strategies.

The care of agricultural crops has undergone a transformation thanks to artificial intelligence. Farmers may access real-time data, make educated decisions, and enhance resource allocation for action-oriented sustainable agriculture by leveraging the potential of AI-powered analytics tools. This section discusses the applications of AI in agriculture, such as soil health monitoring, water quality management, crop disease testing, insect control, and yield calculation. Artificial intelligence will play a significant role in revolutionizing agriculture and assuring global food security as technology develops further. Figure 1 shows the role of AI in keeping an eye on important agricultural challenges.

2. ARTIFICIAL INTELLIGENCE BASED SOIL HEALTH MONITORING

Artificial intelligence (AI) has the potential to significantly contribute to soil health monitoring by delivering precise and timely data on the fertility, condition, and overall health of the soil. By giving farmers and academics invaluable insights into more effective and sustainable practices, this technology has the potential to revolutionize agricultural and environmental management. Here's how AI can be applied to soil health monitoring:

1. Remote Sensing and Imaging: Using satellite and drone images analysis, AI can evaluate different soil factors like moisture content, organic matter, and nutrient levels. Farmers may keep an eye on wide tracts of land and decide on irrigation, fertilisation, and other agricultural practises while using these real-time data streams.
2. Soil testing and analysis: Artificial intelligence (AI) can analyse and analyse data from soil samples, cutting down on the time and expense of conventional laboratory testing. It can find patterns and links between soil parameters and crop health by incorporating machine learning algorithms, which improves soil management techniques.Predictive Modeling: AI can create predictive models using historical data and other environmental factors to forecast future soil health conditions. These models can help farmers anticipate potential issues and optimize their management practices accordingly.
3. 3. Internet of Things (IoT) sensors and data collection: IoT devices with sensors may continuously monitor soil conditions and communicate real-time data to AI systems. AI is able to provide real-time recommendations and modifications to farming practises thanks to these data streams.
4. Precision Agriculture: By offering individualized advice for individual plots or even certain locations within a field, AI can enable precision agriculture practices. This focused strategy improves crop yields while reducing environmental impact and maximizing resource use.
5. Disease and Pest Detection: By examining photos of plants and leaves, AI-powered image recognition systems can spot early indications of illnesses, pests, or nutrient deficits in crops. Early detection of issues enables prompt response and reduces crop losses.
6. Soil Erosion and Degradation Monitoring: Using historical data and satellite pictures, AI can monitor the trends of soil erosion and degradation over time. This knowledge aids in maintaining soil health and carrying out erosion control procedures.
7. Decision Support Systems: AI can help farmers by making tailored recommendations based on the facts about their particular soil health and the surrounding environment. The crop rotation, planting times, and fertilization schedules can all be improved by these decision support systems.

Overall, AI-based soil health monitoring provides farmers and agricultural professionals with insightful data that improves productivity, sustainability, and environmental preservation. To guarantee widespread acceptance and benefit, it is imperative to solve issues including data quality, privacy concerns, and farmer access to technology.

2.1 Remote Sensing and Imaging

Powerful technologies like remote sensing and imaging are used to obtain data about the Earth's surface and atmosphere from a distance. They entail gathering data without coming into contact with the target region directly by using a variety of sensors and devices installed on satellites, aircraft, drones, or other platforms. These technologies have several uses in a variety of industries, including urban planning, disaster management, agriculture, forestry, and environmental monitoring (Luciani et al., 2019 ; Cillis et al., 2018).

Here are some key aspects of remote sensing and imaging:

1. Types of Sensors: Data can be collected using remote sensing equipment from various electromagnetic spectrum regions. sensors typical types of include:
   - Optical sensors: These sensors record visible light and infrared radiation, collecting data on surface features including plant and land cover.
   - Thermal sensors: These devices gauge the surface temperature of the planet to look for anomalies and analyse how heat is distributed.
   - Radar sensors: Radar collects information about surface roughness, moisture content, and elevation by sending microwave signals through clouds.
   - • LiDAR (Light Detection and Ranging): LiDAR sensors emit laser pulses to detect distances and produce accurate three-dimensional (3D) models of the Earth's surface.
2. Satellite-based Remote Sensing: Satellites equipped with remote sensing instruments orbit the Earth and continuously collect data over large areas. They can cover remote or inaccessible regions, making them valuable tools for global-scale monitoring and research.
3. Aerial and Drone-based Imaging: Aircraft and drones are used for high-resolution imaging and data collection at regional or local scales. They are particularly useful for precise and detailed assessments of specific areas.
4. Image Interpretation and Analysis: Remote sensing imagery requires interpretation and analysis to extract valuable information. This involves the identification and classification of features such as land use, vegetation types, water bodies, and other geographical elements.
5. Time Series Analysis: Repeated imaging of the same area over time allows researchers to monitor changes and trends in land cover, vegetation growth, and other dynamic processes. Time series analysis is vital for understanding long-term environmental changes and detecting anomalies.
6. Applications:
   - Agriculture: Monitoring crop health, assessing soil moisture, and optimizing irrigation practices.
   - Environmental Monitoring: Tracking deforestation, studying urbanization, monitoring water quality, and detecting natural disasters.
   - Climate Studies: Studying the Earth's energy balance, monitoring glaciers, and assessing changes in sea ice extent.

  - Disaster Management: Assisting in disaster response and damage assessment after events like wildfires, floods, and earthquakes.

Our understanding of Earth's processes has substantially improved because to remote sensing and imaging, which are now indispensable tools for making well-informed decisions in a variety of fields. The ability to analyse and comprehend enormous volumes of data has been further improved by the integration of remote sensing data with geographic information systems (GIS) and artificial intelligence (AI), providing fresh insights into the intricate interconnections that make up our environment.

2.2 Soil Testing and Analysis

Soil testing and analysis is a critical process used to assess the health and fertility of soil for agricultural, environmental, and construction purposes. By analyzing soil samples, valuable information about its nutrient content, pH levels, organic matter, texture, and other properties can be obtained. This data helps in making informed decisions about appropriate land management practices, crop selection, and the application of fertilizers or soil amendments. Here's an overview of the soil testing and analysis process:

1. Soil Sampling: Gathering representative soil samples from the region of interest is the first stage in soil testing. In order to guarantee that the collected samples accurately reflect the overall features of the soil, proper soil sampling methods must be used.
2. Laboratory Analysis: Following collection, soil samples are forwarded to a soil testing laboratory for evaluation. Several tests are carried out in the lab to ascertain the physical and chemical characteristics of the soil. Common tests consist of:
   - Soil pH: Determines how acidic or alkaline the soil is, which has an impact on the nutrients that are available to plants.
   - Nutrient analysis: Identifies the concentrations of micronutrients (such as iron, zinc, copper, and phosphorus) and important nutrients (such as nitrogen, phosphorus, potassium, calcium, magnesium, and sulphur).
   - Organic Matter Content: Indicates the quantity of organic matter in the soil, which affects nutrient absorption and soil structure.
   - Soil Texture: Controls how much sand, silt, and clay is present in the soil, which affects its ability to retain water and facilitate drainage.
3. Interpretation and Recommendations: Following laboratory analysis, the results of the soil test are analyzed to determine the fertility level and any potential drawbacks of the soil. Experts in soil testing or agricultural extension agencies make recommendations for soil management practices based on the analysis.
4. Recommendations for Fertilizer and Lime: Soil testing aids in determining the proper amounts and types of Fertilizer and Lime (if required) to get the best plant development. This careful application of inputs encourages cost effectiveness and reduces negative effects on the environment.
5. Sustainable Practices: To maximize resource usage and lessen nutrient runoff, soil testing promotes the adoption of sustainable agricultural practices such site-specific nutrient management and precision farming.

6. 6. Environmental Applications: Soil testing is essential for identifying the viability of sites for construction projects, evaluating the possible dangers connected with land use changes, and detecting the extent of soil contamination.

Benefits of Soil Testing and Analysis:

- Increased Crop Yield: By understanding the soil's nutrient status, farmers can apply fertilizers and other amendments in the right amounts, leading to improved crop productivity.
- Cost Savings: Soil testing helps avoid overuse of fertilizers and lime, preventing unnecessary expenses and reducing environmental pollution.
- Soil Health Improvement: With regular testing and appropriate management, soil health can be enhanced over time, leading to sustainable agricultural practices.
- Environmental Protection: By optimizing fertilizer use, soil testing minimizes the risk of nutrient leaching into water bodies, reducing potential environmental harm.

Overall, soil testing and analysis are essential tools for promoting sustainable agriculture, responsible land management, and environmental conservation. By understanding the soil's characteristics and needs, we can make informed decisions to ensure the long-term productivity and health of our lands.

2.3 Predictive Modeling

Data science and machine learning use the technique of predictive modeling to build models that can forecast future results or make wise judgments based on historical data. Predictive modeling aims to produce precise forecasts about future events or trends by utilizing patterns and linkages revealed in the data. These models can anticipate the results of fresh, unforeseen data because they were created using algorithms and trained on historical data with known outcomes.

The following steps are often included in the process of creating predictive models:

1. Data Collection: Gathering pertinent data from diverse sources is the initial stage. This information could include historical records, sensor data, user interactions, or any other details connected to the issue at hand.
2. Data Preprocessing: After data has been gathered, it must be cleaned and prepared. This includes dealing with missing numbers, eliminating outliers, and formatting the data for modeling.
3. Feature Selection: Predictions may not be possible for all features (variables) in the data. Choosing the most significant and informative features to be utilized as inputs to the prediction model is known as feature selection.
4. Model selection: There are many different types of predictive models, including neural networks, support vector machines, decision trees, random forests, and more. The type of the problem and the qualities of the data will determine which model is most appropriate.
5. Model Training: Using the preprocessed data, the chosen model is trained. The model learns the underlying patterns and connections between the input data and the target variable (the variable to be predicted) during the training process.
6. Model Evaluation: After the model has been trained, it is assessed using a different set of data (the test set) that was not utilized during training. Depending on the type of task (for example, classifi-

cation or regression), the model's performance is assessed using a variety of measures, including accuracy, precision, recall, F1 score, and others.

7. Model tuning: If the model's performance is unsatisfactory, model hyperparameters may be adjusted to improve it even more. To enhance the model's predictions, this procedure entails changing parameters that are not learned during training.
8. Prediction and Deployment: The predictive model is prepared to be used for making predictions on fresh, unforeseen data following successful training and evaluation. The model can be used to support decision-making procedures in practical situations.

Numerous industries and applications, such as banking, healthcare, marketing, customer churn prediction, fraud detection, demand forecasting, and others, heavily rely on predictive modeling. Making data-driven forecasts enables businesses and organizations to make more knowledgeable and effective decisions, improving results and better allocating resources.

2.4 IoT Sensors and Data Collection

With the help of real-time monitoring and data-driven insights, IoT (Internet of Things) sensors and data collecting have revolutionized the agriculture sector. To enable precision farming and enhance all agricultural practices, IoT in agriculture integrates sensor devices, communication networks, and data analytics. The usage of IoT sensors and data collection in agriculture is demonstrated here:

1. Sensor Technology: In order to gather information on numerous aspects linked to soil, weather, crop health, and environmental conditions, IoT sensors are used in agricultural fields. These sensors are available in various varieties, including:
    - Soil Moisture Sensors: These devices detect the amount of soil moisture present in order to optimize irrigation and avoid over- or under-watering.
    - Temperature and Humidity Sensors: Keep an eye on the weather so that farmers can choose wisely when to grow, harvest, and safeguard their crops.
    - Crop health sensors are devices that can quickly identify disease, pests, or nutritional deficits.
    - Nutrient Sensors: These devices measure soil nutrient levels to allow for exact fertiliser application for the best plant growth.
    - Meteorological stations: For precise forecasting and risk assessment, gather complete meteorological data, including temperature, humidity, wind speed, and precipitation.
2. Data Collection and Communication: IoT sensors continuously gather data from the field, and this data is communicated to central servers or cloud-based platforms for storage and analysis over wireless communication networks (e.g., cellular, Wi-Fi, LoRa, or NB-IoT).
3. Cloud-Based Data Analytics: Edge computing systems or cloud-based data analytics platforms are used to process and analyze the gathered data. The data is analyzed using cutting-edge algorithms and machine learning techniques to spot patterns and produce insights that can be put to use.
4. Decision Support Systems: Through user-friendly interfaces or mobile applications, IoT data analytics platforms give farmers real-time information and actionable advice. This aids farmers in making decisions on timing of irrigation, insect management, fertilization, and other farming tasks.

5. Precision Farming: Farmers can embrace precision agriculture practices by using the knowledge collected from IoT sensors and data analysis. Based on differences in soil and crop conditions, this entails adapting agricultural practices to particular fields. Precision farming improves resource efficiency, lowers waste, and boosts overall output.
6. Automated Systems: IoT sensors can be integrated with automated irrigation systems, drones, and robotic machinery to carry out tasks efficiently and autonomously based on real-time data.
7. Remote Monitoring: Remotely monitoring fields and crops is one of the major benefits of IoT in agriculture. On their cellphones or PCs, farmers may access data and receive notifications, enabling them to react quickly to any changes or emergencies.
8. Environmental Sustainability: IoT sensors facilitate more sustainable agricultural practices by reducing water consumption, minimizing the use of chemical inputs, and optimizing energy usage.

IoT sensors and data collection have revolutionized agriculture by enabling data-driven decision-making and transforming traditional farming practices into smart, efficient, and sustainable systems. As the technology continues to evolve, it is expected to play an increasingly vital role in addressing the challenges of food security and resource management in a rapidly changing world.

2.5 Precision Agriculture

Precision agriculture, commonly referred to as smart farming or precision farming, is a cutting-edge farming technique that optimizes agricultural practices by using technology, data analytics, and information management. Precision agriculture aims to improve farming operations' productivity, sustainability, and efficiency while reducing waste and adverse environmental effects. This is accomplished by adjusting farm management practices to the unique requirements of various fields rather than employing a one-size-fits-all strategy.

Key components and techniques involved in precision agriculture include:

1. IoT Sensors: Using different sensors, such as soil moisture sensors, temperature sensors, and crop health sensors, to deploy in the field to gather real-time data. These sensors offer useful data on crop health, weather patterns, and soil characteristics.
2. GIS and GPS Technology: To precisely map the spatial variability of soil qualities, crop health, and other factors, geographic information systems (GIS) and Global Positioning System (GPS) technology are utilized. Farmers may make decisions that are site-specific thanks to this geographical data.
3. Remote Sensing: This technique makes use of satellite pictures, drones, and other remote sensing technology to monitor crop growth, identify stress, and evaluate general field conditions across wide regions.
4. Variable Rate Technology (VRT): Based on the variability in soil and crop conditions, VRT enables farmers to alter the application rate of inputs (such as fertilizers, herbicides, and water). As a result, resources are used more effectively, and inputs are only used when and where they are required.
5. Automated Machinery: Using self-driving tractors and drones, as well as IoT and GPS technologies, automated machinery can execute precise operations like planting, spraying, and harvesting.

Benefits of Precision Agriculture:

1. Enhanced production: Precision agriculture maximizes crop yields and total farm production by adjusting agricultural practices to local conditions.
2. Resource Efficiency: Precision agriculture makes the most efficient use of resources like water, fertilizer, and pesticides, which lowers production costs and waste.
3. Sustainability: Lessening the negative effects of agricultural practices on the environment by using fewer chemicals and practicing better resource management.
4. Improved Decision-Making: Data-driven insights and suggestions assist farmers in making wise choices in the present, enhancing overall farm management.
5. Risk management: Timely interventions can be made in the event of early identification of crop stress, diseases, or insect infestations, minimizing crop losses.
6. Economic Benefits: Farmers may see cost savings and higher profitability as a result of precision agriculture.

Overall, precision agriculture is a key development in contemporary agricultural methods that addresses the growing requirement for sustainable and effective food production to fulfill rising worldwide population demands while protecting natural resources.

2.6 AI-Based Disease and Pest Detection

Artificial intelligence-based disease and pest detection in agriculture is a cutting-edge method of identifying and diagnosing plant diseases and insect infestations in crops. With early identification and prompt action, this technology has the potential to completely change how farmers manage crop health, preventing large output losses and reducing the need for chemical pesticides.

Here's how AI-based disease and pest detection in agriculture works:

1. Data Collection: The process begins with the collection of various types of data relevant to crop health. This data may include images of plants and leaves, weather conditions, soil moisture levels, and historical disease and pest outbreak data.
2. Image Processing: High-resolution images of plants are captured using cameras or drones equipped with sophisticated sensors. These images are then processed to extract relevant features and information related to plant health.
3. Training Data Preparation: To build an AI model, a labeled dataset is created, where each image is annotated with information about whether the plant is healthy or infected by a specific disease or pest.
4. Development of AI models: The labeled dataset is used to train machine learning techniques like convolutional neural networks (CNNs). The AI model gets better at identifying the patterns and characteristics that indicate healthy plants, illnesses, and pest damage.
5. Disease and Pest Detection: After the AI model has been trained, it may be used to instantly examine fresh plant photos. Based on the patterns it has discovered throughout training, the model is able to detect the presence of illnesses or pests with great accuracy.
6. Decision Support System: A decision support system can incorporate the AI-based illness and pest detection system. This method notifies farmers of probable disease or insect outbreaks and gives them with real-time information about the health state of their crops.

Benefits of AI-based Disease and Pest Detection in Agriculture:

1. Early Detection: AI systems can identify disease outbreaks and insect infestations before they spread, giving farmers the opportunity to act quickly to stop the spread.
2. Precision Treatment: By locating the problem regions, farmers may use focused treatments rather than broad-spectrum pesticides, minimizing any negative effects on the environment.
3. Increased Yields: Timely disease and pest treatment can result in healthier crops and higher yields.
4. Lower expenses: Farmers may cut expenses by forgoing needless pesticide and fungicide treatments through early identification and accurate treatment.
5. Sustainable Farming: By reducing chemical usage and optimizing resource allocation, AI-based disease and pest detection aligns with sustainable farming practices.

Challenges:

- Robustness: The AI model must be trained on diverse datasets to ensure robustness in detecting various diseases and pests accurately.
- Data Availability: Access to high-quality, labeled datasets can be a challenge, especially for less common or region-specific crop diseases.
- Interpretability: It can be difficult to interpret judgments made by sophisticated AI models, raising questions regarding model responsibility and transparency.

AI-based disease and pest detection in agriculture holds great promise in improving crop management and sustainability. As technology and data availability continue to improve, these systems are expected to become increasingly valuable tools for modern farmers seeking to optimize crop health and productivity.

2.7 AI-Based Soil Erosion and Degradation Monitoring

AI-based soil erosion and degradation monitoring is a novel use of machine learning and artificial intelligence to assess and reduce the detrimental effects of soil erosion and degradation on agricultural areas. Soil erosion and degradation are significant challenges that can lead to reduced soil fertility, decreased crop productivity, and environmental issues. By leveraging AI and data analytics, farmers and researchers can monitor soil erosion and degradation more effectively, enabling them to implement targeted conservation measures and sustainable land management practices. Here's how AI is applied in this context:

1. Data Collection: Various data sources are used to monitor soil erosion and degradation. These include topographic maps, satellite imagery, drone imagery, weather data, soil moisture data, and historical land use records.
2. Remote Sensing and Imaging: Satellite imagery and drones equipped with high-resolution sensors are used to capture data on soil erosion and land cover changes over time. This data provides valuable insights into erosion hotspots and vulnerable areas.
3. Image Analysis: AI algorithms, such as convolutional neural networks (CNNs), are employed to analyze the remote sensing and imaging data. The AI model can automatically detect land cover changes, erosion features, and other indicators of soil degradation.

4. GIS Integration: Geographic Information Systems (GIS) are used to integrate and analyze the spatial data collected from various sources. This enables the creation of detailed soil erosion maps and facilitates better decision-making.
5. Machine Learning Models: Machine learning algorithms are trained on historical erosion data to predict future erosion trends and identify factors contributing to soil degradation. These models can provide early warnings of erosion-prone areas and guide land management strategies.
6. Decision Support Systems: AI-powered soil erosion and degradation monitoring systems can be integrated into decision support tools. These tools provide farmers and land managers with real-time information, erosion risk assessments, and conservation recommendations.
7. Precision Conservation: By identifying areas prone to erosion and degradation, AI allows for precision conservation efforts, including the implementation of contour farming, terracing, cover cropping, and other erosion control measures.
8. Long- Term Impact Assessment: AI models may be used to assess the long-term impacts of land management practices on soil health and erosion rates in order to establish sustainable land use strategies.

Benefits of AI-based Soil Erosion and Degradation Monitoring:

- Prompt Intervention: Farmers may act quickly to stop future soil loss and deterioration by detecting erosion and degradation early.
- Targeted Conservation: AI aids in focusing conservation efforts on particular regions, optimising the use of available resources and enhancing the efficiency of conservation techniques.
- Sustainable Agriculture: AI helps to promote sustainable agricultural practises and maintains soil fertility for future generations by preventing soil erosion and degradation.
- Data-Driven Decision Making: AI-powered monitoring offers data-driven insights that help farmers decide on land management strategies.

An effective technique for improving sustainable agriculture and reducing the effects of soil degradation on crop yield and the environment is AI-based soil erosion and degradation monitoring. Technology has the ability to greatly enhance global soil health and land management practises as it develops and becomes more widely available.

2.8 AI-Based Decision Support Systems

Advanced software programs called AI-based Decision Support Systems (DSS) in agriculture use artificial intelligence, machine learning, and data analytics to help farmers and agronomists make data-driven choices. These systems use real-time data, historical data, and predictive modeling to increase resource efficiency, optimize agricultural practices, and improve overall farm management. The following are some salient features of agriculture-related AI-based decision support systems:

1. Data gathering: Weather stations, soil sensors, satellite imaging, drones, crop health monitors, and historical records are just a few of the sources that decision support systems use to collect data. The system incorporates the data for analysis and updates it continually.

2. Data Analysis and AI Modeling: Patterns, correlations, and trends are found in the acquired data by applying AI algorithms, such as machine learning, deep learning, and statistical techniques. These models have the capacity to learn from the past and forecast future events.
3. Crop Monitoring and Management: AI-based DSS continuously track the health, development, and growth of crops. They let farmers to use precision agricultural techniques and take prompt action by spotting early indications of illnesses, pests, or nutrient deficits..
4. Irrigation Optimization: To schedule irrigation more effectively, AI-powered DSS utilises information on soil moisture, weather, and crop water needs. This aids in reducing water waste, preventing over- or under-watering, and increasing crop output.
5. Nutrient Management: To provide suggestions for accurate and balanced fertilization, AI-based DSS analyzes soil nutrient levels and crop nutrient requirements. This guarantees optimal plant nutrition and reduces wasting of nutrients.
6. Pest and Disease Prediction: AI algorithms can forecast the possibility of pest outbreaks or disease epidemics by examining historical data and present circumstances. This aids farmers in proactively putting pest control measures into place and minimizing crop losses..
7. Market Analysis and pricing Prediction: Some AI-based DSS combine pricing trends and market data to help farmers decide on crops, harvest dates, and marketing tactics.
8. Weather Forecasting and Risk Assessment: AI-powered DSS use weather data and predictive models to assess weather-related risks, such as droughts, floods, or extreme temperatures. This information helps farmers plan and adapt their farming practices accordingly.
9. Personalized Recommendations: AI-based DSS can provide personalized recommendations tailored to specific farm conditions, soil types, and crop varieties, ensuring that decision-making is context-specific and optimal.
10. Mobile and Cloud Integration: Many AI-powered DSS offer mobile applications and cloud-based platforms, allowing farmers to access real-time data and receive recommendations on their smartphones or computers from anywhere.

Benefits of AI-based Decision Support Systems in Agriculture:

- Increased Productivity: AI-powered DSS optimize farming practices, leading to improved crop yields and higher farm productivity.
- Resource Efficiency: Precise management of water, fertilizers, and pesticides reduces waste and conserves resources.
- Sustainability: AI-based DSS promote sustainable farming practices, minimizing environmental impacts.
- Risk Mitigation: Early warning systems and risk assessments help farmers mitigate potential risks and losses.
- Data-Driven Decisions: AI-powered DSS enable data-driven decision-making, enhancing the effectiveness of farm management.

Overall, AI-based Decision Support Systems are valuable tools for modern agriculture, helping farmers navigate the complexities of farming and achieve more efficient, profitable, and sustainable outcomes.

3. PRECISION WATER MANAGEMENT

Artificial Intelligence (AI) can be a game-changer when it comes to precision water management in agriculture. It enables farmers to optimize water usage, increase crop yield, and conserve water resources effectively (Dela Cruz et al., 2017). Here's how AI can be applied to precision water management:

1. Smart Irrigation: To calculate the correct quantity of water required for irrigation, AI can analyze a variety of data sources, including weather patterns, soil moisture levels, crop water requirements, and evapotranspiration rates. AI guarantees that plants receive the proper quantity of water at the right time, decreasing water waste and improving crop health by automating irrigation plans based on real-time data (Kim et al., 2008).
2. Soil Moisture Monitoring: AI can integrate data from soil sensors and IoT devices to continuously monitor soil moisture levels. This information is crucial for determining when and where irrigation is required, preventing overwatering and underwatering (Kuyper and Balendonck, 2001).
3. Weather Prediction and Forecasting: AI can process weather data from various sources to predict rainfall patterns and drought conditions. By considering this information alongside soil moisture levels, AI can proactively adjust irrigation schedules to compensate for changing weather conditions.
4. Crop Water Stress Detection: The earliest indications of water stress in crops may be identified using AI-powered picture analysis and sensor data. This enables farmers to take prompt action to lessen the effect on crop productivity, such as modifying irrigation or introducing drought-resistant types.
5. Water Quality Monitoring: AI can also help in monitoring water quality in irrigation systems, ensuring that the water used for crops is free from contaminants that could harm plant health.
6. Water Resource Allocation: AI can assist in optimizing water allocation across different fields and crops based on their water needs, soil conditions, and expected weather patterns. This ensures that water resources are distributed efficiently, maximizing crop productivity.
7. AI-powered decision support systems can give farmers instant advice on irrigation plans while taking into consideration a variety of factors, such as crop type, soil quality, weather, and water availability.
8. Data Integration and Analysis: AI is capable of handling enormous volumes of data from several sources, including sensors, satellites, and archived documents. Artificial intelligence (AI) may use this data to find trends, patterns, and correlations that can guide improved water management strategies.
9. Autonomous Irrigation Systems: AI may be used in autonomous irrigation systems to allow for automated water flow and distribution adjustments depending on real-time inputs. As a result, less manual intervention is required, saving both time and money on labor.

Precision water management powered by AI not only enhances agricultural productivity but also promotes sustainable water use, conserving this valuable natural resource for future generations. However, it's essential to ensure that farmers have access to AI-based technologies and are properly trained to leverage the full potential of precision water management systems (Lamm et al., 2002).

4. DISEASE DETECTION AND CONTROL

Artificial Intelligence (AI) can significantly improve crop disease detection and control in agriculture, helping farmers identify and manage plant diseases more efficiently. Here's how AI can be applied to crop disease detection and control:

1. Image Recognition and Analysis: Images of plants and leaves may be analyzed by AI-powered image recognition systems to look for early indicators of disease or insect infestations. AI systems can pinpoint the precise illness impacting the crops by comparing these photos to a sizable database of recognized diseases.
2. Sensor-based Disease Monitoring: IoT devices can continually monitor the environmental conditions in the fields thanks to their different sensors, including temperature, humidity, and leaf wetness sensors. Real-time alerts and early warnings about prospective epidemics may be provided by AI, which can evaluate this data and link it with disease incidences.
3. Disease Pattern Recognition: In order to locate recurrent disease hotspots and forecast future outbreaks, AI can analyse historical data on crop diseases and weather trends. This knowledge aids farmers in taking preventative actions to stop or slow the spread of illness.
4. Precision Application of Pesticides: AI can optimize the use of pesticides by precisely identifying the areas affected by diseases and providing targeted recommendations for pesticide application. This reduces chemical usage, minimizes environmental impact, and saves costs for farmers.
5. Disease Diagnosis and Recommendations: AI-powered chatbots or mobile applications can assist farmers in diagnosing crop diseases by analyzing photos of affected plants. The AI system can then recommend suitable treatments or management strategies based on the identified disease.
6. Early Warning Systems: In order to create early warning systems for certain diseases, AI may incorporate data from a variety of sources, such as disease databases, weather predictions, and remote sensing. Farmers may get alerts and advice on disease prevention or control measures in a timely manner.
7. 7. Development of disease-resistant crops: AI can help with the analysis of massive genomic databases to find genes linked to disease resistance in crops. This knowledge may be utilized to develop new crop types that are more resistant to common illnesses, hence lowering the demand for chemical treatments.
8. 8. Disease Spread Prediction: AI can forecast how illnesses can spread throughout an area by taking into account elements like climate, wind patterns, and crop type. This makes it possible for farmers to successfully plan and carry out preventive actions.
9. Data Sharing and Collaboration: AI facilitates the sharing of disease-related data between farmers, researchers, and agricultural experts. This collaborative approach helps build a comprehensive understanding of disease dynamics and enables more informed decisions for disease control.

AI-based methods for agricultural disease identification and control improve crop output while also lowering the negative environmental effects of chemical use (Lee et al., 2017). However, for these AI-driven solutions to be fully utilized, successful deployment calls on readily available and trustworthy data, the incorporation of AI technology into current agricultural practices, and farmer training.

5. PEST MONITORING AND CONTROL

1. Pest Identification: Artificial intelligence (AI) can successfully recognize and identify numerous pests and insects from crop photos. AI systems can swiftly identify the individual bug harming the crops by comparing these photographs with a sizable database of known pests.
2. Sensor-based Pest Monitoring: IoT devices equipped with sensors, such as traps, cameras, and environmental sensors, can continuously monitor pest activities in the fields. These devices collect data on pest populations, movement patterns, and environmental conditions. AI processes this data to provide real-time insights into pest presence and behavior.
3. Early Warning Systems: AI can integrate data from multiple sources, including weather forecasts, pest databases, and historical data, to create early warning systems for specific pests. Farmers receive timely alerts and recommendations on pest prevention or control measures.
4. Precision Pest Control: AI enables precision application of pesticides by identifying the areas with high pest density and providing targeted recommendations for pesticide use. This reduces chemical usage, minimizes environmental impact, and optimizes pest control effectiveness.
5. Pest Lifecycle Prediction: AI can forecast the lifespan of pests by examining past data and environmental factors. Farmers may use this knowledge to better successfully plan their pest management plans and implement control measures at the pests' most susceptible life cycle stages.
6. Data-driven Pest Management: AI facilitates the analysis of vast datasets, including pest population dynamics, crop health, and weather conditions. Farmers can make informed decisions based on this data, leading to more efficient and proactive pest management practices.
7. Decision Support Systems: Farmers may obtain personalized pest management recommendations from AI-powered decision support systems based on their specific crop type, location, and bug problems.
8. Disease-Pest Interaction Analysis: AI can analyze data on crop diseases and pest infestations to identify potential interactions between pests and diseases. Understanding these interactions allows farmers to implement integrated pest management strategies that address multiple threats simultaneously.
9. Drone-based Pest Surveillance: Drones with AI capabilities and high-resolution cameras and sensors can swiftly survey enormous tracts of agriculture. They can spot early indications of insect infestations and assist farmers in quickly focusing on afflicted regions.
10. Biological Pest Control: AI can assist in identifying natural predators and beneficial insects that can control pest populations. By promoting natural pest control methods, farmers can reduce their reliance on chemical pesticides.

AI-based crop pest monitoring and control offer numerous benefits, including increased crop yields, reduced pesticide usage, and minimized environmental impacts. To fully leverage these advantages, it's crucial to integrate AI technologies into existing agricultural practices and provide farmers with the necessary training and support. To guarantee the acceptable and ethical use of AI in agriculture, data privacy and security should also be addressed (Liakos et al., 2018).

6. GROWTH AND YIELD FORECASTING

Artificial Intelligence (AI) can significantly improve crop growth and yield forecasting in agriculture by leveraging data analytics, machine learning, and predictive modeling techniques. Here's how AI can be applied to crop growth and yield forecasting:

1. Data Collection: Crop forecasting with AI begins with gathering pertinent data from multiple sources. This includes past weather information, soil characteristics, crop management techniques, and satellite pictures. Drones and IoT sensors may also deliver real-time information on crop health and environmental variables.
2. Machine Learning Models: Machine learning algorithms are used by AI to process and analyze the gathered data. In order to find patterns and links between crop growth, environmental conditions, and management practices, supervised learning models can be trained on historical data.
3. Weather Data Integration: Weather conditions have a significant impact on crop growth and yield. AI can integrate weather data from meteorological stations or satellites to assess how temperature, rainfall, humidity, and other weather variables influence crop development.
4. Remote Sensing and Imaging: In order to track crop health, development phases, and possible stresses like pests, illnesses, or nutritional deficits, AI may analyze satellite and drone footage. This real-time information improves the precision of yield forecasts.
5. Crop Growth Stage Recognition: AI-powered image recognition systems can identify specific growth stages of crops, enabling more accurate forecasting based on the current developmental phase.
6. Yield Prediction Models: AI can develop predictive models that consider multiple variables, including historical yield data, environmental conditions, and management practices, to forecast crop yields for the current season.
7. 7. Real-Time Monitoring and Updates: AI enables ongoing observation of crop development and environmental factors. The predictive algorithms may be adjusted to improve yield projections during the growing season as new data becomes available.
8. 8. Decision Support Systems: Based on the anticipated yield and different situations, AI-based decision support systems may provide farmers with insights and suggestions. This facilitates the improvement of crop management techniques and the making of knowledgeable decisions on irrigation, fertilization, and harvest time.
9. Crop Insurance and Risk Management: AI-based yield forecasting can be valuable for crop insurance companies and policymakers. Accurate yield predictions help assess risk and determine appropriate insurance coverage or allocate resources for support and disaster management.
10. Precision Agriculture Implementation: AI-driven yield forecasting enables farmers to adopt precision agriculture techniques. They can optimize resource usage, adjust planting densities, and implement targeted interventions based on specific areas' yield potential.

Farmers and governments may use AI-based crop growth and yield predictions to make data-driven choices, better resource allocation, and prepare for unforeseen issues. Accurate production forecasts also aid in reducing waste, promoting more sustainable farming practices, and boosting food security. To encourage the use of AI across various farming groups, it is crucial to assure data quality, accessibility, and privacy.

7. CONCLUSION AND FUTURE SCOPE

Artificial intelligence (AI) has made significant strides in agricultural surveillance and is redefining how farming is done. By integrating AI with various data sources, such as IoT sensors, remote sensing technologies, and historical records, farmers can access real-time, precise, and data-driven insights into crucial aspects of agriculture. AI-based systems enable early detection of crop health issues, soil conditions, weather patterns, and pest infestations, empowering farmers to make informed decisions, optimize resource utilization, and implement sustainable farming practices. The benefits of AI for monitoring agricultural essentials include increased productivity, resource efficiency, environmental sustainability, risk mitigation, and data-driven decision-making, fostering a more resilient and prosperous agriculture sector.

The future of AI for monitoring agricultural essentials looks promising, with several exciting developments on the horizon:

1. Advanced Sensing Technologies: Artificial intelligence (AI) is revolutionizing farming by making substantial advancements in agricultural surveillance.
2. AI-Driven Automation: Integration of AI with automated machinery and robotic systems will lead to autonomous farm operations, such as precision planting, harvesting, and weeding, further optimizing agricultural processes.
3. Edge Computing: Farmers will be able to react quickly to changing conditions if AI processing is done at the network's edge, closer to the data source. This will minimize latency and enable real-time decision-making.
4. Blockchain Integration: By combining AI and blockchain technology, the agricultural supply chain can be made more transparent and traceable, preserving the integrity and quality of the crop.
5. Multi-Modal Data Fusion: AI will keep enhancing the fusion of data from many sources, including satellite imaging, drones, weather stations, and ground-based sensors, to give a thorough understanding of agricultural ecosystems..
6. Personalized Recommendations: AI-based systems will offer personalized recommendations tailored to the specific needs of each farm, considering its unique characteristics and challenges.
7. Collaborative AI: AI systems that encourage data sharing and collaboration among farmers, researchers, and agricultural stakeholders will foster collective learning and accelerate agricultural innovation.
8. Climate Change Adaptation: AI will be essential in assisting farmers in adapting to climate change by advising climate-smart farming practises and offering climate-resilient methods.
9. Global Accessibility: By making AI technology more inexpensive and available to smallholder farmers in underdeveloped nations, the advantages of AI in agriculture will become more widely shared.

In conclusion, the future of Artificial Intelligence for monitoring agricultural essentials holds immense potential to revolutionize the agriculture sector further. As AI technologies continue to evolve and become more widespread, they will continue to drive sustainable and efficient agricultural practices, ensuring food security, environmental stewardship, and economic prosperity for farmers and the global community.

REFERENCES

Cillis, D., Pezzuolo, A., Marinello, F., & Sartori, L. (2018). Field-scale electrical resistivity profiling mapping for delineating soil condition in a nitrate vulnerable zone. *Applied Soil Ecology*, *123*, 780–786. doi:10.1016/j.apsoil.2017.06.025

Dela Cruz, J. R., Baldovino, R. G., Bandala, A. A., & Dadios, E. P. (2017, May). Water usage optimization of Smart Farm Automated Irrigation System using artificial neural network. In *2017 5th International Conference on Information and Communication Technology (ICoIC7)* (pp. 1-5). IEEE. 10.1109/ICoICT.2017.8074668

Gandhi, M., Kamdar, J., & Shah, M. (2020). Preprocessing of non-symmetrical images for edge detection. *Augmented Human Research*, *5*(1), 1–10. doi:10.100741133-019-0030-5

Garre, P., & Harish, A. (2018, December). Autonomous agricultural pesticide spraying uav. *IOP Conference Series. Materials Science and Engineering*, *455*, 012030. doi:10.1088/1757-899X/455/1/012030

Hanson, B., Orloff, S., & Sanden, B. (2007). *Monitoring soil moisture for irrigation water management* (Vol. 21635). University of California, Agriculture and Natural Resources.

Kim, Y., Evans, R. G., & Iversen, W. M. (2008). Remote sensing and control of an irrigation system using a distributed wireless sensor network. *IEEE Transactions on Instrumentation and Measurement*, *57*(7), 1379–1387. doi:10.1109/TIM.2008.917198

Kuyper, M. C., & Balendonck, J. (1997, August). Application of dielectric soil moisture sensors for real-time automated irrigation control. In *III International Symposium on Sensors in Horticulture 562* (pp. 71-79). Academic Press.

Lamm, R. D., Slaughter, D. C., & Giles, D. K. (2002). Precision weed control system for cotton. *Transactions of the ASAE. American Society of Agricultural Engineers*, *45*(1), 231.

Lee, J., Wang, J., Crandall, D., Šabanović, S., & Fox, G. (2017, April). Real-time, cloud-based object detection for unmanned aerial vehicles. In *2017 First IEEE International Conference on Robotic Computing (IRC)* (pp. 36-43). IEEE. 10.1109/IRC.2017.77

Liakos, K. G., Busato, P., Moshou, D., Pearson, S., & Bochits, D. (2018). Machine Learning in Agriculture: A Review. *Sensors*.

Luciani, R., Laneve, G., & JahJah, M. (2019). Agricultural monitoring, an automatic procedure for crop mapping and yield estimation: The great rift valley of Kenya case. *IEEE Journal of Selected Topics in Applied Earth Observations and Remote Sensing*, *12*(7), 2196–2208. doi:10.1109/JSTARS.2019.2921437

Schmid, T., Rodríguez-Rastrero, M., Escribano, P., Palacios-Orueta, A., Ben-Dor, E., Plaza, A., Milewski, R., Huesca, M., Bracken, A., Cicuendez, V., Pelayo, M., & Chabrillat, S. (2015). Characterization of soil erosion indicators using hyperspectral data from a Mediterranean rainfed cultivated region. *IEEE Journal of Selected Topics in Applied Earth Observations and Remote Sensing*, *9*(2), 845–860. doi:10.1109/JSTARS.2015.2462125

Chapter 8
Challenges and Solutions in the Agricultural Sector of India

Pratibha Verma
Constituent Government Degree College, Puranpur, India

Pratyush Bibhakar
Galgotias University, India

ABSTRACT

In this chapter, agriculture stands as the foundational pillar of the Indian economy, serving as a vital catalyst for growth and yielding substantial dividends in terms of revenue. Despite the commendable growth rate of the Indian economy, the state of agriculture remains less than satisfactory, marred by a multitude of disregarded challenges. India holds a prominent position as a leading supplier of essential commodities such as spices, pulses, saffron, and milk, significantly contributing to the nation's GDP and profoundly impacting the lives of the majority of its populace. Regrettably, these contributions have witnessed a decline. This study aims to elucidate the primary concerns and obstacles confronting Indian agriculture while also proposing potential remedies.

1. INTRODUCTION

The Indian economy draws partial reliance from agriculture, with a substantial portion of the population engaged in various forms of farming in F.R. (1971). These activities encompass horticulture, pisciculture, floriculture, and animal husbandry. It is undeniable that the implementation of the Green Revolution has led to a significant surge in crop production in Jha and Kumar (2006). Despite this, a considerable segment of the Indian populace continues to reside below the poverty threshold, grappling with issues of malnutrition. Regions characterized by limited rainfall have yet to witness substantial enhancements in both productivity and rural income. Those residing in remote and underdeveloped areas continue to bear the brunt of these challenges. Due to a lack of access to fundamental services such as finance, extension services, and agricultural inputs, a sense of distress has taken hold within the rural population. Farmers find themselves confined to small plots of land, contributing to the rise of economically

DOI: 10.4018/979-8-3693-0782-3.ch008

unviable holdings. This situation has led to the necessity of generating greater food output from limited resources. Addressing these gaps is imperative to alleviate the strain on scarce and strained natural resources. Furthermore, the existing infrastructure does not adequately support increased agricultural yields. Despite recent efforts to diversify into high-value fruit and vegetable varieties, the escalating demand for food products has contributed to a phenomenon of food inflation. Compounding the challenges, Indian farmers now face an increasing trend of trade deregulation, which compels them to compromise on both product quality and pricing before exporting to domestic and international markets. In many instances, farmers are coerced into tailoring their production to meet specific demands and quality standards, further exacerbating the complexities they already encounter. This has led to a proliferation of farming agreements. However, it is crucial to ensure that the interests of growers are not exploited. One effective approach is the advancement of innovative crop technologies. Pal and Jha (2007), in their research, unveiled a growing involvement of the private sector in research and development (R&D), with substantial investments primarily directed toward chemicals and food processing. According to the planning commission (2011), certain private sector research and industrial efforts prioritize crops with higher market value, inevitably resulting in a lack of attention for many other crops.

The Government of India (2011) reported that the rapid expansion of supermarkets and the agricultural sector has heightened the demand for food items, subsequently leading to sluggish supply in various commodities.

The main issues and challenges in Indian agriculture discussed in Jibran, and Mufti (2019).

2. CHALLENGES FACING INDIAN AGRICULTURE AND POTENTIAL SOLUTIONS

2.1 Small and Fragmented Land Holdings

One of the most pressing issues undermining Indian agriculture is the prevalence of small and unorganized land holdings. These fragmented plots of land, dispersed across the landscape, present a significant challenge in terms of effective management. This problem is particularly pronounced in densely populated states such as Kerala, West Bengal, Uttar Pradesh, and Bihar. The root cause of this issue lies in inheritance laws that mandate the division of land among multiple heirs, resulting in smaller and less productive plots. Consequently, this division hampers overall agricultural productivity and operational efficiency. Goyal etl. (2016) the logistical challenges associated with these small land parcels further compound the problem. The need to transport resources like manure, fertilizers, seeds, and livestock between these scattered plots consumes valuable time and resources. Additionally, the limited space makes implementing efficient irrigation systems a complicated endeavor, leading to wastage of fertile land. To address this, a potential solution involves the consolidation of these small land holdings into larger, more cohesive farms. By encouraging farmers to pool their land resources and collaborate, a shift from fragmented patches to more substantial and manageable agricultural units is possible. Such an approach could lead to increased productivity and shared profits among the participating farmers.

The challenge of small and fragmented land holdings in Indian agriculture can be effectively addressed through a combination of policy reforms, technological interventions, and community-driven initiatives in Kamble et al. (2018). Consolidating land holdings, promoting sustainable land use practices, and empowering small farmers are essential for enhancing agricultural productivity and ensuring rural livelihoods. Here are potential solutions to tackle the small and fragmented land holdings challenge:

1. **Land Consolidation and Land Banks:** Establishing land consolidation programs and land banks can facilitate the pooling of small land holdings. This approach allows for the creation of larger, more productive plots, making mechanization and modern farming practices more feasible.
2. **Promotion of Cooperative Farming:** Encouraging small farmers to form cooperatives can enable joint cultivation of land. This collaborative approach enhances resource-sharing, access to credit, and adoption of modern agricultural technologies.
3. **Leasing and Rental Arrangements:** Encouraging leasing or rental agreements among farmers can lead to larger, consolidated land holdings. Legal frameworks and awareness campaigns can ensure transparent and mutually beneficial arrangements.
4. **Technology Adoption:** Promoting mechanization, precision farming, and technology-driven practices can enhance productivity even on small land holdings. Mechanized equipment and advanced tools optimize resource utilization and improve crop yield.
5. **Promotion of High-Value Crops:** Encouraging the cultivation of high-value crops, such as fruits, vegetables, and horticultural products, on small plots can increase income for farmers. These crops have shorter cultivation cycles and can generate better returns.
6. **Crop Diversification:** Encouraging diversification of crops based on agro-climatic conditions can optimize land use and reduce risk. Crop rotation and mixed cropping can maximize yield from limited land resources.
7. **Integrated Farming Systems:** Implementing integrated farming systems that combine crops, livestock, and agroforestry can optimize land use and enhance farm income. This holistic approach improves resource efficiency.
8. **Access to Credit and Financial Services:** Facilitating access to credit for small farmers enables investment in productivity-enhancing measures. Microfinance institutions, farmer producer organizations, and government-sponsored schemes can provide financial support.
9. **Extension Services and Training:** Providing extension services and training to small farmers on modern farming practices, sustainable techniques, and crop management can enhance their skills and productivity.
10. **Land Tenure Reforms:** Implementing land tenure reforms that ensure secure land rights can empower small farmers. Legal recognition of land ownership provides a foundation for investment and growth.
11. **Promotion of Agribusiness and Value Addition:** Encouraging small farmers to engage in agribusiness activities such as processing, packaging, and marketing adds value to their produce. This generates additional income and reduces post-harvest losses.
12. **Digital Platforms for Market Access:** Developing digital platforms that connect small farmers directly to markets can eliminate intermediaries and ensure better prices for their produce.
13. **Promotion of Farmer Producers Organizations (FPOs):** Forming FPOs allows small farmers to collectively negotiate prices, access technology, and improve market linkages.
14. **Government Support and Policy Reforms:** Implementing policies that incentivize land consolidation, provide access to resources, and support technological adoption can empower small farmers.
15. **Community-Based Approaches:** Promoting community initiatives, knowledge-sharing platforms, and farmer collectives can empower small farmers and amplify their voices.

In conclusion, addressing the challenge of small and fragmented land holdings requires a comprehensive strategy that leverages policy interventions, technology adoption, and community participation. By enabling farmers to overcome resource limitations, adopt modern practices, and enhance their income, Indian agriculture can achieve sustainable growth and improve rural livelihoods.

2.2 Seeds Quality and Accessibility

The availability of high-quality seeds is a cornerstone of successful crop production. Regrettably, many farmers lack access to reliable sources of quality seeds, which subsequently affects the overall yield in Baruah etl. (2020). The high cost of these seeds places them out of reach for marginal farmers, leading to compromised crop quality. To mitigate this challenge, the Indian Government has established entities like the National Seeds Corporation (NSC) and the State Farms Corporation of India (SFCI) to promote the distribution of quality seeds among farmers. Furthermore, initiatives like the High Yielding Variety Programme (HYVP) launched in 1966 seek to enhance food grain production by introducing improved seed varieties. Enhancing seed quality and accessibility can substantially elevate agricultural productivity and diversify crop varieties to better suit India's diverse climatic conditions. Seed enhancement procedure performs as shown in Fig.1.

The challenges related to seeds quality and accessibility in Indian agriculture can be effectively tackled through a combination of policy measures, technological advancements, and community engagement. Ensuring reliable access to high-quality seeds is essential to enhance crop yield, improve agricultural productivity, and contribute to food security. Here are potential solutions to address these challenges:

Figure 1. Seed enhancement procedure

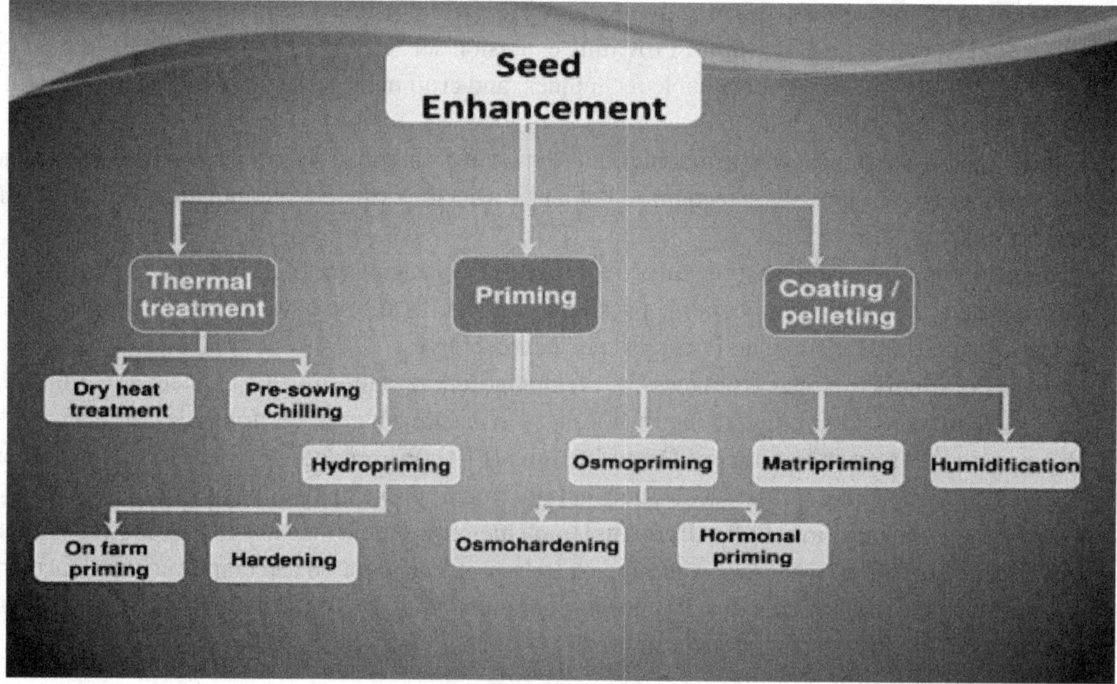

1. **Establishment of Seed Banks and Storage Facilities:** Creating seed banks at regional and local levels can ensure the availability of quality seeds to farmers. These banks can store diverse seed varieties, including traditional and climate-resilient ones, to meet varying agricultural needs. Adequate storage facilities with controlled conditions can maintain seed viability over extended periods.
2. **Public-Private Partnerships for Seed Production:** Collaborations between public institutions and private seed companies can increase the production of quality seeds. This partnership can ensure the availability of certified seeds to farmers while also promoting the development of improved seed varieties through research and development efforts.
3. **Strengthening National Seed Corporations:** Supporting and enhancing the role of organizations like the National Seeds Corporation (NSC) can bolster the supply of quality seeds. NSC can focus on producing and distributing certified seeds, including high-yielding and disease-resistant varieties, to farmers across different regions.
4. **Promotion of Farmer-Managed Seed Systems:** Encouraging farmers to engage in seed saving, selection, and exchange can enhance seed diversity and accessibility. Training farmers in seed management techniques, along with creating community seed banks, can empower them to take control of their seed resources.
5. **Introduction of Seed Vouchers and Subsidies:** Providing farmers with seed vouchers or subsidies for purchasing quality seeds can improve accessibility, especially for marginalized farmers. This approach ensures that cost is not a barrier to accessing seeds that can lead to higher crop productivity.
6. **Seed Quality Certification and Testing:** Establishing stringent quality control measures, including seed testing and certification, can ensure that seeds meet specified standards. Regular quality checks by dedicated agencies can prevent the distribution of subpar seeds in the market.
7. **Promotion of High-Yielding Varieties:** Building on the success of initiatives like the High Yielding Variety Programme (HYVP), continued research and promotion of high-yielding and climate-resilient seed varieties are essential. These varieties can better adapt to changing climatic conditions and contribute to improved crop yields.
8. **Capacity Building for Farmers:** Conducting training sessions, workshops, and demonstrations on seed selection, storage, and management can educate farmers about the importance of using quality seeds. Empowering farmers with knowledge can lead to better decision-making regarding seed choices.
9. **Digital Platforms for Seed Information:** Developing digital platforms or mobile applications that provide information about seed availability, prices, and quality can empower farmers to make informed seed purchasing decisions. These platforms can also facilitate direct interactions between farmers and seed suppliers.
10. **Community Seed Fairs and Exchanges:** Organizing community-based seed fairs and exchanges can promote the sharing of traditional and locally adapted seed varieties. This fosters a culture of seed diversity conservation and strengthens the resilience of farming systems.
11. **Investment in Research and Development:** Allocating resources to research and development can lead to the development of improved seed varieties tailored to local agro-climatic conditions. Public and private investment can drive innovation in seed technology.
12. **Regulation and Oversight:** Government regulatory bodies should enforce stringent quality standards for seed production, packaging, and labeling. This ensures that farmers receive seeds that are accurately represented and of the promised quality.

In conclusion, addressing the challenges related to seeds quality and accessibility requires a multi-pronged approach involving collaboration between government agencies, research institutions, seed companies, and the farming community. By promoting quality seed production, distribution, and education, India's agricultural sector can ensure a steady supply of seeds that contribute to increased agricultural productivity and food security.

2.3 Manures and Fertilizers

Sustainable soil health is imperative for consistent agricultural productivity. However, due to historical practices, soil quality has been depleted over time. This degradation has led to decreased productivity and crop yield. Implementing effective manure and fertilizer application can help rejuvenate soil health. Manures, akin to nutrition for humans, play a vital role in soil enrichment. Nutrient-rich soils yield enhanced productivity, contributing to up to 70% growth in overall production. While traditional sources of manure like cow dung remain popular among Indian farmers, chemical fertilizers are often expensive, posing affordability issues for resource-constrained farmers. The establishment of 52 fertilizer quality control laboratories across the country, alongside the Central Fertilizer Quality Control and Training Institute, demonstrates the government's commitment to maintaining the quality of available fertilizers.

The challenge of inadequate use of manures and fertilizers in Indian agriculture can be effectively addressed through a combination of sustainable practices, technological advancements, and policy reforms. Enhancing soil fertility and nutrient management is crucial for improving crop productivity and ensuring long-term agricultural sustainability. Here are potential solutions to tackle the manures and fertilizers challenge:

1. **Promotion of Organic Farming:** Encouraging the adoption of organic farming practices can reduce the reliance on chemical fertilizers. Organic methods, such as composting, green manuring, and crop rotation, enhance soil fertility naturally while minimizing environmental impacts as shown in Fig.1.
2. **Integrated Nutrient Management:** Implementing integrated nutrient management involves combining organic sources of nutrients with judicious use of chemical fertilizers. This approach optimizes nutrient availability to crops, reduces dependency on external inputs, and improves soil health.
3. **Soil Testing and Nutrient Balancing:** Conducting regular soil testing to assess nutrient levels helps farmers apply fertilizers more precisely. Providing farmers with personalized nutrient management recommendations based on soil tests can prevent overuse of fertilizers and minimize nutrient imbalances.
4. **Site-Specific Nutrient Application:** Utilizing precision agriculture techniques, such as variable rate fertilization, ensures that fertilizers are applied where they are needed most. This reduces wastage, improves nutrient efficiency, and minimizes environmental pollution.
5. **Fertilizer Quality Control:** Strengthening the regulatory framework for fertilizer quality control can ensure that farmers receive fertilizers with accurate nutrient content. Regular monitoring and testing of fertilizers in laboratories can maintain their quality standards.
6. **Promotion of Biofertilizers and Microbial Inoculants:** Encouraging the use of biofertilizers and microbial inoculants enhances soil biological activity and nutrient availability. These natural

Figure 2. Promotion of organic farming

products can improve nutrient uptake, reduce reliance on chemical fertilizers, and contribute to sustainable agriculture.

7. **Nutrient Recycling and Resource Management:** Promoting the recycling of organic waste, such as crop residues and livestock manure, can provide valuable nutrients to the soil. Efficient resource management through composting and vermicomposting reduces waste and enriches soil fertility.
8. **Public-Private Partnerships for Fertilizer Distribution:** Collaborations between public institutions and private fertilizer companies can ensure the availability of quality fertilizers to farmers. Government oversight can help prevent adulteration and ensure fair pricing.
9. **Education and Training for Farmers:** Conducting training programs and workshops on nutrient management can educate farmers about the importance of balanced fertilization. Teaching them about the 4Rs (Right Source, Right Rate, Right Time, Right Place) of nutrient management can guide their fertilizer application practices.
10. **Digital Tools for Fertilizer Management:** Developing mobile apps or online platforms that provide fertilizer recommendations based on crop type, soil type, and local conditions can empower farmers with accurate information for efficient fertilizer use.
11. **Government Subsidies and Incentives:** Providing subsidies and incentives for organic fertilizers, biofertilizers, and micro-nutrients can encourage farmers to adopt more sustainable nutrient management practices.
12. **Research and Development:** Investing in research to develop crop varieties that are more responsive to nutrient-efficient practices can lead to improved crop yield with reduced fertilizer input.
13. **Awareness Campaigns:** Launching public awareness campaigns about the benefits of balanced fertilization, environmental conservation, and sustainable farming practices can encourage a mindset shift among farmers.

In conclusion, addressing the challenge of manures and fertilizers requires a holistic approach that combines scientific knowledge, technological innovations, and policy support. By adopting sustainable nutrient management practices and fostering responsible fertilizer use, Indian agriculture can achieve higher productivity, enhance soil health, and contribute to long-term food security.

2.4 Irrigation Challenges

In a tropical country like India, irrigation holds paramount importance due to the erratic nature of rainfall. India stands as the second-largest irrigated country globally, second only to China. Ensuring an effective irrigation system is crucial for sustaining agricultural growth, as heavy reliance on rainfall proves insufficient. However, a balance must be struck to prevent issues of over-irrigation and the potential consequences it brings. In conclusion, Indian agriculture grapples with several complex challenges, including fragmented land holdings, inadequate access to quality seeds, soil degradation, and irrigation issues. Addressing these challenges requires a comprehensive approach involving policy changes, technological advancements, and collaborative efforts among farmers and relevant authorities in Tripathi etl. (2023). By collectively addressing these issues, Indian agriculture can move towards sustainable growth and enhanced food security.

The challenges related to irrigation in Indian agriculture can be effectively mitigated through a combination of water management strategies, technological innovations, and policy reforms. Ensuring efficient water use, expanding irrigation infrastructure, and promoting sustainable practices are essential to enhance crop yield, water availability, and agricultural productivity as shown in Fig.2. Here are potential solutions to tackle the irrigation challenges:

1. **Diverse Irrigation Techniques:** Promoting a mix of irrigation techniques such as drip, sprinkler, and micro-irrigation systems can optimize water usage by delivering water directly to plant roots. These methods reduce water wastage and enhance crop water efficiency.
2. **Rainwater Harvesting:** Encouraging the construction of rainwater harvesting structures, such as ponds, tanks, and check dams, can capture and store rainwater for agricultural use. This technique enhances groundwater recharge and provides a supplementary water source.
3. **Modernization of Canal Systems:** Upgrading traditional canal systems with efficient water distribution mechanisms and sensor-based monitoring can prevent water losses due to seepage and evaporation.
4. **Groundwater Management:** Implementing groundwater management practices, including aquifer recharge and regulation of excessive extraction, can ensure sustainable use of this vital resource.
5. **Smart Irrigation Technologies:** Adopting smart irrigation technologies that utilize weather forecasts and soil moisture sensors can enable precise irrigation scheduling. This minimizes water wastage and optimizes crop water requirements.
6. **Micro-Level Water Budgeting:** Implementing micro-level water budgeting at the farm level helps farmers estimate water needs for each crop cycle. This empowers them to manage water efficiently and avoid over-irrigation.
7. **Promotion of Drought-Resistant Crops:** Researching and promoting drought-resistant crop varieties can reduce water dependency in regions prone to water scarcity. These varieties require less water while maintaining yield potential.
8. **Efficient Use of Greywater:** Encouraging the use of treated greywater from households and industries for irrigation purposes can alleviate pressure on freshwater sources.
9. **Floodplain Management:** Implementing floodplain management techniques, such as flood-resistant crops and contour bunding, can harness excess monsoon water for irrigation while preventing soil erosion.
10. **Community-Based Water Management:** Establishing water user associations and community-driven water management systems can enhance equitable water distribution and promote collective responsibility.
11. **Policy Support for Water Conservation:** Implementing policies that incentivize water-efficient practices, provide subsidies for efficient irrigation systems, and penalize wasteful water use can drive adoption of sustainable irrigation techniques.

Figure 3. Smart irrigation technologies

12. **Reforming Water Pricing:** Pricing water based on its scarcity value and agricultural water needs can encourage efficient use and prevent over-exploitation.
13. **Awareness and Training:** Conducting workshops, training programs, and awareness campaigns for farmers on efficient irrigation practices and water conservation can foster behavior change.
14. **Remote Sensing and GIS Technologies:** Utilizing remote sensing and Geographic Information System (GIS) technologies can help monitor and manage irrigation efficiency across larger landscapes.
15. **Research and Development:** Investing in research to develop crop varieties with higher water-use efficiency can lead to better crop yields with reduced water input.

In conclusion, addressing the irrigation challenges in Indian agriculture requires a multi-dimensional approach that combines technological innovation, sustainable practices, and policy reforms. By optimizing water usage, improving irrigation infrastructure, and fostering responsible water management practices, India's agricultural sector can achieve higher productivity, enhance food security, and contribute to water resource sustainability.

2.5. Dependence on Manual Labour

A significant obstacle hindering the progress of Indian agriculture is the prevalent reliance on manual labor for various tasks such as plowing, irrigation, weeding, harvesting, and threshing. This dependence on human labor not only leads to inefficiencies but also results in the underutilization of human resources that could be better directed towards more skilled and productive activities. The integration of mechanization into these processes holds the key to achieving enhanced outcomes. Mechanized operations not only optimize labor utilization but also enhance the overall efficiency of farming activities. In the years following independence, mechanization in agriculture has made noteworthy strides, with the increasing adoption of tools like tractors, power tillers, combine harvesters, and irrigation pumps. The proportion of mechanical and electrical power in agricultural operations has grown significantly from 40% in 1971 to 84% in 2003-04. To further this progress, continuous efforts are required to encourage farmers to embrace advanced agricultural equipment that enables timely and precise farming activities, thus streamlining the production process.

The challenge of heavy dependence on manual labor in Indian agriculture can be effectively addressed through a combination of mechanization, skill development, policy reforms, and community participation. Enhancing the efficiency of farming operations, reducing labor-intensive tasks, and improving rural livelihoods are essential to modernize the agricultural sector. Here are potential solutions to tackle the dependence on manual labor:

1. **Promotion of Mechanized Farming:** Encouraging the adoption of modern machinery such as tractors, power tillers, and combine harvesters can reduce the reliance on manual labor for plowing, planting, and harvesting. Government subsidies and credit facilities can make these machines accessible to small farmers.
2. **Custom Hiring Centers:** Establishing custom hiring centers where farmers can rent machinery for specific tasks can make mechanization affordable and accessible to smallholders who may not be able to purchase expensive equipment.
3. **Training and Capacity Building:** Offering training programs on the operation and maintenance of agricultural machinery can empower farmers with the skills needed to effectively use modern equipment.
4. **Women Empowerment in Agriculture:** Promoting the involvement of women in agriculture by providing training and access to modern tools can reduce the drudgery of labor-intensive tasks traditionally performed by women.
5. **Innovative Tools and Technologies:** Introducing labor-saving tools such as weeders, transplanters, and seeders can significantly reduce manual labor requirements and enhance productivity.
6. **Promotion of Agroecological Practices:** Encouraging agroecological practices like cover cropping and intercropping can minimize the need for labor-intensive activities such as weeding and pest control.
7. **Digital Platforms for Labor Sharing:** Developing digital platforms where farmers can find and share labor for specific tasks can efficiently allocate human resources and reduce labor demand during peak seasons.
8. **Government Subsidies and Incentives:** Providing subsidies for machinery purchase, along with tax incentives for equipment manufacturers, can encourage mechanization and technological advancements.
9. **Collaboration With Research Institutions:** Partnering with agricultural research institutions can lead to the development of specialized machinery suited to the needs of Indian farmers.
10. **Community-Based Mechanization:** Establishing farmer cooperatives that collectively own and operate machinery can reduce costs and increase access to mechanization.
11. **Credit Facilities:** Facilitating easy access to credit for purchasing machinery through formal financial institutions can alleviate the financial burden on farmers.
12. **Extension Services and Awareness Campaigns:** Conducting extension services and awareness campaigns can educate farmers about the benefits of mechanization, dispelling myths and misconceptions.
13. **Promotion of Startups and Innovation:** Encouraging startups to develop affordable and adaptable agricultural machinery tailored to the needs of small farmers can drive innovation in the sector.
14. **Policy Reforms:** Implementing policies that prioritize mechanization, streamline equipment import and manufacturing processes, and provide regulatory support can foster a conducive environment for modernization.
15. **Research and Development:** Investing in research and development to create energy-efficient and environment-friendly machinery can make mechanization sustainable in the long run.

In conclusion, addressing the challenge of dependence on manual labor in Indian agriculture requires a multi-pronged approach that leverages mechanization, skill development, policy support, and technological innovation. By adopting modern farming practices, farmers can enhance productivity, reduce drudgery, and contribute to the overall growth and sustainability of the agricultural sector.

2.6. Marketing of Agricultural Produce

The marketing of agricultural produce remains a critical challenge for rural India. The absence of robust marketing infrastructure forces farmers to rely on local traders and intermediaries, often leading to exploitative pricing. This situation, compounded by debt pressures, compels farmers to sell their produce at suboptimal rates. Establishing organized marketing systems can help mitigate these challenges. Government intervention to regulate markets, standardize weights and measures, and prevent malpractices is essential to create a fair market environment that benefits both producers and consumers.

The challenge of marketing agricultural produce in India can be effectively addressed through a combination of infrastructure development, market reforms, technological advancements, and capacity building. Ensuring efficient and transparent market access, minimizing intermediaries, and enhancing farmer incomes are essential to modernize the agricultural marketing system in Rajput etl. (2023). Here are potential solutions to tackle the marketing of agricultural produce challenge:

1. **Strengthening Market Infrastructure:** Investing in the development of modern market infrastructure, including cold storage facilities, warehouses, and transportation networks, can reduce post-harvest losses and improve the quality of produce.
2. **Establishment of Agri-Marketing Cooperatives:** Encouraging the formation of farmer producer organizations (FPOs) and cooperatives can empower farmers to collectively market their produce, negotiate better prices, and access bulk buyers.
3. **Electronic Mandis and Online Platforms:** Developing online marketplaces and mobile applications can enable direct interaction between farmers and buyers, reducing the need for intermediaries and ensuring fair prices for produce.
4. **Contract Farming:** Promoting contract farming agreements between farmers and agribusinesses can provide a guaranteed market for produce and reduce price fluctuations.
5. **Price Information Systems:** Setting up real-time price information systems through mobile apps, SMS services, and digital kiosks can help farmers make informed decisions about when and where to sell their produce.
6. **Removal of Intermediaries:** Streamlining the marketing chain by reducing the number of intermediaries can lead to better price realization for farmers. This can be achieved through direct marketing and improved market linkages.
7. **Market Intelligence and Research:** Conducting market research and analysis to identify demand trends and consumer preferences can guide farmers in producing crops that are in demand.
8. **Value Addition and Processing:** Encouraging value addition through processing, packaging, and branding can open up new market opportunities and increase the shelf life of produce.
9. **Promotion of Export:** Supporting the export of agricultural products by providing export incentives, facilitating quality certification, and ensuring compliance with international standards can diversify markets for farmers.

10. **Standardization and Grading:** Implementing quality standards and grading systems can enhance the marketability of produce, ensuring that buyers receive consistent quality.
11. **Market Access for Smallholders:** Facilitating market access for small and marginal farmers through aggregation centers, collective selling, and access to transportation can level the playing field.
12. **Awareness and Capacity Building:** Conducting training programs and workshops on modern marketing techniques, negotiation skills, and understanding market dynamics can empower farmers to navigate the market more effectively.
13. **Policy Reforms:** Implementing policies that deregulate markets, encourage private investment, and ensure fair pricing can create a conducive environment for agricultural marketing.
14. **Supply Chain Management:** Integrating supply chains through collaboration between farmers, processors, distributors, and retailers can streamline the movement of produce from farm to market.
15. **Government Support:** Providing financial assistance, subsidies, and incentives for setting up farmer-friendly market infrastructure can improve market access for rural farmers.

In conclusion, addressing the challenge of marketing agricultural produce requires a holistic approach that combines market reforms, technology adoption, capacity building, and policy support. By creating efficient and transparent market systems, Indian agriculture can achieve higher income for farmers, minimize wastage, and contribute to overall sectoral growth.

2.7. Inadequate Storage Facilities

A lack of adequate storage facilities predominantly plagues rural India. Farmers, constrained by inadequate storage options, are compelled to sell their produce immediately after harvest, often at unfavorable prices. Developing efficient storage infrastructure is crucial to prevent post-harvest losses and enable farmers to capitalize on better market conditions. Proper storage facilities not only safeguard farmers' interests but also contribute to stable pricing for consumers.

The challenge of inadequate storage facilities in Indian agriculture can be effectively addressed through a combination of infrastructure development, technology adoption, policy reforms, and community-driven initiatives. Improving storage capabilities, reducing post-harvest losses, and ensuring food security are crucial for modernizing the agricultural sector. Here are potential solutions to tackle the inadequate storage facilities challenge:

1. **Expansion of Cold Storage and Warehousing:** Investing in the establishment and expansion of cold storage facilities, warehouses, and silos can provide farmers with the means to store perishable and non-perishable produce for extended periods.
2. **Community-Based Storage Solutions:** Encouraging communities to set up decentralized storage units and community grain banks can help small farmers store their produce collectively and reduce individual costs.
3. **Modern Storage Technologies:** Introducing innovative storage technologies such as controlled atmosphere storage, vacuum packaging, and modified atmosphere packaging can extend the shelf life of produce.
4. **Capacity Building for Farmers:** Providing training to farmers on proper storage techniques, pest management, and quality preservation can empower them to store their produce effectively.

5. **Promotion of On-Farm Storage:** Encouraging farmers to invest in on-farm storage structures such as granaries, bins, and small storage sheds can reduce post-harvest losses and provide immediate access to produce.
6. **Public-Private Partnerships:** Collaborations between the government, private sector, and farmer cooperatives can lead to the construction of storage facilities, ensuring widespread coverage.
7. **Mobile Storage Units:** Introducing mobile storage units that can be moved to areas with surplus produce can prevent wastage and ensure the efficient utilization of storage resources.
8. **Solar-Powered Cold Storage:** Implementing solar-powered cold storage units in areas with unreliable electricity can provide farmers with a reliable means of preserving their produce.
9. **Financial Support and Subsidies:** Providing financial assistance and subsidies to farmers and cooperatives for constructing storage facilities can lower the financial burden on farmers.
10. **Integration With Supply Chains:** Integrating storage facilities with supply chains, markets, and transportation networks can streamline the movement of produce and reduce bottlenecks.
11. **Real-Time Monitoring:** Implementing real-time monitoring systems that track temperature, humidity, and inventory levels can ensure that stored produce remains in optimal condition.
12. **Climate-Resilient Storage:** Designing storage structures that can withstand extreme weather conditions, such as floods and cyclones, can prevent damage to stored produce.
13. **Policy Support:** Implementing policies that incentivize private investment in storage infrastructure and provide tax benefits can encourage the growth of storage facilities.
14. **Research and Innovation:** Supporting research and development to create cost-effective and sustainable storage technologies can drive innovation in the sector.
15. **Public Awareness and Education:** Conducting awareness campaigns about the importance of proper storage practices can educate farmers and reduce post-harvest losses.

In conclusion, addressing the challenge of inadequate storage facilities requires a comprehensive approach that leverages technology, infrastructure development, policy support, and community participation. By enhancing storage capabilities, India's agricultural sector can minimize post-harvest losses, ensure food security, and contribute to overall rural development.

2.8. Transportation Challenges

The absence of efficient and affordable transportation options presents a significant hurdle for Indian agriculture. Many villages lack proper connectivity to highways or main roads, limiting farmers' access to broader markets with better pricing prospects. Enhancing transportation infrastructure is vital to enable farmers to tap into previously inaccessible markets and fetch competitive prices for their produce.

The challenge of transportation in Indian agriculture can be effectively tackled through a combination of infrastructure development, technological interventions, policy reforms, and community-driven initiatives. Enhancing transportation networks, reducing logistical constraints, and improving market access are crucial for modernizing the agricultural supply chain. Here are potential solutions to address the transportation challenges:

1. **Rural Road Infrastructure Development:** Investing in the development of rural road networks connecting agricultural areas to markets and transportation hubs can improve accessibility and reduce transportation costs.

2. **Last-Mile Connectivity:** Focusing on improving last-mile connectivity through feeder roads and village linkages can facilitate the efficient movement of agricultural produce.
3. **Transportation Hubs and Aggregation Centers:** Establishing transportation hubs and aggregation centers near production clusters can facilitate the consolidation of produce, reducing the number of trips and optimizing transportation.
4. **Cold Chain Infrastructure:** Expanding the cold chain infrastructure, including refrigerated trucks and storage facilities, can help transport perishable goods over longer distances without quality deterioration.
5. **Use of Information Technology:** Implementing digital platforms and mobile applications for real-time tracking of transportation routes and schedules can enhance logistical efficiency.
6. **Shared Transport Models:** Promoting shared transport models where multiple farmers share a single vehicle for transporting produce can reduce transportation costs for smallholders.
7. **Rail and Water Transport:** Exploring the use of rail and waterways for bulk transportation of agricultural produce can be more cost-effective for long distances.
8. **Public-Private Partnerships:** Collaborating with private transport companies to provide efficient and reliable transportation services for farmers can bridge the transportation gap.
9. **Capacity Building for Farmers:** Providing training to farmers on proper packaging, loading, and unloading techniques can minimize damage during transportation and improve produce quality.
10. **Time-Sensitive Transportation:** Promoting time-sensitive transportation for perishable goods through dedicated routes and fast-track procedures can reduce spoilage.
11. **Weather-Resilient Transport:** Upgrading transportation infrastructure to withstand adverse weather conditions, such as floods and monsoons, can prevent disruptions in supply chains.
12. **Warehousing and Storage at Transportation Hubs:** Establishing temporary storage facilities at transportation hubs can enable farmers to store their produce safely before it is transported to markets.
13. **Route Optimization:** Implementing route optimization algorithms and software can help identify the most efficient routes for transportation, reducing travel time and costs.
14. **Multi-Modal Transport Integration:** Integrating multiple modes of transport, such as road, rail, and waterways, can create a seamless transportation network that maximizes efficiency.
15. **Policy Reforms:** Implementing policies that incentivize private investment in transportation infrastructure, promote inter-state movement, and simplify transport regulations can enhance the transportation landscape.

In conclusion, addressing the transportation challenges in Indian agriculture requires a comprehensive strategy that combines infrastructure development, technological innovation, policy support, and community involvement. By improving transportation networks and reducing logistical bottlenecks, India's agricultural sector can enhance market access, reduce post-harvest losses, and contribute to overall rural development.

2.9 Emerging Challenges in Indian Agriculture

Indian agriculture continues to grapple with multifaceted challenges, despite substantial efforts and investments. Rapid climate changes and the need to meet growing demands within shorter timeframes present formidable challenges. The prevalence of traditional practices and reluctance to adopt new technologies hinder progress in rural India. Additionally, the high population density exerts pressure on limited arable land, contributing to further fragmentation of holdings. Addressing challenges in land tenure systems, such as exploitative rent and ownership practices, is also essential to empower farmers and foster equitable growth.

3. SOLUTION TO EMERGING CHALLENGES IN INDIAN AGRICULTURE

To overcome the emerging challenges facing Indian agriculture, a comprehensive and strategic approach is required. The integration of technological advancements, policy reforms, and community participation is essential to ensure the sustainability and growth of the agricultural sector. Here are potential solutions to address these challenges:

1. **Climate-Resilient Farming Practices:** Implementing climate-resilient agricultural practices is crucial to mitigate the impact of changing weather patterns. This involves promoting drought-resistant and flood-tolerant crop varieties, improving soil health, and adopting water-efficient irrigation techniques.
2. **Promotion of Agricultural Technology:** Raising awareness and providing training to farmers about modern agricultural technologies can help overcome technological reluctance. Government initiatives, farmer education programs, and collaborations with research institutions can facilitate the adoption of advanced farming practices and tools.
3. **Land Consolidation and Cooperative Farming:** Encouraging land consolidation through incentives and promoting cooperative farming can address land fragmentation. This allows for efficient use of resources, sharing of machinery, and better access to markets.
4. **Land Tenure Reforms:** Enacting land tenure reforms that ensure secure land rights for farmers can empower them and enable long-term planning. This can be achieved through legal changes and community awareness campaigns.
5. **Sustainable Soil Management:** Implementing soil conservation techniques such as crop rotation, cover cropping, and organic farming can restore soil health. Government support in providing training, subsidies, and incentives can encourage farmers to adopt sustainable practices.
6. **Efficient Water Management:** Investing in efficient irrigation systems like drip and sprinkler irrigation, alongside rainwater harvesting, can alleviate water scarcity issues. Awareness campaigns and subsidies for adopting water-saving technologies can be beneficial.
7. **Agricultural Marketing Reforms:** Strengthening agricultural marketing infrastructure, establishing farmer-producer organizations, and promoting direct market access can help farmers fetch fair prices for their produce. Technology-enabled platforms can bridge the gap between farmers and consumers.
8. **Rural Development Initiatives:** Creating non-farm employment opportunities in rural areas through agribusiness, agro-processing, and value addition can curb rural-urban migration. This requires investments in skill development and entrepreneurship training.
9. **Diversification and Nutrition Enhancement:** Encouraging diversification of crops to include nutritious and climate-resilient varieties can address changing dietary patterns. Government subsidies and awareness campaigns can promote the cultivation and consumption of these crops.

10. **Sustainable Resource Management:** Implementing sustainable agricultural practices that conserve resources and reduce environmental impact is essential. This involves promoting integrated pest management, organic farming, and agroforestry.
11. **Research and Development Investment:** Allocating resources to agricultural research and development can lead to innovations tailored to address specific challenges. This includes developing disease-resistant crop varieties, climate-adaptive technologies, and precision farming techniques.
12. **Policy and Institutional Reforms:** Policy changes that prioritize agriculture, provide financial incentives, and streamline regulations can create an enabling environment for growth. Strengthening agricultural extension services and promoting farmer cooperatives can enhance support structures.
13. **Community Engagement and Empowerment:** Involving local communities in decision-making processes, providing them with access to knowledge and resources, and promoting participatory approaches can lead to sustainable agricultural development.

In conclusion, addressing the emerging challenges in Indian agriculture requires a multi-faceted approach that leverages technology, policy reforms, and community participation. Collaboration between government agencies, research institutions, NGOs, and the farming community is essential for effective implementation. By embracing innovation and sustainable practices, India's agricultural sector can navigate these challenges and thrive in an evolving landscape.

4. CONCLUSION

The Indian agricultural sector faces a range of challenges that necessitate comprehensive and coordinated efforts from both the government and the agricultural community. Mechanization, marketing reforms, storage enhancement, transportation infrastructure development, accessible capital, and streamlined supply chains are crucial areas that require attention. While challenges persist, a concerted approach toward innovative solutions holds the potential to transform Indian agriculture and ensure sustainable growth.

REFERENCES

Baruah, I., & Baruah, G. (2020). Improvement of Seed Quality: A Biotechnological Approach. In A. K. Tiwari (Ed.), *Advances in Seed Production and Management*. Springer. doi:10.1007/978-981-15-4198-8_26

Frankel, F.R. (1971). *India's Green Revolution. Economic Gains and Politir*. I Costs Bombay.

Government of India. (2011). *State of the Economy and Prospects Economic Survey*. http:// exim.indiamart.com/economic-survey10-11/pdfs/ echap-01.pdf

Goyal. (2016). Indian Agriculture and Farmers-Problems and Reforms. Academic Press.

Jha, D., & Kumar, S. (2006). *Research resource allocation in Indian agriculture, Policy Paper 23*. National Centre for Agricultural Economics and Policy Research.

Jibran, S., & Mufti, A. (2019). Issues and challenges in Indian agriculture. *International Journal of Commerce and Business Management*, *12*(2), 85–88. doi:10.15740/HAS/IJCBM/12.2/85-88

Kamble, P. (2018). *Sustainability of Indian Agriculture of Indian Agriculture: Challenges and Opportunities*. Academic Press.

Pal, S., & Jha, D. (2007). Public-private partnerships in Agricultural R&D: Challenges and Prospects. In V. Ballabh (Ed.), *Institutional Alternatives and Governance of Agriculture*. Academic Foundation.

Planning Commission. (2011). *Draft on Faster, Sustainable and More Inclusive Growth – An approach to Twelfth Five Year Plan*. Author.

Rajput, S., Khanna, L., & Kumari, P. (2023). *Artificial Intelligence and Machine Learning-Based Agriculture*. doi:10.4018/978-1-6684-6418-2.ch002

Sharma V. (2020). Major Agricultural Problems of India and Various Government Initiatives. doi:10.13140/RG.2.2.33455.76968

Tripathi, P., Kumar, N., Rai, M., Shukla, P. K., & Verma, K. N. (2023). *Applications of Machine Learning in Agriculture*. doi:10.4018/978-1-6684-6418-2.ch006

Chapter 9
Conventional to Modern Agriculture Using Artificial Intelligence

Smrati Tripathi
Ajay Kumar Garg Engineering College, India

Prasen Jeet
Ajay Kumar Garg Engineering College, India

Rijwan Khan
https://orcid.org/0000-0003-3354-3047
School of Computing, Galgotias University, India

Akhilesh Kumar Srivastava
ABES Engineering College, India

ABSTRACT

Because of increased pollution and population, the agriculture system is being affected drastically. New technologies should be implemented in this sector. The author discussed irrigation techniques and patterns using artificial intelligence that can change our irrigation process and minimize water consumption. Artificial intelligence approaches for soil fertility properties, the connection between soil quality, effect of fertiliser on soil possession and crop fecundity, season's growth, modelling of some soil properties, and estimation of polluted soil are being discussed. One more essential factor that reduces crop growth, quality, and yield is called weeds. Different types of weed control applications using ANN and ANFIS model have been discussed. To minimize monetary losses for farmers, disease control becomes a necessary parameter for better crop production. Image-based strategies using machine learning and deep learning for exact location, order of the disease, for precise and accurate identification have been considered.

DOI: 10.4018/979-8-3693-0782-3.ch009

INTRODUCTION

The day the human race exit, it has three basic requirements, food, cloth, and shelter. Among all three foods is the most essential requirement, in its absence, we cannot even imagine our beautiful human race. With the growing population of the world, we must propagate our agriculture system for better yield, operationally efficient, strong enough to deal with changing climate situations, with decreasing size of cultivable land area, enhanced quality, and various food options for fulfilling the needs of upcoming generations and it should be increased by 150% by 2050 (Chukwu et al., 2019). In this scenario, when the population is increasing drastically, the refurbishment of agriculture in India, which is based on agriculture even its economy depends on it, becomes very crucial. Our economy will become stronger if there is a large production and this target can be achieved only if we introduce smart agriculture in our farming system. Smart agriculture can unbelievably enhance the production of our farms and this way there will be a significant increase in the economy of our country and can be a reason of increased growth in the world economy because it is an industry of 5 trillion dollars all over world.

Growth per capita income has been declining in the past few years. There will be an estimated increase in population by 10% and a decrease in food production by 20% as per United States Department of Agriculture. There is a positive indication in demand of food by 58 to 98% but a negative indication in supply. This negative indication will break the food supply chain in future. These are the adverse effects of increasing population and lesser involvement of technology in the agriculture field.

Artificial intelligence is nothing but a machine vision available for humans in which the machine recognises its habitat and takes necessary measures that maximize its chances of attaining its goals. Its main goals are attaining, knowledge, mining, representing, planning, processing, and perceiving.

LITERATURE REVIEW

AI Simulation

AI simulates machines with a person's intellect, and these machines are set in such a way that they can give judgment and act like humans. In other words, we can say it is a machine that thinks and acts like humans. Computer systems are cloned to work as specialists, capable of processing the native tongue and recognizing oral expression in artificial intelligence. Artificial Intelligence (AI) and new inventions in Deep Learning are strengthening the class of artificial speech. The use of these implementations is very common.

To increase food security for future generations, effective and smart techniques like precision farming and smart agriculture must be adopted. Despite the availability of various advancements in technology, agriculture systems have not yet fully embraced the use of AI and different IoT platforms and devices. AI systems are skilled at coming up with prognostic vision regarding which crop to plant in a particular year and when the required dates to sow and harvest are in a particular land area, thus enhancing crop yields and minimizing water requirements, crop yield enhancers, and weed killers. With the help of the implementation of AI technologies, the effect on the ecosystem can be minimized, and employee safety may improve, which will, in turn, keep food prices lower and ensure that food production remains balanced with the increasing population. AI sets the parameters required to provide more suitable climate-smart agricultural systems for healthy food and a comfortable living for the farmers.

The implementation of AI in agriculture will change the entire growth of the economy, which, in turn, affects our world socially, economically, and politically. Artificial intelligence is for humans; without humans, this intelligence has no meaning, and this implies human intervention will always be there, so being jobless is not a big concern.

Working of AI

As it is highly hyped that AI should be used for smart work in various domains, traders dealing with brands and benevolence in AI have a tussle related to their publicity. They believe that AI is one component, just like machine learning, but it cannot be reduced to a single feature since human intelligence cannot be defined by a single trait. AI combines hardware and software, and it also involves the use of various computer languages. The most supported languages for AI are Java, Python, and R.

According to Katariya et al. (2015), in AI, we convert a machine with human characteristics and competence into sophisticated technical variations with planned obsolescence. This way, we enhance computational prowess. Jha et al. (2019), explain the three paramount technologies being used in AI: expert systems, artificial neural networks, and fuzzy systems. In this paper, we will explore how AI techniques can be utilized to address challenges in agriculture.

Figure 1. AI-based smart agriculture control device
Source: Syers et al. (1997)

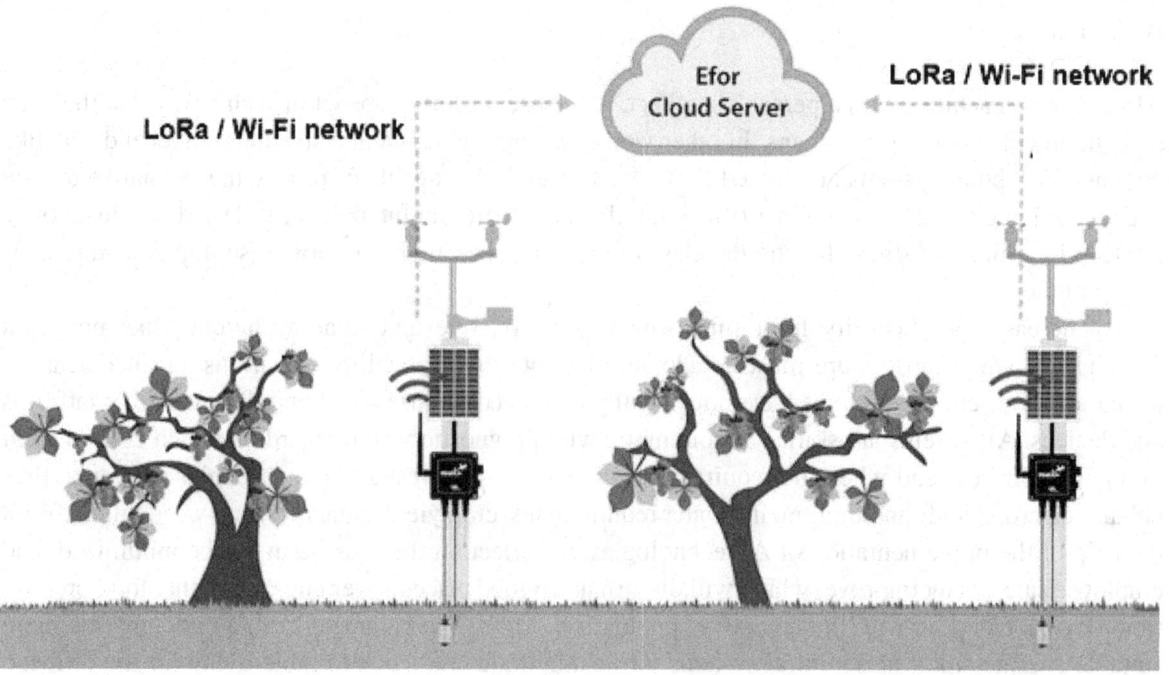

Soil and Irrigation Water Management

Irrigation water management, which extensively affects the season's growth and requirement of water, can be identified and supervised by supervising the tension and content in soil moisture. Determination of a suitable gap in sprinkling water, dampness, water content absorbed by roots, and enough moistening can be done by hygroscope by Hanson et al. (2000), examined various batches to explore the execution of these devices. It is being found that these devices are reasonably accurate and can check the dampness. Continuous super vision of wetness helps farmers in irrigating their crops even if these devices are not very accurate. Soil moisture sensor is shown in Figure 2.

A plethora of control devices is currently employed in agriculture. Figure 1 depicts one of the AI-based control devices. According to Syers et al. 1997, two models using artificial neural networks and adaptive neuro-fuzzy inference systems have been proposed to determine contamination in tainted soil. Contamination may cause changes in soil color, density, hardness, flammability, oxidation, corrosion, and other essential nutrients. It has been observed that the ANN model, with a higher R2 and lower RMSE, is more efficient than the ANFIS and ML models. Bazoobandi et al. (2022) compared the models using sensitivity analysis and found that unprocessed material was the most influential parameter in predicting cadmium and lead as contaminants. Manek et al. 2016, compared several neural network architectures for anticipating cloud bursts and discovered that radial basis function neural networks outperformed other models. Choudhary et al. (2019) designed a system comprising two crucial units, with one module providing information on precipitation quantity, rainfall, and wetness index using expert systems. The second module involves the deployment of profit-making equipment based on an internet link, which, compared to current flood management practices, requires less labor and conserves water. These modules currently rely on anemometer data and perform calculations based on it. To deploy these modules in rural areas with limited water sources, on-site sensors must be used. Table 1 was used for the comparative analysis.

Weed Management

For achieving high-yielding and top-quality crops, effective weed management plays a pivotal role. Unfortunately, many farmers find themselves with no choice but to resort to pesticides to safeguard their plants against various diseases, pests, and weeds, despite being well aware of the adverse effects on the environment and human health. Gai et al, 2020, reported that the treatment of crops with insecticides and herbicides encompasses over 90% of the cultivated land area in the United States of America. The use of weed killers not only helps farmers in conserving manual labor but also ensures the health and vitality of their crops.

Partel et al. (2019) have devised an innovative solution to reduce the use of herbicides through the development of an economical intelligent sprayer, as depicted in Figure 3. This intelligent sprayer employs expert systems and AI-based software, utilizing deep learning technology to accurately differentiate between unwanted weeds and desirable crop plants based on hyperspectral imaging. By leveraging semantic networks for identification and differentiation of weeds, this technology informs the precise amount of weed killer to be applied, effectively minimizing herbicide usage. Notably, the integration of both color and depth images enables this system to detect and locate plants at various growth stages, particularly in conditions with high weed density.

Table 1. Comparison with different technologies for soil management

S.NO.	Reference	Work Area	Problem Description	AI Model Used	Data Used	Advantages	Limitation
1	Hanson et al, 2000	Irrigation water management	Soil moisture content	Dielectric soil moisture sensors	Retiff at all 1983,taylor 1965,doorenboss and pruitt1973	Better than Tensiometers and electrical resistance blocks.	Provided high accuracy is available
2	Bazoobandi et al, 2022	Soil management	Cadmium and lead content in the polluted soil	Sensitivity analysis for Compared ANN & ANFIS	250 soil samples collected in the Province of Gilan in Iran	ANN is better than all other models, organic carbon was the most effective parameter in predicting both parameters.	Study is limited to cadmium and lead
3	Manek et al, 2016	Irrigation management	Prediction of rainfall	Several neural network architectures is been compared	Test data	RBNN gives best result for prediction.	Area for pridiction is limited
4	Choudhary et al, 2019	Soil and Iirrigation management	Prediction of amount of rainfall and the soil moisture content	Machine learning and predictive algorithms (internet of things)	Government of India Portal for local regions	decreased efforts required and aids water conservation compared to current irrigation method	system currently depends on weather station information for its calculation
5	Sarmadian et al, 2010	Analysing soil nutrients	Develop a Ped transfer function	Multiple linear regression and neural network model	125 soil samples collected from different horizons of 32 soil profiles located in the Gorgan Province, North of Iran	ANN with two neurons in hidden layers is better than multivariate regression soil nutrients	Measurement errors might cause the poor prediction of the parameters
6	Stine et al, 2002	Soil management	Relationship between soil quality and crop productivity	Back Propagation Neural Network.	Test data	Highly correlated	Lack of replication

Christensen et al. (2003) propose the implementation of specialized machinery equipped with weed control technologies tailored to specific locations. This machinery identifies weeds and optimizes weed control based on local factors provided as input. The system integrates semantic systems and introduces an image filtering method called the 'greenness method.' This method involves pixel-wise differentiation of color intensity for comprehensive image recognition, reduced processing time, and memory efficiency. By capturing images between rows of corn, the system calculates the green color intensity and its percentage in each image, providing insights into the weed-covered area and weed plant locations across the entire field. Utilizing this information, the system calculates the precise requirement for weed killers. Notably, the support vector machine (SVM) demonstrates a high degree of accuracy in clas-

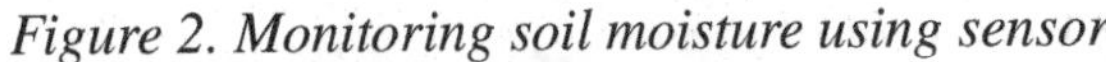

Figure 2. Monitoring soil moisture using sensor

Figure 3. Weed based on hyperspectral images. Note: The four weed species (a): Thistle (TT), (b): Yellow bristle grass (YBG), (c): Buttercup (BC), and (d): wind grass (WG)
Source: Partel et al. (2019)

Table 2. Comparison between different technologies for weed control

S.No.	Reference	Work Area	Problem Description	AI Model Used	Data Used	Advantages	Limitation
1	Partel et al, 2019	Development and evaluation of a low-cost and smart technology	Precision weed control	Deep learning	Experimental data	Accurately differentiate weed and sprays, useful in labor shortage and climate change	amount of chemicals applied based on the weed canopy size and vehicle speed is unavailable
2	Gai et al, 2020	robotic weed control	Automated crop plant detection based on the fusion of color and depth images	Fusion-based algorithm	USDA, (2016,17) Experimental data	More robust to high weed density, extreme lighting condition	Lower success rate at early growth stages
3	Yang et al, 2003	Development of a herbicide application map	Develop a precision herbicide-spraying system in a corn field	Artificial neural networks and fuzzy logic	Experimental data	Create weed coverage and weed patchiness maps herbicide application rates can be implemented	Lower success recognition rates
4	Monteiro, A. L et al, 2021	A new alternative to determine weed control in agricultural	Beginning of weed control	Artificial neural networks (ANNs)	Experimental data	Beginning of weed control, degree of weed interference in sesame and melon.	Complex data
5	Karimi et al, 2006	Application of support vector machine technology tool for classifying hyperspectral images	Weed and nitrogen stress detection in corn	SVM method evapotranspiration (ETo)	Field experiment data	Classifies with high accuracy	Done only for corn
6	Lamm et al, 2002	Real-time robotic weed control system	Precision weed control system for cotton	Areal-time machine vision system, a controlled illumination chamber, and a precision chemical applicator	Experimental data	The system correctly sprayed 88.8% of the weeds while correctly identifying and not spraying 78.7% of the cotton plants while traveling at 0.45 m/s.	Not done for other crops
7	Tobal et al, 2014	Evolutionary Artificial Intelligence Algorithm	Recognizing weed and plants other than crop plants	Genetic ANN algorithm which works on consecutive iterations.	Experimental data	Online decision supportive system for controlled and accurate spray of herbicides.	Hardware answers software design for real time are not available to use directly in the field

sification compared to artificial neural network (ANN) models. Lamm et al. (2002) developed a practical automated weed control system comprising an intelligent recognition system, a controlled lighting chamber, and a precise weed killer sprayer. This system effectively distinguishes unwanted weeds from crop plants and selectively applies weed killer to weeds with a moving speed of 0.45 m/s. The system achieved recognition and targeted chemical application to over 80% of weeds while preserving more than 78% of crop plants. However, it's important to note that there has been limited consideration of robotic weed killers utilizing ordered labeling and genetic engineering to distinguish crops from weeds. Table 2 has comparison between different technologies for weed control.

Disease Management

Plant diseases pose a significant threat to crop yields and agricultural profitability. Jha et al. (2012) highlighted that crop yield losses due to plant pathogens can reach up to 16% annually. Traditional farming practices often rely on the extensive use of chemical treatments to combat plant diseases, despite the well-established risks these chemicals pose to human health. Amara et al. (2017) introduced an innovative tool for the automated classification and identification of banana diseases, with a particular focus on diseases like banana sigatoka and banana speckle. This model serves as a decision support tool for disease detection in banana leaves. The author has devised a convolutional neural network (CNN) model featuring a three-layered structure, including convolutional, pooling, and fully connected layers. The convolutional layer extracts images from pre-stored data, followed by size reduction through pooling, and ultimately classification into predetermined disease categories using the SoftMax activation function. For the detection of soybean diseases, Bajwa et al. (2017) presented a method that leverages spectroscopic remote sensing to relate leaf reflectance to disease conditions. The system distinguishes disease-afflicted plants through wave band identification. Notably, the green and red wave bands showed the most significant correlations with disease. Early-stage disease detection is crucial for optimizing crop yields. The author conducted a comparative analysis of various neural network models, such as AlexNet, GoogleNet, Inception v3, ResNet-50, ResNet-101, and SqueezeNet. The study revealed that ResNet-101 is the most effective, although it's complex. As a practical alternative, ResNet-50 is recommended for leaf prognosis models. In the context of Fox grape disease detection, a system without expert systems and trainers exhibited a significantly lower error rate of 64.23% to 77.12%. This system eliminates the need for trained users, expensive sensing equipment, and input images while delivering high accuracy in disease identification, not limited to GY (Grapevine Yellow). The SoftMax aggregation method is recommended for improving the accuracy of disease recognition and localization. Yang et al. (2003) proposed a CNN-based model for the classification of ten common rice diseases from a dataset of 500 images using image recognition. This approach significantly enhances speed and accuracy. Figure 4 has an idea about banana leaves disease. The CNN method relies on stochastic pooling and scattered robotic encoding for image feature extraction. For maize disease information gathering, Sannakki et al 2011, employed a mobile terminal and the SVM classification method, which has demonstrated effectiveness even with limited sample data. Table 3 has data for different technologies on diseases control.

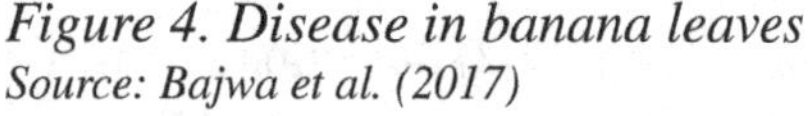

Figure 4. Disease in banana leaves
Source: Bajwa et al. (2017)

Crop Management

Crop management is nothing but precision agriculture, which can recognize and maintain the balance between soil–plant–atmosphere systems. Precision agriculture works toward increased yield, profitability, and quality of the crop and minimizes the use of fertilisers, seeds, chemicals, and water to reduce climate hazards. This starts with sowing the seed and keeping a record of process, harvesting, and crop storage and circulation. It is a plan which has two main objectives with four steps, growth improvement, yield of agricultural products, and data addition, analysis, decision, and control, respectively, which minimises pollination, providing excess flow of minerals and other micronutrients which ends up contaminating the water. Since climate conditions and its uncertainty affects agricultural crop production and its conservation, the prior information about crop yield and its harvest time will help farmers to plan accordingly and take proper decision for selling and conservation. These predictions not only help farmers but also help allied industries, national food policies, and international crop trade in making strategies.

The authors found that CNN model has the highest accuracy of 97.59% for training images and 92.45% for testing images. Opportunities and limitations have been discussed by Moran et al. 1997 for image-based remote sensing in precision crop management. Different types of crop management practices should be adopted by farmers to handle water-related issues. Barnes et al. 2000, detected desiccation, nitrogen percentage and canopy density coincidentally using image data within specific wavelength ranges across the electromagnetic spectrum. Results shows that remotely sensed data can be used to detect areas of desiccation, nitrogen percentage, and canopy density besides all parameters are changing together. Because of this model, it is easy to manage the food supply chain when a large variety of

Table 3. Table for different technologies on disease control

S.No.	Reference	Work Area	Problem Description	AI Model Used	Data Used	Advantages	Limitation
1	Amara et al, 2017	Banana leaf diseases classification	Classify and identify diseases automatically for banana leafs	Deep learning technology	Experimental data	Capable in classifying disease.	Automation is not available
2	Apan et al, 2004	EO-1 Hyperion hyperspectral imagery	Detecting sugarcane 'orange rust' disease	Discriminant function analysis		Superior ability to discriminate sugarcane areas affected by orange rust disease.	Disease severity level and crop protection have not been discussed.
3	Bajwa et al, 2017	Relating leaf reluctance to disease condition and differentiating disease plant by wave band identification	Soybean Disease Monitoring	Spectroscopic remote sensing	Crop data were collected on a weekly basis over a 10-week period, starting from 71 days after planting	Able to detect 97% of healthy plants and 58% of the diseased plants.	Limited to certain parameters related to soil and weather for disease discrimination
4	Moshou et al, 2004	Automatic detection of 'yellow rust' in wheat	Simple and cost-effective optical remote disease detection device	Canopy reflectance in several wavebands and neural network	5137 leaf spectra were used	Performance increased up to 99%	Limited to crop
5	Cruz et al, 2019	Detection of grapevine yellows symptoms in Vitis viniferous L.	Leaf discoloration and irregular wood ripening,	Convolutional neural networks, machine learning	GitHub data	ResNet-50 better than all other neural networks. the system has a sensitivity of 98.96% and a specificity of 99.40%.	Uses complex deep learning algorithms leaf images are not taken in field conditions
6	Hu G., et al. 2019	Low shot learning method	Tea leaf's disease identification	Support vector machine and deep learning networks,C-DCGNA,VGG-16	Mandalay data	90% accuracy in disease spot detection	Generalized performance is not available
7	Lu Y., et al 2017	Rice diseases identification method	Identification of rice diseases	Deep convolutional neural networks	Experimental data	Identifies disease with 95.48% accuracy	Depends on experimental and large set of data, optimal no of layers and neurons
8	Zhang et al, 2014	Recognition of Maize Disease Based on Mobile Internet and Support Vector Machine Technique	Recognition of Maize Disease	Support Vector Machine Technique	Experimental data	Improves the efficiency of agricultural producers and technical staff	Works very well with small sample data
9	Balleda et al, 2014	Efficient rule-based expert system to prevent pest diseases of rice & wheat crops	Crop disease management	Fuzzy logic based interference	Experimental data	provides extensively-effective interactive user interface on web for live interactions	Limited to certain crops

Table 4. Table for different technologies on crop management

S.No.	Reference	Work Area	Problem	AI Model Used	Data Used	Advantages	Limitation
1	Paymode et al, 2021	Onion model sorting and grading onions quality	Vegetables Crop Onion Sorting and Grading	Deep learning based Convolution Neural Network (CNN) architecture	Experimental data & deep learning neural network keras library	Accuracy on training datasets is 97.59% and 92.45%	Crop type is limited
2	Moran et al, 1997	Opportunities and limitations precision crop management	Image-based remote sensing	Near-term implementation with available remote sensing technology and instrumentation	HyperSpectral data	Provide valuable information for PCM	Presently limited applications are available
3	Barnes et al, 2000	Precision crop management	Coincident Detection of Crop Water Stress, Nitrogen Status and Canopy Density	Analysis of Ground-Based Multispectral Data	Ground-Based Multispectral Data	Remotely sensed data can be used to prepare maps of water stress, N status and canopy density simultaneously	Difficult to ralate until the crop reaches full cover changes in canopy density
4	Rahutomo et al, 2019	Precision crop management	Time efficiency and minimize human error	Web-based application with back-end and front-end side	Crowd sourcing annotation system	Minizes the use of resources such as water, fertilizer, insecticides, and packaging materials.	Crop limited
5	PILARSKI et al, 2002	The Demeter System for Automated Harvesting	Automation of agricultural harvesting equipment	GPS and vision system	Computer controlled system	Planning harvesting, executing its plan by cutting crop rows, repositioning itself in the field, detecting unexpected obstacles	Obstacle detection capability needs to be improved
6	Khaki et al, 2020	Crop yield prediction	A CNN-RNN Framework for Crop Yield Prediction	CNN-RNN	USDA-NASS, 2019). (Thornton et al., 2018) (Thornton et al., 2018)	Best compared to all other data available	Hybrid, low yielding, high yielding performances are not discussed
7	Lal et al, 1992	FARMSYS—A whole-farm machinery management decision support system	Precision crop management	A PROLOG (Programming in Logics) using object-oriented data structures. VGG-CNN-S	*Wheat Disease Database 2017* (WDD2017),	Easily adaptable for all types of agricultural and industrial applications.	Presently limited application s available
8	Ramos et al, 2017	Precision crop management	Automatic fruit count on coffee branches using computer vision	Machine Vision System with an image processing algorithm	Experimental data	An efficient, non-destructive, and low-cost method	For small coffee growers
9	Barbosa et al, 2020	Modeling yield response to crop management	Crop management	Convolutional neural networks	DIFM, experimental data	Better than available models	Limited to a particular season
10	Nyéki et al, 2021	Application of spatio-temporal data in site-specific maize yield prediction	Yield prediction	Machine learning	Soil data, meteorological data and satellite-based data.	Results obtained with 95.38% for accuracy, 91.3% for sensitivity and 97.62% for specificity	Specific to site

food items are available in a large amount. Moore et al 1993, discussed terrain analysis which gives clarity of soil maps and other data sources used for soil-specific crop management. Khaki et al. (2020) proposed a CNN-RNN framework capable to detect genetic improvement of seeds by time dependency and environmental factors, accurately predicting yield variation in crop yield due to weather predictions, soil conditions, and management practices. Table 4 define different technologies on crop management.

Demeter plans the harvest procedure by identifying the abrupt hurdle and doing harvesting row by row successively. A cost-effective robot for harvesting cucumber, which is an autonomous vehicle having a 4-DOF hinged operator, an end effector a recorder, and 4 DC servo drive system, was developed. This robot works with CAN -bus transmission between the recorder and controller. It identifies and takes 3D images of the fruit and finally with the help of collision free motions harvesting is being done. Ramos et al, 2017, have proposed a machine vision system for numbering and detecting collectable and not collectable fruits in a specific coffee branch. It is a well-organized, non-harmful, and cost-effective model for small coffee growers. To capture suitable dimensional structures of various properties and predict yield with reference to nutrient and seed rate management, Barbosa et al. (2020) proposed a CNN based model. This model reduces the dataset up to 68%by comparing with multiple linear regression. Which is very much demanding since due to the agriculture machinery a lot of data is being created, we need sophisticated models to deal with this type of data.

Yield Prediction

Crop yield prediction is highly depended on many parameters such as genetic variety of crops, atmospheric parameters, administration methods, and their reciprocal actions. For deciding marketing strategies and crop cost estimation, crop yield prediction is very beneficial. Selection of every crop becomes essential for crop yield prediction which directly affects financial system as yield increases as it increases. This method can represent that it can anticipate the production of yield in those areas for which data is not available without affecting precision. Moreover, when precision agriculture is very much in trend, prediction models can be used for the analysis of these dependent parameters on which crop yield depends. Nyeki et al. (2021) presented a ML model for the calculation of the effect of geographical -terrestrial data on soil fertility in the prediction of particular terrestrial maize field. Bhojani et al. (2020) presented a mechanism which gives valuable and correct results in minimum time because it uses a function that delivers an output based on the input from an artificial neuron known as activation function. The prime purpose of the suggested method is to deliver a more accurate multiple linear programming neural network and get crop prediction using various environmental variable datasets by revising arbitrary weights and partial values. Figure 5 has equipment for spray on crop so that crop can grow properly.

Purview of Automation

Agriculture system is working on conventional practices, such as growing population, water shortage, weather conditions, soil pollution, etc. are continuously demanding for change, and this change can be made by simply implementing new technologies at each step. According to Behrens et al. (2005), the most crucial part of agriculture where automation required is irrigation for minimising water wastage. For automation in irrigation, sensors can be used for assessing the suitable time intervals for watering, dampness depth, extent of uprooting by roots and sufficiency of moisture and automatically watering the farm as soon as it goes down to the baseline. For estimating the wetness of soil in arable fields,

Figure 5. UAV spraying herbicide
Source: Brazeau (2018)

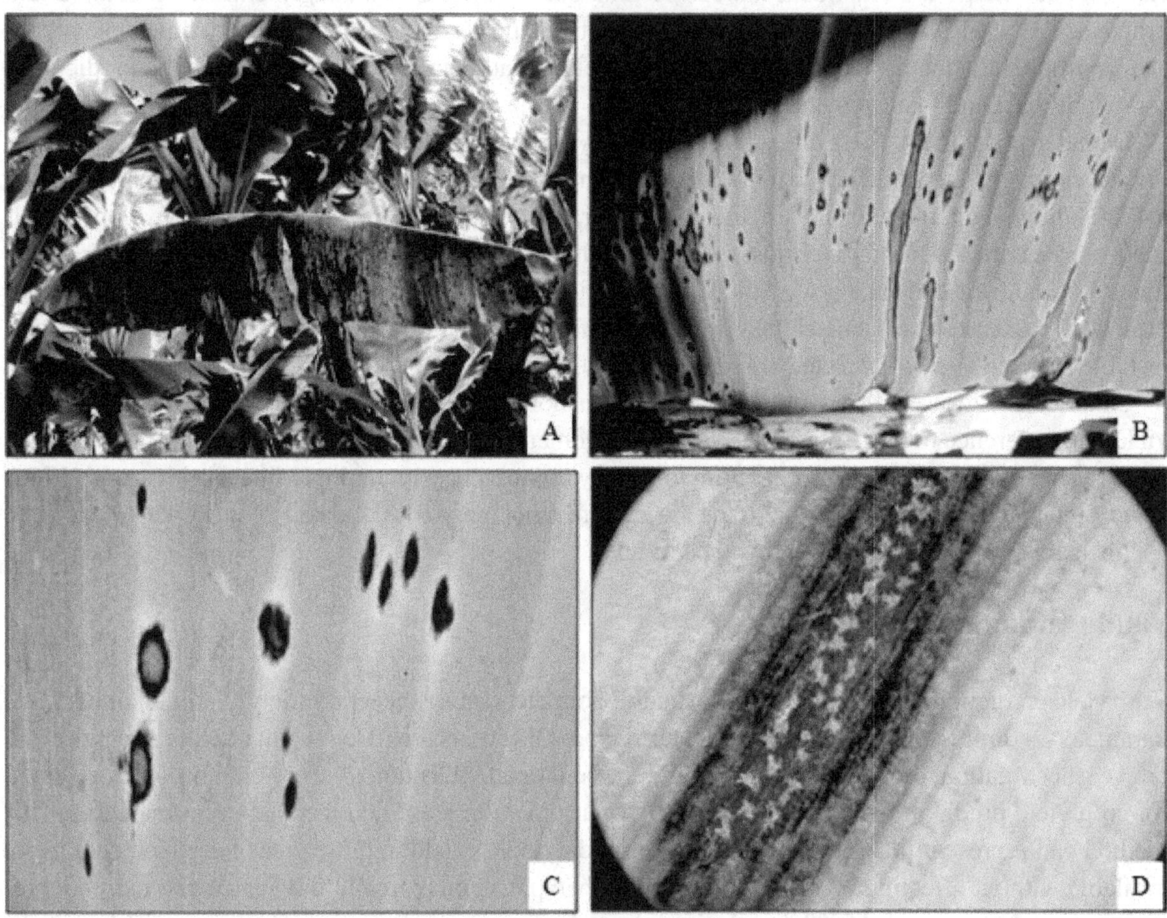

various ANN models are available based on restricted environmental data. Liakos et al. (2018) says a close-packed system can be developed for administering moisture content without any human interference. Machine learning, Artificial Intelligence, Deep learning, Neural network, fuzzy logic are available methods which work very efficiently and effectively in robotization which reduces time and labour wastage. Deep structured learning improves the correctness and stratification potential of the model for the identification of plant numbered fruit, and prediction of crop yield. The training of these models is being done by large image data set whereas other models with other techniques use text data. In Future agriculture, DL has many dimensions which will provide acceleration to agribusiness.

Farmers can get the information of ripening of fruit or crops by categorising crops or fruit. During the cultivation period, areas can be identified where water, fertilizer, or pesticides are required by real-time estimates using drones, copters, or other airborne systems says Rajotte et al. (1992). Drones have the capability to face major challenges in agribusiness like investigation of soil and field, sowing, administering crop progress, application of fertilizer, control of weed, water management and disease control. Drones can provide early soil analysis using precise 3D maps and accordingly plan for seed planting. It has been seen that the drone planting system has decreased the planting cost by 85%. Harker et al. found that devices which can administer crop health, water management, weed detection, etc. can be conveyed by drones. Jiayu et al. (2015) found that the aeronautical method of application of weed killer and fertiliser using drones multiplies by five.

LIMITATIONS OF AI ON AGRICULTURE BASED TECHNIQUES

The advancement in artificial intelligent is very recent, even though the solution for almost all problems related to agriculture is being proposed by AI. Kothari et al. (2018) observed that in the next 50 yrs. These problems will be eclipsed by AI and we will approach to a new era of evolution. However, still a large population of the world does not yet have access to basic requirements such as water, food, shelter, etc. Moreover, a large population will go jobless since AI will save time and labour. The majority of the repetitive works and tasks can be replaced by smart machines and robots._Agriculture is an industry of $3 trillion in which over one and half bill. Khoa et al. (2019) said that water management system and platform in smart agriculture are overpriced Amalgamation of various technologies leads to complexity, almost all IoT based agricultural devices are based on manual system. Extra sensors are required to administer more than two farms on the same mobile application because of the inflexibility of the devices.

Since the repairing and maintenance cost is very high, that will increase the cost of yield and hence the final product will be very costly. AI based models are based on internet connections which are very difficult to manage all time. According to Kang et al. (2020), proposed CNN based models time may take so long to train their models However, in comparison to other available methods like Support vector machine or kernel neural network time effectiveness increases hugely. Another disadvantage is the necessity of big data sets of different types of images and their proper notations which is occasionally tactful method and can be performed by the experts only. A. Kamilaris et al. (2018) found that the problem of poor data labelling can give notable depletion in performance and exactness achieved. Waleed et al. (2020) suggested optimization point for previously skilled models on analogous and small-scale data sets is a difficulty because of hardware limitations and convolution of the model. Development of wireless sensor networks creates a large quantity of statistics. Smartphone connectivity creates indeterminable facts and needs promising warehouses for a big amount of data. Khoa et al. (2019) discussed that Smart Farming uses Internet of things-based multisensory which needs promising warehouses for big quantity of facts (data). Authors presented the detection of solo leaf, fronting, on an entire field which tells that DL models can conclude up to a certain limit beyond that it does not work because many diseases does not appear on the upper side of the leaf as explained by Kim et al. (2019), that plants can also be affected by environmental factors example puckered surface or insect damage. Rahutomo et al. (2019) presented a web-based AI model which on the implementation of Pineapple Object Counting performs better than other models, but the recognition and counting of pineapple objects is limited up to 70 metres for achieving the best results by the system.

FUTURE RESEARCH DIRECTIONS

Sweeping transformation with advanced approaches can change the traditional ways and limits of agriculture using AI. The future generation farmers will slowly adopt this new automation approach by investing more. In a very short time, an agricultural revolution occurs with more yield using less resources.

Precision farming will occur over conventional farming using CNN, ANN, DL, and other technologies. Organic farming and green house farming soon will be in trend because a particular environment can be provided with human interventions. With the help of AI, vertical cropping can be done in buildings of urban areas where AI can lessen the worry of man power. AI not only gives information regarding next year crop seasons/weather/climate/rainfall etc. but it also precisely suggests the amount of water requirement, fertilizer, pesticides timely before the large incidence of loss or disease. AI powered chat boxes navigated with administered and fortified by both technologies of ML for regular and conditions specific learning will work as a virtual assistant and answer queries and provide recommendations based on specific problems in their native languages. The UAV, which is commonly known as unmanned Ariel vehicle, is the combination of various technologies, AI, informatics, communication, robotics, big data analytics, and IOT. 2 D and 3D map provided by UAV's draws a lot of attention as it gives information about the dimensions of the farm area, soil conditions, and crop condition, which can be highly beneficial for making models and strategies said by Sánchez et al. (2016). UAVs showed unlimited potential in mapping, spraying, planting, crop monitoring, irrigation, artificial pollination, and diagnosis of insect pests. UAV can minimize pesticide use by measuring pesticides per hectare in a lesser time on a larger area which reduces man power requirements as well as worker malady and environmental pollution. According to Shengde et al. (2017), damage can be minimised or even eliminated by early diagnosis by using high-spectral RGB image capturing devices and various continuum sensors fitted on unmanned Arial vehicles. Various inconsistent amalgamated ideas are possible to apply for imparting a feasible atmosphere and enhanced productivity. Administering a large farm requires a lot of time and labour, usually it is done by satellite, but in precision farming it is not appropriate.

In the future, farmers and breeders can be benefitted by the implementation of AI in effective livestock management, automated video monitoring of animal and human habits such as calving, heat intervals, or various physiological phenomena. With this labour-saving idea, a 24 hrs help benefit can be obtained with alerts and reports. This diminishes a lot of burden on the farmers and they can concentrate on the animal health.

CONCLUSION

Growth per capita income has been declining in the past few years. There will be an estimated increase in population by 10% and a decrease in food production by 20% as per United States Department of Agriculture. There is a positive indication in demand of food by 58 to 98%but a negative indication in supply. This negative indication will break the food supply chain in future. These are the adverse effects of increasing population and lesser involvement of technology in the agriculture field. To increase the food security for future generations, effective and smart techniques like precision farming and smart agriculture must be adopted. Despite the availability of various advancements of technology, agriculture systems have not yet accepted the use of AI and different IOT platforms and devices. Smart methods primarily work with reduced labour and increased production of food. Country Reference networks classification dataset helps farmers in selecting crops and fertilizers with crop no. fruit name with several other factors. With the current application of AI in agricultural sector as well as future prospectus of AI is also been analysed. The implementation of AI in agriculture will change the entire growth of economy which in turn affects our world socially, economically, and politically.

Data Availability

This is a review paper so no data is needed.

Competing Interests

The authors of this publication declare there are no competing interests.

Funding Statement

This research received no specific grant from any funding agency in the public, commercial, or not-for-profit sectors. Funding for this research was covered by the author(s) of the article.

REFERENCES

Amara, J., Bouaziz, B., & Algergawy, A. (2017). A deep learning-based approach for banana leaf diseases classification. *Datenbanksysteme für Business, Technologie und Web (BTW 2017)-Workshopband.*

Apan, A., Held, A., Phinn, S., & Markley, J. (2004). Detecting sugarcane 'orange rust'disease using EO-1 Hyperion hyperspectral imagery. *International Journal of Remote Sensing, 25*(2), 489–498. doi:10.1080/01431160310001618031

Bajwa, S. G., Rupe, J. C., & Mason, J. (2017). Soybean disease monitoring with leaf reflectance. *Remote Sensing (Basel), 9*(2), 127. doi:10.3390/rs9020127

Balleda, K., Satyanvesh, D., Sampath, N. V. S. S. P., Varma, K. T. N., & Baruah, P. K. (2014, January). Agpest: An efficient rule-based expert system to prevent pest diseases of rice & wheat crops. In *2014 IEEE 8th International Conference on Intelligent Systems and Control (ISCO)* (pp. 262-268). IEEE.

Barbosa, A., Trevisan, R., Hovakimyan, N., & Martin, N. F. (2020). Modeling yield response to crop management using convolutional neural networks. *Computers and Electronics in Agriculture, 170*, 105197. doi:10.1016/j.compag.2019.105197

Barnes, E. M., Clarke, T. R., Richards, S. E., Colaizzi, P. D., Haberland, J., Kostrzewski, M., . . . Moran, M. S. (2000, July). Coincident detection of crop water stress, nitrogen status and canopy density using ground based multispectral data. In *Proceedings of the Fifth International Conference on Precision Agriculture, Bloomington, MN, USA* (*Vol. 1619*, p. 6). Academic Press.

Bazoobandi, A., Emamgholizadeh, S., & Ghorbani, H. (2022). Estimating the amount of cadmium and lead in the polluted soil using artificial intelligence models. *European Journal of Environmental and Civil Engineering, 26*(3), 933–951. doi:10.1080/19648189.2019.1686429

Behrens, T., Förster, H., Scholten, T., Steinrücken, U., Spies, E. D., & Goldschmitt, M. (2005). Digital soil mapping using artificial neural networks. *Journal of Plant Nutrition and Soil Science, 168*(1), 21–33. doi:10.1002/jpln.200421414

Bhojani, S. H., & Bhatt, N. (2020). Wheat crop yield prediction using new activation functions in neural network. *Neural Computing & Applications*, *32*(17), 13941–13951. doi:10.100700521-020-04797-8

Brazeau, M. (2018). *Fighting weeds: Can we reduce, or even eliminate, herbicides by utilizing robotics and AI*. Genetic Literacy Project, North Wales.

Choudhary, S., Gaurav, V., Singh, A., & Agarwal, S. (2019). Autonomous crop irrigation system using artificial intelligence. *International Journal of Engineering and Advanced Technology*, *8*(5), 46–51. doi:10.35940/ijeat.E1010.0585S19

Cruz, A., Ampatzidis, Y., Pierro, R., Materazzi, A., Panattoni, A., De Bellis, L., & Luvisi, A. (2019). Detection of grapevine yellows symptoms in Vitis vinifera L. with artificial intelligence. *Computers and Electronics in Agriculture*, *157*, 63–76. doi:10.1016/j.compag.2018.12.028

Eli-Chukwu, N. C. (2019). Applications of artificial intelligence in agriculture: A review. *Engineering, Technology &. Applied Scientific Research*, *9*(4), 4377–4383.

Fan, M., Shen, J., Yuan, L., Jiang, R., Chen, X., Davies, W. J., & Zhang, F. (2012). Improving crop productivity and resource use efficiency to ensure food security and environmental quality in China. *Journal of Experimental Botany*, *63*(1), 13–24. doi:10.1093/jxb/err248 PMID:21963614

Gai, J., Tang, L., & Steward, B. L. (2020). Automated crop plant detection based on the fusion of color and depth images for robotic weed control. *Journal of Field Robotics*, *37*(1), 35–52. doi:10.1002/rob.21897

Gerhards, R., & Christensen, S. (2003). Real-time weed detection, decision making and patch spraying in maize, sugarbeet, winter wheat and winter barley. *Weed Research*, *43*(6), 385–392. doi:10.1046/j.1365-3180.2003.00349.x

Hanson, B., Orloff, S., & Peters, D. (2000). Monitoring soil moisture helps refine irrigation management. *California Agriculture*, *54*(3), 38–42. doi:10.3733/ca.v054n03p38

Hu, G., Wu, H., Zhang, Y., & Wan, M. (2019). A low shot learning method for tea leaf's disease identification. *Computers and Electronics in Agriculture*, *163*, 104852. doi:10.1016/j.compag.2019.104852

Jha, K., Doshi, A., Patel, P., & Shah, M. (2019). A comprehensive review on automation in agriculture using artificial intelligence. *Artificial Intelligence in Agriculture*, *2*, 1–12. doi:10.1016/j.aiia.2019.05.004

Jha, P., Kumar, V., Godara, R. K., & Chauhan, B. S. (2017). Weed management using crop competition in the United States: A review. *Crop Protection (Guildford, Surrey)*, *95*, 31–37. doi:10.1016/j.cropro.2016.06.021

Jiayu, Z., Shiwei, X., Zhemin, L., Wei, C., & Dongjie, W. (2015, May). Application of intelligence information fusion technology in agriculture monitoring and early-warning research. In *2015 International Conference on Control, Automation and Robotics* (pp. 114-117). IEEE. 10.1109/ICCAR.2015.7166013

Kamilaris, A., & Prenafeta-Boldú, F. X. (2018). Deep learning in agriculture: A survey. *Computers and Electronics in Agriculture*, *147*, 70–90. doi:10.1016/j.compag.2018.02.016

Kang, H., & Chen, C. (2020). Fast implementation of real-time fruit detection in apple orchards using deep learning. *Computers and Electronics in Agriculture*, *168*, 105108. doi:10.1016/j.compag.2019.105108

Karimi, Y., Prasher, S. O., Patel, R. M., & Kim, S. H. (2006). Application of support vector machine technology for weed and nitrogen stress detection in corn. *Computers and Electronics in Agriculture*, *51*(1-2), 99–109. doi:10.1016/j.compag.2005.12.001

Katariya, S. S., Gundal, S. S., Kanawade, M. T., & Mazhar, K. (2015). Automation in agriculture. *International Journal of Recent Scientific Research*, *6*(6), 4453–4456.

Khaki, S., Wang, L., & Archontoulis, S. V. (2020). A cnn-rnn framework for crop yield prediction. *Frontiers in Plant Science*, *10*, 1750. doi:10.3389/fpls.2019.01750 PMID:32038699

Khoa, T. A., Man, M. M., Nguyen, T. Y., Nguyen, V., & Nam, N. H. (2019). Smart agriculture using IoT multi-sensors: A novel watering management system. *Journal of Sensor and Actuator Networks*, *8*(3), 45. doi:10.3390/jsan8030045

Kim, J., Kim, S., Ju, C., & Son, H. I. (2019). Unmanned aerial vehicles in agriculture: A review of perspective of platform, control, and applications. *IEEE Access: Practical Innovations, Open Solutions*, *7*, 105100–105115. doi:10.1109/ACCESS.2019.2932119

Kim, N., Ha, K. J., Park, N. W., Cho, J., Hong, S., & Lee, Y. W. (2019). A comparison between major artificial intelligence models for crop yield prediction: Case study of the midwestern United States, 2006–2015. *ISPRS International Journal of Geo-Information*, *8*(5), 240. doi:10.3390/ijgi8050240

Kothari, J. D. (2018). Plant Disease Identification using Artificial Intelligence: Machine Learning Approach. *International Journal of Innovative Research in Computer and Communication Engineering*, *7*(11), 11082–11085.

Lal, H., Jones, J. W., Peart, R. M., & Shoup, W. D. (1992). FARMSYS—A whole-farm machinery management decision support system. *Agricultural Systems*, *38*(3), 257–273. doi:10.1016/0308-521X(92)90069-Z

Lamm, R. D., Slaughter, D. C., & Giles, D. K. (2002). Precision weed control system for cotton. *Transactions of the ASAE. American Society of Agricultural Engineers*, *45*(1), 231.

Liakos, K. G., Busato, P., Moshou, D., Pearson, S., & Bochtis, D. (2018). Machine learning in agriculture: A review. *Sensors (Basel)*, *18*(8), 2674. doi:10.339018082674 PMID:30110960

Lu, Y., Yi, S., Zeng, N., Liu, Y., & Zhang, Y. (2017). Identification of rice diseases using deep convolutional neural networks. *Neurocomputing*, *267*, 378–384. doi:10.1016/j.neucom.2017.06.023

Manek, A. H., & Singh, P. K. (2016, July). Comparative study of neural network architectures for rainfall prediction. In 2016 IEEE Technological Innovations in ICT for Agriculture and Rural Development (TIAR) (pp. 171-174). IEEE doi:10.1109/TIAR.2016.7801233

Mohanty, S. P., Hughes, D. P., & Salathé, M. (2016). Using deep learning for image-based plant disease detection. *Frontiers in Plant Science*, *7*, 1419. doi:10.3389/fpls.2016.01419 PMID:27713752

Monteiro, A. L., de Freitas Souza, M., Lins, H. A., da Silva Teófilo, T. M., Júnior, A. P. B., Silva, D. V., & Mendonça, V. (2021). A new alternative to determine weed control in agricultural systems based on artificial neural networks (ANNs). *Field Crops Research*, *263*, 108075. doi:10.1016/j.fcr.2021.108075

Moore, I. D., Gessler, P. E., Nielsen, G. A., & Peterson, G. A. (1993, May). Terrain analysis for soil specific crop management. In *Proceedings of Soil Specific Crop Management: A Workshop on Research and Development Issues* (pp. 27-55). Madison, WI, USA: American Society of Agronomy, Crop Science Society of America, Soil Science Society of America. 10.2134/1993.soilspecificcrop.c3

Moran, M. S., Inoue, Y., & Barnes, E. M. (1997). Opportunities and limitations for image-based remote sensing in precision crop management. *Remote Sensing of Environment, 61*(3), 319–346. doi:10.1016/S0034-4257(97)00045-X

Moshou, D., Bravo, C., West, J., Wahlen, S., McCartney, A., & Ramon, H. (2004). Automatic detection of 'yellow rust'in wheat using reflectance measurements and neural networks. *Computers and Electronics in Agriculture, 44*(3), 173–188. doi:10.1016/j.compag.2004.04.003

Nyéki, A., Kerepesi, C., Daróczy, B., Benczúr, A., Milics, G., Nagy, J., Harsányi, E., Kovács, A. J., & Neményi, M. (2021). Application of spatio-temporal data in site-specific maize yield prediction with machine learning methods. *Precision Agriculture, 22*(5), 1397–1415. doi:10.100711119-021-09833-8

Partel, V., Kakarla, S. C., & Ampatzidis, Y. (2019). Development and evaluation of a low-cost and smart technology for precision weed management utilizing artificial intelligence. *Computers and Electronics in Agriculture, 157*, 339–350. doi:10.1016/j.compag.2018.12.048

Paymode, A. S., Shinde, J. N. M. U. B., & Malode, V. B. (2021). Artificial intelligence for agriculture: A technique of vegetables crop onion sorting and grading using deep learning. *International Journal, 6*(4).

Pilarski, T., Happold, M., Pangels, H., Ollis, M., Fitzpatrick, K., & Stentz, A. (2002). The demeter system for automated harvesting. *Autonomous Robots, 13*(1), 9–20. doi:10.1023/A:1015622020131

Rahutomo, R., Perbangsa, A. S., Lie, Y., Cenggoro, T. W., & Pardamean, B. (2019, August). Artificial intelligence model implementation in web-based application for pineapple object counting. In *2019 International Conference on Information Management and Technology (ICIMTech)* (Vol. 1, pp. 525-530). IEEE. 10.1109/ICIMTech.2019.8843741

Rajotte, E. G., Bowser, T., Travis, J. W., Crassweller, R. M., Musser, W., Laughland, D., & Sachs, C. (1992, February). Implementation and adoption of an agricultural expert system: The Penn State Apple Orchard Consultant. In *III International Symposium on Computer Modelling in Fruit Research and Orchard Management 313* (pp. 227-232). 10.17660/ActaHortic.1992.313.28

Ramos, P. J., Prieto, F. A., Montoya, E. C., & Oliveros, C. E. (2017). Automatic fruit count on coffee branches using computer vision. *Computers and Electronics in Agriculture, 137*, 9–22. doi:10.1016/j.compag.2017.03.010

Sannakki, S. S., Rajpurohit, V. S., Nargund, V. B., Kumar, A., & Yallur, P. S. (2011). Leaf disease grading by machine vision and fuzzy logic. *International Journal (Toronto, Ont.), 2*(5), 1709–1716.

Sarmadian, F., & Keshavarzi, A. (2010). Developing pedotransfer functions for estimating some soil properties using artificial neural network and multivariate regression approaches. *Int. J. Environ. Earth Sci, 1*, 31–37.

Shengde, C., Lan, Y., Jiyu, L., Zhiyan, Z., Aimin, L., & Yuedong, M. (2017). Effect of wind field below unmanned helicopter on droplet deposition distribution of aerial spraying. *International Journal of Agricultural and Biological Engineering*, *10*(3), 67–77.

Stine, M. A., & Weil, R. R. (2002). The relationship between soil quality and crop productivity across three tillage systems in south central Honduras. *American Journal of Alternative Agriculture*, *17*(1), 2–8.

Syers, J. K. (1997). Managing soils for long-term productivity. *Philosophical Transactions of the Royal Society of London. Series B, Biological Sciences*, *352*(1356), 1011–1021. doi:10.1098/rstb.1997.0079

Teimouri, N., Dyrmann, M., & Jørgensen, R. N. (2019). A novel spatio-temporal FCN-LSTM network for recognizing various crop types using multi-temporal radar images. *Remote Sensing (Basel)*, *11*(8), 990. doi:10.3390/rs11080990

Tobal, A., & Mokhtar, S. A. (2014). Weeds identification using Evolutionary Artificial Intelligence Algorithm. *Journal of Computational Science*, *10*(8), 1355–1361. doi:10.3844/jcssp.2014.1355.1361

Torres-Sánchez, J., Peña, J. M., de Castro, A. I., & López-Granados, F. (2014). Multi-temporal mapping of the vegetation fraction in early-season wheat fields using images from UAV. *Computers and Electronics in Agriculture*, *103*, 104–113. doi:10.1016/j.compag.2014.02.009

Waleed, M., Um, T. W., Kamal, T., Khan, A., & Iqbal, A. (2020). Determining the precise work area of agriculture machinery using internet of things and artificial intelligence. *Applied Sciences (Basel, Switzerland)*, *10*(10), 3365. doi:10.3390/app10103365

Yang, C. C., Prasher, S. O., Landry, J. A., & Ramaswamy, H. S. (2003). Development of a herbicide application map using artificial neural networks and fuzzy logic. *Agricultural Systems*, *76*(2), 561–574. doi:10.1016/S0308-521X(01)00106-8

Zhang, L. N., & Yang, B. (2014). Research on recognition of maize disease based on mobile internet and support vector machine technique. *Advanced Materials Research*, *905*, 659–662. doi:10.4028/www.scientific.net/AMR.905.659

Chapter 10
Image Processing Applications in Agriculture With the Help of AI

Raj Kishor Verma
https://orcid.org/0009-0005-7216-7752
ABES Institute of Technology, Ghaziabad, India

Kaushal Kishor
https://orcid.org/0000-0002-7131-1389
ABES Institute of Technology, Ghaziabad, India

ABSTRACT

Agriculture has significant challenges in the 21st century, which is why cutting-edge technologies like artificial intelligence (AI) and image processing are being employed to alter current farming operations. The primary study concern is the increased availability of visual data from a variety of sources, such as drones, satellites, and ground-based imaging equipment, which give critical information on crop and soil health as well as animal monitoring. The authors look at how AI algorithms can analyze this rich data and detect illnesses, pests, and nutritional deficits early on to help precision agriculture and reduce chemical use. In addition, they explore how image processing is used in automated irrigation management, where artificial intelligence (AI) uses meteorological data and satellite photos to optimize water use and protect resources. They also explore mechanized harvesting with AI-controlled robots that not only improve efficiency but also increase productivity.

INTRODUCTION

John McCarthy first proposed a study based on the idea that "every aspect of learning or any other feature of intelligence can, in principle, be so precisely described that a machine can be made to simulate it" at the 1955 Dartmouth Conference, which is where the term "Artificial Intelligence" was first used (McCarthy Shannon et al 1955). Agriculture, the backbone of human civilization, is undergoing a trans-

DOI: 10.4018/979-8-3693-0782-3.ch010

formative phase with the integration of cutting-edge technologies. Among these, image processing and artificial intelligence (AI) are playing a pivotal role in revolutionizing the way farming and agricultural practices are carried out. Image processing applications in agriculture harness the power of AI-driven computer vision to extract valuable insights from visual data, enabling farmers to make more informed decisions, increase efficiency, and optimize crop yield. In this era of data-driven agriculture, drones equipped with advanced cameras, satellites, and ground-based imaging systems capture vast amounts of visual data from agricultural fields. These images contain a wealth of information about crop health, pest infestations, nutrient deficiencies, and other vital factors that influence agricultural productivity. However, analysing this massive volume of data manually would be impractical and time-consuming. Here is where AI-driven image processing steps in, leveraging its ability to process and interpret visual information at an unprecedented speed and accuracy (R. Khan et al 2019).

In this article, we will explore the diverse applications of image processing in agriculture, facilitated by AI algorithms. From crop monitoring and disease detection to precision herbicide application and automated irrigation management, we will delve into how these technologies are transforming farming practices. Furthermore, we will uncover how AI assists in automated harvesting, soil health assessment, livestock monitoring, and quality grading and sorting. Additionally, we will explore how AI-driven image processing aids in land use planning, crop rotation strategies, and even farm equipment automation (Gupta S. et al 2022).

The convergence of AI and image processing in agriculture holds the promise of more sustainable, efficient, and economically viable farming practices (R. Khan, et al 2022). By enabling early detection of crop diseases, reducing chemical usage, optimizing water resources, and enhancing overall productivity, these technologies pave the way for a greener and more food-secure future. As we journey through the various applications, it becomes evident that the fusion of AI and image processing is empowering farmers and stakeholders across the agricultural landscape, fostering innovation and growth in this critical sector (Rai, B. K et al 2022).

IMPORTANCE OF ARTIFICIAL INTELLIGENCE

Digital assistance: The current trend is employing machines to engage with people using "avatars" on behalf of humans. Replicas or digital assistants contribute to a decrease in the demand for human resources. Robots are designed to reason rationally and make the best judgments possible based on the experience that has been imparted to the machine (K Kishor et al., 2023). In Figure 1 show the Farming Industry Machine Learning Model.

Constant Accessibility: Machines don't need regular breaks from work as people and other living things do. They may be designed to operate constantly for a lengthy period of time without being distracted or even weary. With machines, we also anticipate consistent outcomes devoid of bugs regardless of time, season, or weather.

Medical Applications: The use of artificial intelligence in the medical area is one of its many benefits. Doctors use artificial machine intelligence to evaluate the patient's health-related data. It functions as personal digital care and aids the patient in understanding the adverse effects of various medications. We currently have extensive tools to identify and track neurological problems. It has the ability to mimic how the human brain works. A modern use of AI in medicine is radio surgery, which helps remove tumours without harming healthy surrounding tissues.

Figure 1. Farming industry machine learning model

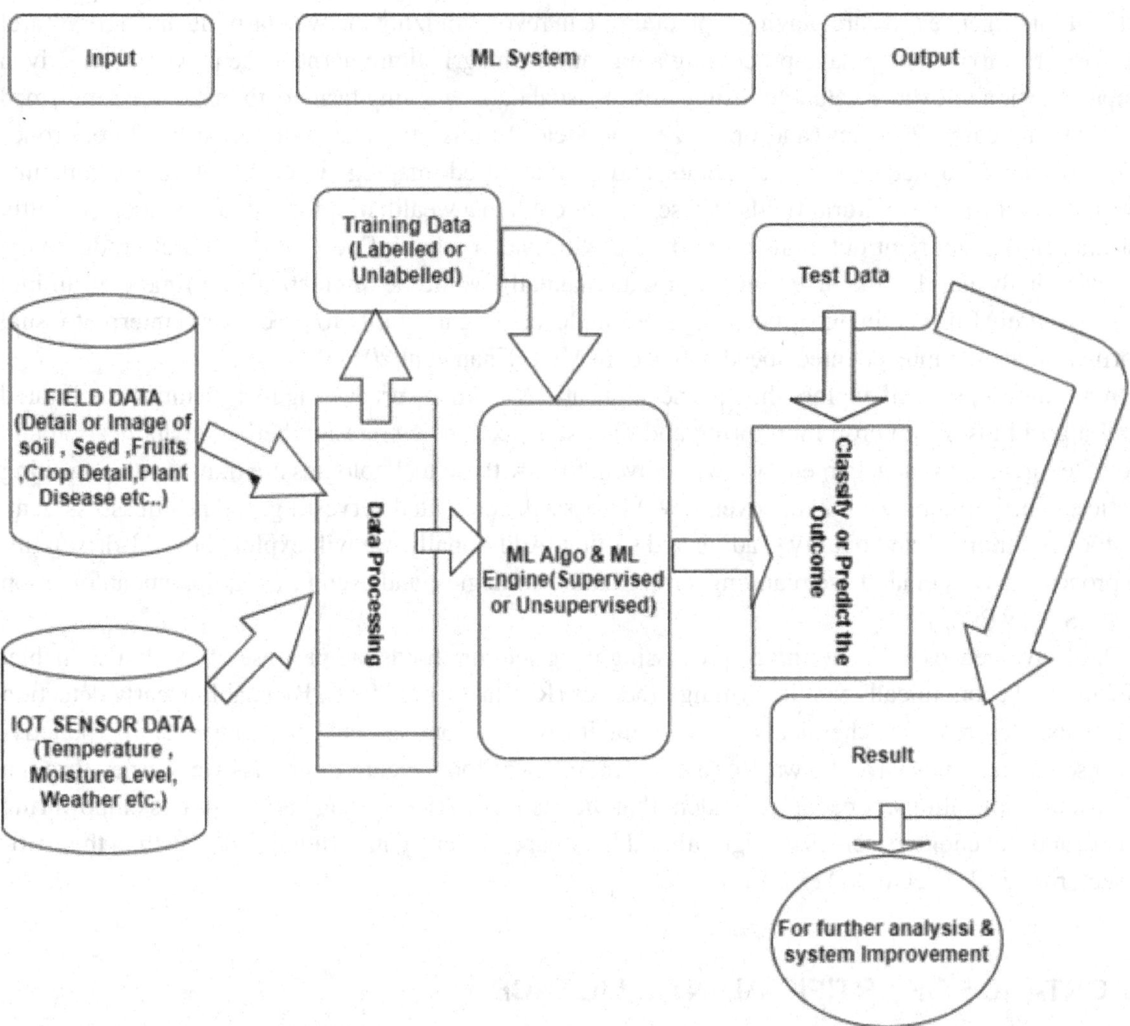

Managing Repeated Tasks: AI algorithms may be used to tackle repetitive tasks with ease. Between the procedures, these tasks don't demand much intellect. Machine learning may be used to carry out

Dangerous Exploration: By using AI, we can manage the enormous volume of data that has to be processed and stored. The sophisticated machines can be used to get around human limitations. Wherever we believe a human-powered procedure to be dangerous, we may use these machines as an alternative.

Error Reduction: Using artificial intelligence has the benefit of reducing mistakes and increasing the likelihood of achieving better accuracy with more precision.

Agriculture: One of the oldest, most significant and well-known vocations in the world is agriculture. People need to get inventive, more efficient, and imaginative about farming in order to increase agricultural output and yield while utilising less land overall. This is necessary due to the growing global population and decreasing availability of land. The global agricultural sector is now relying on AI technologies to increase crop yields and crop quality, manage pests, track soil and growing conditions, organise data for farmers, lessen their workload, and improve a wider range of agriculture-related tasks throughout the entire food supply chain. There are several areas of difficulty where research and development of AI in agriculture may produce excellent outcomes (R. Khan, et al, 2021). They are Lack of assured irrigation, indiscriminate use of agrochemicals, insufficient demand forecasting, usage of real-time advice systems to upgrade and increase crop yields early identification of pest assaults Analysis of crop damage Weed management and weed crop distinction; Market forecast for premium crop practises.

SENSING TECHNOLOGIES

Sensing technologies are an essential part of precision agriculture because they make it easier to gather and analyse data for wise decision-making about numerous facets of crop management (Lee et al., 2010). Modern information and communication systems, together with cutting-edge sensing and actuation technologies, have the potential to significantly enhance agricultural operations (Bochtis et al., 2014).Sensing technologies cover a wide range of technologies, including ground-based sensors like weather stations, soil moisture sensors, and nutrient sensors, as well as remote sensing methods like satellite images, aerial photography, and drones (Sishodia, et al., 2020). Additionally, the use of GPS and GIS applications in precision agriculture improves data collection and analysis for informed crop management decisions as well as precise location tracking, mapping, and informed crop management decisions, optimising resource utilisation (Tagung et al., 2022; Tayari et al., 2015). According to a different research, GPS and GIS technology, when used in conjunction with other digital tools, are vital for optimising agricultural practises, enabling precision farming and the production of sustainable food (Ghosh and Kumpatla, 2022).

The use of precision agriculture technologies, such as GPS, GIS, remote sensing, active canopy crop sensors, and sensor-based N optimisation algorithms, can optimise crop yields while requiring the least amount of input (Farid et al., 2023). As a result, these technologies are important tools for maximising profitability and sustainability in developing countries. Both passive and active sensors are used in remote sensing in precision agriculture. Active sensors produce internal stimuli to gather information about the planet, whereas passive sensors detect and record natural energy, mainly reflected sunlight from the planet's surface.

VARIABLE RATE SEEDING (VRS)

The accurate agricultural method known as variable rate seeding (VRS) increases crop yields and boosts farm profitability by accurately adjusting seed quantity based on soil variability and other factors. According to (Bullock et al. 1998), the potential economic gain of VRS in maize production over uniform rate seeding (URS) ranged from \$0.15 to \$12.83 per hectare, depending on the accessibility of information and the uptake of new technology. Additionally, compared to standard fixed seeding rate selections, variable rate seeding utilizing the soybean VRS simulator could increase profits by \$5 to \$57 per hectare (Correndo et al., 2022). This illustrates the possible economic gains from implementing VRS in various crop production systems. Variable rate seeding (VRS) hasn't been widely used in agriculture, although interest is growing as a result of technological developments and easier application. Farmers are given the ability to create and apply personalised strategies for exact seed planting through the integration of high-precision satellite systems with Variable rate seeding (VRS) technologies, leading to improved yield outcomes (Kishor K., 2022).

VARIABLE-RATE NUTRIENT

A precision agricultural approach called variable rate nutrient application (VRNA) includes applying different fertiliser or manure rates to various parts of a field that is producing the same crop in accordance with a predetermined prescription. According to (Teoh et al. 2016), the application of variable rate fertiliser management is anticipated to minimise fertiliser consumption, increase or sustain production, improve crop quality, and eventually improve environmental quality. The potential for increased profitability and environmental advantages of variable rate nutrient application. While VRNA for nitrogen (N) has the potential for economic and environmental benefits, it is currently not widely used by producers. VRNA for phosphorus (P) can result in long-term economic gains by increasing soil-test phosphorus. The benefits of variable rate nutrient management can be further enhanced by the use of real-time sensors for VRNA-N, which may allow nitrogen administration to be optimised based on crop needs (K. Kishor, (2023).

IRRIGATION AT A VARIABLE RATE

The ability to distribute water to a field with varied depths in a spatially specified manner, taking into consideration the needs of particular crops and other pertinent circumstances, is known as variable rate irrigation. Depending on the unique field features, VRI technology offers variable degrees of advantages and disadvantages, with results varied from field to field (Pokhrel, et al., 2018). The return on investment associated with implementing VRI depends heavily on factors including crop variety, equipment lifespan, water cost, and pumping costs (Sharma R. et al., 2022).The site-specific Variable Rate Irrigation system used in the state has the potential to increase crop water production, encourage energy and water conservation, and promote environmental sustainability (Kishor, K et al., 2023). According to a study done in a vineyard, water use was significantly reduced by 18% without affecting yield or product quality (Ortuani et al., 2019).

The study also showed that vine cultivation may efficiently use variable-rate irrigation to improve crop uniformity and water management. Variable rate irrigation (VRI) was shown to result in a 25% reduction in irrigation water usage and a 2.8% increase in soybean production in a study comparing VRI management to uniform rate irrigation (URI) management (Kishor K., (2022). Although variable rate irrigation technology has the potential to improve water management and boost crop yields, its implementation may face difficulties due to insufficient salt leaching, unintentional yield reductions in large fields, and the need to address the owner/operators attitude and philosophy (Oshaughnessy, et al., 2019).

PRECISE CROP MANAGEMENT

Through the incorporation of various data sets and precision agriculture practises, Precision Crop Management (PCM), a data-driven strategy in agriculture, maximises profitability, sustainability, and environmental protection. During the growing season, various types of data, such as yield distribution, soil characteristics, remote sensing, crop scouting observations, and weather data, can be gathered at a site-specific level to help farm managers optimise crop management practises (Kishor, Kaushal, 2023). As a result, agricultural productivity is increased while resource waste and environmental effects are reduced (Jones and Barnes, 2000). Farmers are able to make well-informed decisions based on accurate information. Precision agriculture enables the definition of management zones and the optimisation of crop management practises through the analysis and interpretation of obtained data (Arno et al., 2009). Precision agriculture uses management zones created through data analysis and geostatistics to apply variable rate input application, increasing crop management effectiveness (R. Khan et al., 2021). Applications for GIS include biotic and abiotic damage assessment and intervention, crop monitoring and yield prediction, soil health and fertility management, biomass evaluation, and land suitability assessment (Ghosh and Kumpatla, 2022).

In a 10-day period, remote sensing (RS) technology can identify single crop cultivation areas with 90% accuracy and assess the cultivated area with 95% accuracy (P. Nand et al., 2023). This technology can also monitor the dynamic conditions of soil, plants, and the area under cultivation with high accuracy. Drones have become an important tool for agriculture, allowing for smart irrigation, targeted fertilisation, and inspection of huge regions (Gago et al., 2015). Drones equipped with infrared sensors improve crop management and monitoring, resulting in better crop conditions and higher yields as well as useful equipment for power and pipeline inspections (Ahirwar et al., 2019).Drones using infrared cameras may identify the spread of plant diseases and the need for watering, saving a lot of time and money. Agronomists can save money by using less agrochemical (Daponte et al., 2019). With the use of data from soil tests and crop responses, GIS-enabled cloud technology acts as a decision support system for managing soil fertility. It makes precise fertiliser recommendations. This makes it possible for farmers to maximise crop yield and use fertiliser efficiently (Leena et al., 2016).

Additionally, remote sensing allows for accurate weed patch identification and mapping inside fields, enabling targeted herbicide applications and reducing environmental pollution while maximising weed control (Jr et al., 2003).

UAV

A crucial hour has come in the world's race against time for precision farming. The application of innovative agricultural technology has accelerated greatly in an effort to close the demand and supply imbalance between farmers and end consumers. The technology of sensor vision-based robotics (SVR) is currently useful in all field operations. To monitor and detect a field, disease, and insect detection, spraying, harvesting, soil water vegetation, etc., precision agriculture is helpful in many ways. Using this precise technology, more yield and other field operations may be completed in less time. At the moment, drone technology, automated guided vehicles, and agro-robotics are all advancing agriculture technology in different ways. We cannot obtain any data from fieldwork or other operations without sensors. We are unable to collect data from fieldwork or other operations without sensors. This research evaluates UAVs equipped with multispectral imaging sensors for use in precision agriculture. With the aid of numerous reference variables from diverse crops, it has been found that Multispectral camera imaging sensors with UAV (Rotary and Fixed wing) are useful for calculating various vegetation Indices. The linear regression model is frequently used in the literature to get various vegetation indices from diverse crops. Let's hope that everybody who wants to learn about the use of UAV-integrated sensors in precision farming will look to this evaluation as a torchbearer.

VARIETIES OF UAV

UAVs can be broadly categorized into three groups depending on design features, such as fixed wings, rotary wings, hybrid wings, and VTOL wings when applied to precision agriculture (Radoglou-Grammatikis et al., 2020).

A DRONE WITH FIXED WINGS

According to its name, UAV with Fixed Wings has two fixed wings that run the length of the vehicle. Because of its fixed wings, this kind of UAV needs a runway to take off and land. According to (Saeed et al. 2018), it offers some benefits like a relatively basic structure, an easy maintenance or repair process, an aerodynamically stable system, greater energy efficiency, wide area coverage, and high endurance (Kishor K., Pandey D. 2022). There is higher cargo capacity. It has certain advantages, but it also has some drawbacks, such as a higher cost than other types of UAV, poor weather tolerance, and a need for additional floor space for take-off and landing (Tsouros, 2019). Air must flow through the wings of fixed-wing aircraft in order to produce lift. To hover at one place like a multirotor UAV can, constant forward motion is necessary. For stationary applications like inspection work and static video capture, for example, fixed-wing solutions are therefore not the most suitable (K. Kishor, 2023).

DRONE WITH ROTATING WINGS

UAVs with rotating wings are also known as rotorcraft or VTOL (vertical take-off and landing) UAVs. Since these UAVs take off vertically, they needed a very limited area for landing. It has some benefits, such as the ability to conduct operations in a smaller area without the need for a take-off or landing space, the ability to maneuver quickly while flying or hovering, and precision flying. In addition, it will not fly at very high speeds or stay in the air for much extended periods of time, among other disadvantages. Low energy efficiency in comparison to fixed-wing competitors, area coverage, and a very complex structure and programming (Saeed et al. 2018). The number of rotor blades can be used to classify UAVs with rotary wings. Quad rotor and hex rotor drones are the most popular (Kanellakis and Nikolakopoulos 2017). The majority of applications for this kind of UAV are diverse. This particular type of UAV is frequently utilized for precision agriculture (Tsouros et al.2019).

FIXED-WING HYBRID UAV (VTOL: VERTICAL TAKE-OFF AND LANDING)

There are more sophisticated varieties of UAV that have a new design of UAV that has incorporated aspects of both UAV to overcome the drawbacks of the aforementioned two UAV. Vertical takeoff, cruise transition, and landing are all functions performed by hybrid UAVs inside a border spectrum envelope. There are several applications for these UAV kinds (Tsouros et al.2019).

VARIOUS SENSORS ARE UTILISED IN PRECISION AGRICULTURE

By expanding its services in the field of data-driven agriculture, the UAV offers inputs to both small and large farm owners. We won't be able to gather information from the field without sensors. The eyes of UAVs are these sensors. Precision spraying and insect and disease detection are a few of the crucial applications. The usefulness of UAVs in precision agriculture will be increased in a number of ways. To regulate plant nutrition, nitrogen to actual application, UAV-based imagery and spectral reflectance data acquired from these imageries are crucial instruments (Caturegli et al. 2019). UAVs could now be controlled precisely in the field thanks to sensors. The sensors built into UAVs determine how they are used. The integration of different sensors into UAVs is constrained by their payload capacity. Some of the crucial sensors needed by UAVs for operation and control are built in. The flying and operations of the UAV are directly impacted by this sensor. The majority of these sensors include accelerometers, GPS, gyroscopes, rotary encoders, temperature sensors, proximity sensors, voltage sensors, magnetometers, and others (Zhao et al. 2018). Work-specific sensors are spectral sensors, visual sensors, air quality sensors, water quality sensors, etc. External sensors are what we refer to as these. Optical and spectral sensors are typically utilised in agriculture. Each of these sensors, such as cameras, colour sensors, and multispectral imaging, has a particular application in precision farming.

We can observe from this research that comparisons are done between various multispectral UAV cameras utilised for various crops and various estimated and reference variables used for various models.

MULTIPLE-SPECTRAL SENSOR

Measurements of crop water can be made using multispectral sensor cameras to calculate references. Different indices can be calculated, including NDVI, PRI, Fluorescence UAV, TCARI/OSAVI, and Vegetation Indices, Narrow Spectral Bands (Poblete T et al. 2017), CSWI, GNDVI, WDI, and TC-Ta. Crop nutritional status can be determined using a camera with a multispectral sensor.

There are passages that talk about it being monitored. We are able to calculate the NDVI (Benincasa et al. 2018), multiple VIS, a variety of indices, REIP, GNDVI, VIS, Green Pixel Fraction, and vegetation indices. Monitoring of agricultural diseases is demonstrated with cited references. We are able to calculate a variety of indices, including 20 indices and parameters, Image Pixels, Vegetation Indices (Dash et al. 2018, GNDVI, and Image Pixels. The cited references explain the monitoring of pests in crops. Numerous indicators, including NDVI, GNDVI, Vegetation indicators, and DVM, can be computed. The camera can be used in references addressing crop weed monitoring. We can compute various indices.

THERMAL INFORMATION AND NDVI

In order to estimate vegetation indices, multispectral imageries were frequently utilized. These included the Green Normalised Difference Vegetation Index (GNDVI), the Normalised Difference Vegetation

Index (NDVI), the Soil-Adjusted Vegetation Index (SAVI), and others. Comparisons were made between estimated variables from multispectral imageries and ground truth reference variables (Czyba, R et al.2018). Several models, including machine learning, deep learning, multiple linear regression, and linear regression, among others, were employed for this investigation. Crop monitoring, crop nutrition management, crop damage assessment, disease control, etc. are all studied utilizing multispectral sensors.

SCOPE OF AI IN AGRICULTURE

The use of AI in agriculture is quite broad and has the potential to significantly alter the whole agricultural environment (Kishor, K., 2023). AI-driven apps are becoming more widely available, more reasonably priced, and more suited to the particular requirements of farmers and the agricultural sector as technology develops. Here are a few crucial components of the application of AI in Agriculture: -

Precision Agriculture: AI makes it possible for farmers to use real-time data about their crops and fields to make data-driven decisions. Farmers may maximize the use of resources like water, fertilizer, and pesticides by integrating AI, which will enhance productivity, lower costs, and have a smaller negative impact on the environment.

Crop Monitoring and Disease Detection: Drones, sensors, and AI-powered image processing May all be used to continuously check on the health of crops. AI algorithms can identify the first indications of illnesses, nutritional deficits, and pest infestations by analysing visual data and fusing it with additional environmental parameters. Farmers may take quick action to safeguard their crops thanks to this timely information.

Farming tasks that may be automated include planting, harvesting, and weeding. AI makes this possible. Robotic pickers, autonomous tractors, and other AI-driven equipment automate labour-intensive jobs, requiring less physical labour and boosting productivity.

Smart Irrigation Management: To optimize irrigation schedule, AI may examine data from weather predictions, soil moisture sensors, and satellite photography. Artificial intelligence (AI) algorithms-driven intelligent irrigation systems make sure that crops get the proper quantity of water at the right time, preserving water supplies and improving agricultural output.

Crop Yield Prediction: Using historical data, weather patterns, and other environmental parameters, AI can forecast crop yields. Farmers may improve their planning, resource allocation, and marketing techniques with the use of this information.

AI-driven sensors and computer vision may be used to monitor the health, behaviour, and feeding habits of cattle. Farmers can quickly address health concerns and enhance the general welfare of their animals by doing so.

Supply Chain Optimisation: By anticipating demand, enhancing logistics, and lowering post-harvest losses, AI may improve the agricultural supply chain. It makes it possible for distributors, merchants, processors, and farmers to work together more effectively. AI can analyse soil data to determine the health, fertility, and nutrient level of the soil. Farmers may utilise this information to make educated judgements regarding proper fertiliser application and soil management techniques. Market Analysis and Price Prediction: Artificial intelligence (AI) algorithms can analyse market trends, price information, and consumer behaviour to provide farmers insights into the ideal times to sell their goods and help them plan their marketing campaigns.

AI-driven systems may offer farmers personalised farming advice based on their unique farming circumstances, assisting them in adopting the most suited agricultural practises. The applications listed above are just a small portion of the potential uses of AI in agriculture. We may anticipate even more inventive and complex AI-driven solutions to meet the difficulties and possibilities in the agriculture industry as technology improves and data becomes more readily available. AI has the potential to improve farming's resilience, efficiency, and sustainability, hence promoting sustainable agricultural practises and enhancing global food security (K. Kishor et al 2018).

Soil Health Assessment: AI can analyse soil data to assess its health, fertility, and nutrient content. This information helps farmers make informed decisions about soil management practices and the appropriate use of fertilizers.

Market Analysis and Price Prediction: AI algorithms can analyse market trends, pricing data, and consumer behaviour to provide farmers with insights into the best times to sell their produce and make strategic marketing decisions.

Personalized Farming Recommendations: AI-driven platforms can provide personalized recommendations to farmers based on their specific farming conditions, helping them adopt the most suitable agricultural practices.

The scope of AI in agriculture is not limited to the applications mentioned above. As technology advances and data availability increases, we can expect even more innovative and sophisticated AI-driven solutions to address the challenges and opportunities in the agricultural sector. AI has the potential to make farming more sustainable, efficient, and resilient, contributing to global food security and sustainable agricultural practices (Tyagi, R, 2023).

Proposed Work of Image Processing Applications in Agriculture with the Help of AI:

Dataset Collection and Annotation: Gather diverse and representative datasets containing images of crops, weeds, pests, diseases, and soil conditions. Annotate the data to provide ground truth labels for supervised learning.

Algorithm Development and Optimization: Develop and optimize AI algorithms, particularly deep learning models, for image processing tasks in agriculture. Use convolution neural networks (CNNs) and other state-of-the-art architectures to handle complex agricultural data.

Crop Disease and Pest Detection: Implement AI models to accurately detect and classify crop diseases and pest infestations based on leaf or plant images. Evaluate the models on real-world datasets from different regions and crop types.

Weed Detection and Management: Create AI-based systems to identify and distinguish weeds from crops. Develop algorithms that can recommend targeted herbicide application or robotic weeding strategies.

Crop Yield Estimation and Monitoring: Build models to estimate crop yields based on growth stage, plant health, and environmental conditions. Implement real-time monitoring systems using drones or satellite imagery to provide continuous updates on crop status.

Soil Health Assessment: Develop AI algorithms to assess soil health and fertility using soil images. Correlate image-based analysis with soil test results and develop a soil nutrient mapping system.

Precision Irrigation and Fertilization: Implement AI-driven systems that analyse plant water stress and nutrient requirements to optimize irrigation and fertilization schedules. Integrate the system with smart irrigation and fertigation technologies.

Autonomous Farming: Design and deploy AI-controlled agricultural robots equipped with cameras and sensors for autonomous planting, weeding, and harvesting. Ensure seamless integration with existing farming practices.

Figure 2. CNN-based plant stress detection

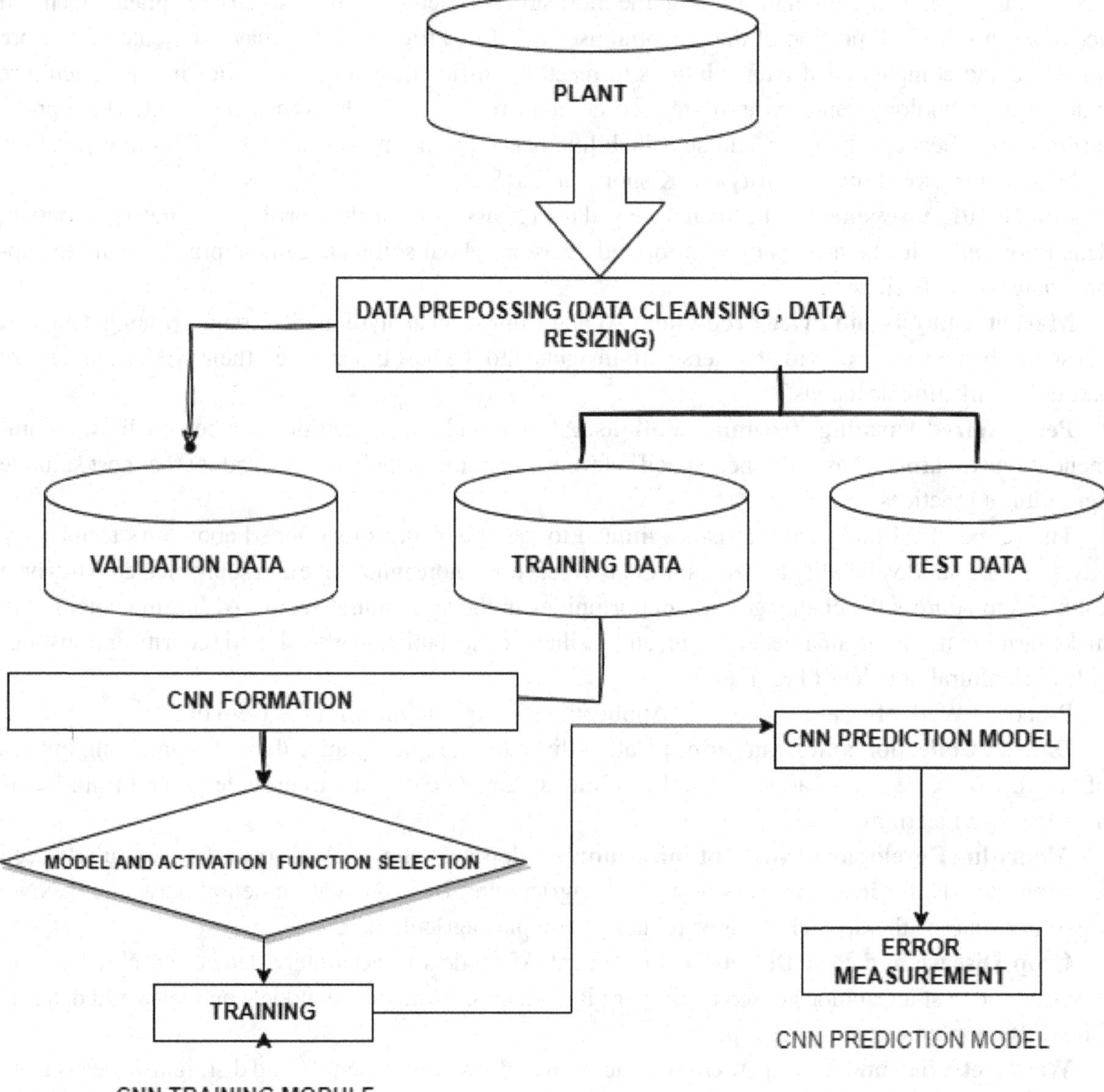

Real-time Crop Monitoring and Alerts: Develop AI-powered applications that offer real-time monitoring and alerts to farmers about crop health, disease outbreaks, and pest infestations. Enable timely interventions for crop protection.

Climate Change Adaptation: Create AI systems that analyse climate data, satellite imagery, and historical trends to predict climate-related risks in specific regions. Offer actionable insights for climate change adaptation in agriculture.

Integration With Farm Management Systems: Integrate AI-powered image processing applications with existing farm management systems, making them accessible and user-friendly for farmers. Provide mobile apps or web interfaces for easy interaction.

Privacy and Data Security Measures: Implement robust data privacy and security measures to protect farmers' sensitive information and ensure ethical use of data collected from farms.

Collaboration With Agricultural Stakeholders: Collaborate with farmers, agricultural researchers, agronomists, and policymakers to understand their specific needs and challenges. Seek feedback and incorporate domain knowledge into the AI applications.

Evaluation and Validation: Thoroughly evaluate the performance and accuracy of the developed AI models and applications in real-world settings. Validate the effectiveness of the proposed work against traditional agricultural practices.

Scalability and Adoption: Design the AI applications with scalability in mind to accommodate various crop types, farming systems, and geographical regions. Promote adoption by providing support, training, and cost-effective solutions for farmers of all scales.

The proposed work aims to harness the power of AI-driven image processing in agriculture to foster sustainable and efficient farming practices, ultimately contributing to global food security and the advancement of agricultural technology.

IMPLICATIONS OF AI IN AGRICULTURE

Image processing applications in agriculture with the help of AI have been gaining traction in recent years due to their potential to revolutionize the way farming and agricultural practices are carried out. Some of the key challenges that these applications aim to address are:

Crop Monitoring and Disease Detection: One of the primary challenges in agriculture is monitoring crops for signs of diseases, pests, and nutrient deficiencies. AI-powered image processing can help by analysing images of crops captured by drones or sensors and identifying early signs of diseases or stress. This allows farmers to take timely action, such as targeted spraying or adjusting fertilizer application, reducing the need for widespread pesticide use and minimizing crop losses.

Weed Detection and Management: Weeds compete with crops for resources and can significantly impact yields. AI-powered image processing can assist in identifying and distinguishing weeds from crops, enabling precise and targeted weed management strategies. This reduces the reliance on herbicides and promotes sustainable farming practices.

Crop Yield Estimation: Estimating crop yield accurately is essential for effective planning, resource management, and marketing. By analysing images of crops throughout the growing season, AI can predict yield potential, enabling farmers to make informed decisions about irrigation, fertilization, and harvest schedules.

Soil Health Assessment: Healthy soil is critical for successful crop growth. AI can be used to process images of soil samples, helping to assess soil health and fertility levels. This information guides farmers in making data-driven decisions about soil amendments and nutrient management.

Irrigation Management: Over-irrigation and under-irrigation can lead to water wastage and reduced crop yields. AI-based image analysis can evaluate crop water stress levels, enabling farmers to optimize irrigation schedules and conserve water while ensuring optimal crop growth.

Precision Agriculture: Precision agriculture involves tailoring farming practices to individual plants or small areas, optimizing resource usage and reducing environmental impact. AI-driven image processing plays a central role in providing real-time data for precision agriculture applications, such as variable rate seeding, fertilization, and pesticide application.

Autonomous Farming: The ultimate goal of AI and image processing in agriculture is to achieve fully autonomous farming systems. Developing reliable algorithms that can interpret images from various sources make decisions, and control agricultural machinery is a significant challenge but holds immense potential for enhancing efficiency and reducing labour requirements (R. Khan.2021).

Data Privacy and Security: As AI applications in agriculture rely heavily on data, ensuring the privacy and security of sensitive information becomes crucial (Kaushal Kishor, 2023). Adequate measures must be implemented to safeguard data from unauthorized access and protect farmers' interests (Kishor, K.et al. 2018).

Integration with Traditional Farming Practices: Introducing AI-based image processing technologies to traditional farming communities may face resistance or challenges in terms of acceptance and adoption. Education and training programs are necessary to familiarize farmers with these new tools and demonstrate their benefits.

Scalability and Affordability: While AI-powered image processing technologies hold tremendous promise, they need to be scalable and affordable for farmers across various economic strata and regions. Ensuring accessibility to these technologies can democratize advanced agricultural practices (R. Khan.2021)

Overcoming these challenges will require collaborative efforts from researchers, engineers, farmers, policymakers, and technology providers. As technology continues to advance and more data becomes available, AI-driven image processing applications in agriculture will continue to evolve, helping address the pressing issues of food security, sustainability, and environmental conservation.

CONCLUSION

In conclusion, Image Processing Applications in Agriculture with the Help of AI offer tremendous opportunities to revolutionize and elevate the agricultural industry. The integration of AI-driven image processing technologies has the potential to address some of the most pressing challenges faced by farmers, researchers, and policymakers, contributing to sustainable and efficient farming practices. Through AI-powered image analysis, farmers can gain valuable insights into their crops' health, enabling timely interventions to combat diseases, pests, and nutrient deficiencies. This proactive approach not only reduces crop losses but also minimizes the need for harmful chemicals, promoting environmentally friendly agriculture.

The future of Image Processing Applications in Agriculture with the Help of AI is marked by advancements in precision farming, where AI algorithms enable data-driven decisions on irrigation, fertilization, and pest control, leading to optimized resource usage and increased crop yields. Additionally, the concept of fully autonomous farming systems is on the horizon, as AI-powered robots equipped with cameras can perform tasks with precision and efficiency, reducing labour requirements and increasing productivity. AI's role in early detection of crop diseases and pests is pivotal in ensuring food security by safeguarding harvests from potential threats. Furthermore, customized crop management based on individual plant needs promises to unlock untapped potential for crop growth and overall agricultural productivity.

Remote sensing and high-throughput phenotyping powered by AI pave the way for large-scale monitoring and accelerated crop breeding, essential components in adapting agriculture to changing climate conditions and global food demand. Nevertheless, implementing Image Processing Applications in Agriculture with the Help of AI comes with challenges. Data privacy, algorithmic biases, and access to comprehensive datasets are critical areas that need attention to ensure responsible and equitable AI implementation in agriculture.

Overall, AI-driven image processing applications offer a paradigm shift in agriculture, empowering farmers, researchers, and policymakers with data-driven insights for informed decision-making. The collaborative efforts of stakeholders, along with ongoing research and development, will be essential to unlock the full potential of AI in agriculture and create a more sustainable, resilient, and productive global food system. By embracing these transformative technologies responsibly, we can work towards a future where technology and agriculture go hand in hand, meeting the challenges of the 21st century and beyond.

FUTURE WORK

The future scope of Image Processing Applications in Agriculture with the help of AI is promising and holds immense potential for transforming the agricultural landscape. As technology continues to advance, and AI algorithms become more sophisticated, the integration of image processing and AI in agriculture is expected to offer various exciting possibilities: Enhanced Precision Farming: AI-driven image processing can lead to more precise and data-driven farming practices. By continuously monitoring crops and fields using drones or satellite imagery, farmers can make real-time decisions regarding irrigation, fertilization, and pest control, optimizing resource usage and maximizing yields.

Autonomous Farming Systems: The development of advanced AI algorithms can enable the creation of fully autonomous farming systems. Robots equipped with cameras and AI can perform tasks like planting, weeding, and harvesting with precision and efficiency, reducing the reliance on human labour. Early Detection of Crop Diseases and Pests: AI-powered image processing can detect early signs of crop diseases and pest infestations, enabling timely interventions. This can significantly reduce crop losses and the need for excessive pesticide use, promoting sustainable and environmentally friendly farming practices.

Customized Crop Management: By analysing images and data from individual crops, AI can provide personalized recommendations for each plant's specific needs. This tailored approach to crop management can optimize growth conditions and improve overall crop health. Climate Change Adaptation: AI-driven image processing can help monitor the effects of climate change on agriculture. By analysing data from various sources, such as satellite imagery and weather sensors, AI can provide insights into adapting farming practices to changing climate conditions.

Improving Soil Health and Fertility: AI can analyse soil images to assess soil health and fertility levels. This information can guide farmers in making informed decisions about soil amendments, improving soil quality, and increasing crop productivity. Smart Irrigation Systems: AI-based image processing can analyse crop water stress levels and help optimize irrigation systems accordingly. This can lead to water conservation and more efficient water usage, especially in regions facing water scarcity.

High-Throughput Phenotyping: AI-powered image processing can enable high-throughput phenotyping, allowing researchers to analyse large sets of plant traits and characteristics rapidly. This can accelerate crop breeding and the development of new, resilient plant varieties. Remote Sensing for Large-Scale Monitoring: AI can process satellite and drone imagery to monitor large agricultural areas efficiently. This capability can aid in assessing crop health, land use, and environmental changes on a regional or global scale. Integration with Farm Management Systems: AI-driven image processing applications can be seamlessly integrated into existing farm management systems, providing farmers with user-friendly interfaces and actionable insights for better decision-making. While the future scope of Image Processing Applications in Agriculture with the Help of AI is promising, it also presents challenges such as data privacy, algorithmic biases, and the need for extensive data sets. Addressing these challenges will require collaborative efforts from researchers, policymakers, and the agricultural community. As advancements in AI and image processing continue, their implementation in agriculture can contribute to increased sustainability, productivity, and food security worldwide.

In conclusion, the integration of image processing and AI technologies into agriculture has opened up a plethora of opportunities for enhancing productivity, sustainability, and efficiency in the industry. By leveraging advanced algorithms and machine learning techniques, various challenges faced by farmers and agriculturalists have been addressed. This has led to improved crop monitoring, disease detection, yield prediction, and resource management. The utilization of drones, satellites, and other imaging devices has further expanded the scope of data collection and analysis, enabling better decision-making processes.

Moreover, the application of AI and image processing has not only benefited large-scale commercial agriculture but has also made strides in supporting small-scale and subsistence farming by providing access to valuable insights and information. The development of user-friendly interfaces and mobile applications has democratized this technology, making it accessible even to those with limited technical knowledge.

While significant progress has been made, there remains ample room for further research and development in the intersection of image processing, AI, and agriculture. Some potential avenues for future work include Multi-Spectral and Hyper spectral Imaging: Incorporating more advanced imaging techniques like multi-spectral and hyper spectral imaging can provide even more detailed information about crops, soil, and environmental conditions, leading to improved decision-making accuracy.

Robotic Farming: Combining image processing with robotics can lead to the development of autonomous farming systems that can perform tasks such as planting, weeding, and harvesting with minimal human intervention.

Climate Resilience: AI and image processing can be employed to predict and mitigate the impacts of climate change on agriculture. This involves developing models that can anticipate extreme weather events and their effects on crops.

Crop Breeding: AI-powered image analysis can aid in the process of crop breeding by identifying desirable traits in plants at an early stage, accelerating the development of new and resilient crop varieties. Data Integration: Integrating image-based data with other sources, such as weather data, soil data, and market trends, can provide a more holistic view of the agricultural ecosystem and enable better decision-making. Data Privacy and Security: As more data is collected and processed, ensuring the privacy and security of sensitive agricultural data becomes increasingly important. Developing robust protocols and encryption methods will be crucial.

Collaborative Platforms: Building platforms that allow farmers, researchers, and stakeholders to collaborate and share insights can facilitate the dissemination of knowledge and best practices.

REFERENCES

Ahirwar, S., Swarnker, R., Bhukya, S., & Namwade, G. (2019). Application of Drone in Agriculture. *International Journal of Current Microbiology and Applied Sciences*, *8*(1), 2500–2505. doi:10.20546/ijcmas.2019.801.264

Arno, J., Casasnovas, M. J., Ribes-Dasi, M., Rosell-Polo, J. R., (2009). Review. Precision Viticulture. Research topics, challenges and opportunitiesin site-specific vineyard management. *Spanish Journal of Agricultural Research, 7*, 779-790.

Benincasa, P., Antognelli, S., Brunetti, L., Fabbri, C. A., Natale, A., Sartoretti, V., Modeo, G., Guiducci, M., Tei, F., & Vizzari, M. (2018). Reliability of NDVI derived by high resolution satelliteand UAV compared to in-field methods for the evaluation of early crop N status and grain yield in wheat. *Experimental Agriculture*, *54*(4), 604–622. doi:10.1017/S0014479717000278

Bochtis, D. D., Sorensen, C. G., & Busato, P. (2014). Advances in agricultural machinery management: A review. *Biosystems Engineering*, *126*, 69–81. doi:10.1016/j.biosystemseng.2014.07.012

Bullock, D. G., Bullock, D. S., Nafziger, E. D., Doerge, T. A., Paszkiewicz, S. R., Carter, P. R., & Peterson, T. A. (1998). Does Variable Rate Seeding of Corn Pay? *Agronomy Journal, 90*(6), 830-836. doi:10.2134/agronj1998.00021962009000060019x

Caturegli, L., Corniglia, M., Gaetani, M., Grossi, N., Magni, S., Migliazzi, M., ... Raffaelli, M. (2016). Unmanned aerial vehicle to estimate nitrogen status ofturfgrasses. *PLoS One*, *11*(6), e0158268. doi:10.1371/journal.pone.0158268 PMID:27341674

Correndo, A., McArtor, B., Prestholt, A., Hernandez, C., Kyveryga, P. M., & Ciampitti, I. A. (2022). Interactive soybean variable-rate seedingsimulator for farmers. *Agronomy Journal*, *114*(6), 3554–3565. doi:10.1002/agj2.21181

Czyba, R., Lemanowicz, M., Gorol, Z., & Kudala, T. (2018). Construction Prototyping, Flight Dynamics Modeling, and Aerodynamic Analysis of Hybrid VTOLUnmanned Aircraft. *Journal of Advanced Transportation*, *2018*, 1–15. doi:10.1155/2018/7040531

Dalamagkidis, K., Valavanis, K. P., & Piegl, L. A. (2012). Aviation history and unmanned flight. In *On integrating unmanned aircraft systems into the nationalairspace system* (pp. 11–42). Springer. doi:10.1007/978-94-007-2479-2_2

Daponte, P., Vito, L. D., Glielmo, L., Iannelli, L., Liuzza, D., Picariello, F., & Silano, G. (2019). A review on the use of drones for precisionagriculture. *IOP Conference Series. Earth and Environmental Science*, *275*(1), 1–10. doi:10.1088/1755-1315/275/1/012022

Farid, H. U., Mustafa, B., Khan, Z. M., Anjum, N. M., Ahmad, I., Mubeen, M., & Shahzad, H. (2023). An Overview of Precision Agricultural Technologies for Crop Yield Enhancement and Environmental Sustainability. In *Climate Change Impacts on Agriculture* (pp. 239–257). Springer. doi:10.1007/978-3-031-26692-8_14

Gago, J., Douthe, C., Coopman, R., Gallego, P., Ribas-Carbo, M., Flexas, J., & Medrando, H. (2015). UAVs challenge to assess water stress forsustainable agriculture. *Agricultural Water Management, 153*, 9–19. doi:10.1016/j.agwat.2015.01.020

Ghosh, P., & Kumpatla, S. P. (2022). GIS Applications in Agriculture. In Geographis Information Systems and Applications in Coastal Studies. Intechopen. doi:10.5772/intechopen.104786

Ghosh, P., & Kumpatla, S. P. (2022). *GIS Applications in Agriculture*. In Geographis Information Systems and Applications in Coastal Studies. Intechopen. doi:10.5772/intechopen.104786

Gupta, S., Tyagi, S., & Kishor, K. (2022). Study and Development of Self Sanitizing Smart Elevator. In D. Gupta, Z. Polkowski, A. Khanna, S. Bhattacharyya, & O. Castillo (Eds.), *Proceedings of Data Analytics and Management. Lecture Notes on Data Engineering and Communications Technologies* (Vol. 90). Springer. doi:10.1007/978-981-16-6289-8_15

Jones, D., & Barnes, E. (2000). Fuzzy composite programming to combineremote sensing and crop models for decision support in precisioncrop management. *Agricultural Systems, 65*(3), 137–158. doi:10.1016/S0308-521X(00)00026-3

Jr, P. P., Hatfield, J. L., Schepers, J. S., Barnes, E. M., Moran, S. M., Daughtry, C. S., & Upchurch, D. R. (2003). Remote Sensing for Crop Management. *Photogrammetric Engineering and Remote Sensing, 18*(6), 647–664. doi:10.14358/PERS.69.6.647

Khan & Amjad. (2019). Mutation Based Genetic Algorithm for Efficiency Optimization of Unit Testing. *International Journal of Advanced Intelligence Paradigms, 12*(3-4), 254-265. . doi:10.1504/IJAIP.2019.098563

Khan, Kumar, Dhingra, & Bhati. (2021). The Use of Different Image Recognition Techniques in Food Safety: A Study. *Journal of Food Quality*. doi:10.1155/2021/7223164

Khan, Kumar, Srivastava, Dhingra, Gupta, Bhati, & Kumari. (2021). Machine Learning and IoT Based Waste Management Model. *Computational Intelligence and Neuroscience*. doi:10.1155/2021/5942574

Khan, R., Shabaz, M., Hussain, S., Ahmed, F., & Mishra, P. (2021). *Early Flood Detection and Rescue Using Bioinformatic devices, Internet of Things (IoT) and Android App. Word Journal of Engineering Journal of Emerald Publishing*. doi:10.1108/WJE-05-2021-0269

Kishor, K. (2022). Communication-Efficient Federated Learning. In S. P. Yadav, B. S. Bhati, D. P. Mahato, & S. Kumar (Eds.), *Federated Learning for IoT Applications. EAI/Springer Innovations in Communication and Computing*. Springer. doi:10.1007/978-3-030-85559-8_9

Kishor, K. (2022). Personalized Federated Learning. In S. P. Yadav, B. S. Bhati, D. P. Mahato, & S. Kumar (Eds.), *Federated Learning for IoT Applications. EAI/Springer Innovations in Communication and Computing*. Springer. doi:10.1007/978-3-030-85559-8_3

Kishor, K. (2023). *10 Study of quantum computing for data analytics of predictive and prescriptive analytics models*. In S. P. Yadav, R. Singh, V. Yadav, F. Al-Turjman, & S. A. Kumar (Eds.), *Quantum-Safe Cryptography Algorithms and Approaches: Impacts of Quantum Computing on Cybersecurity* (pp. 121–146). De Gruyter. doi:10.1515/9783110798159-010

Kishor, K. (2023). *12 Review and significance of cryptography and machine learning in quantum computing*. In S. P. Yadav, R. Singh, V. Yadav, F. Al-Turjman, & S. A. Kumar (Eds.), *Quantum-Safe Cryptography Algorithms and Approaches: Impacts of Quantum Computing on Cybersecurity* (pp. 159–176). De Gruyter. doi:10.1515/9783110798159-012

Kishor, K. (2023). Cloud Computing in Blockchain. In Cloud-based Intelligent Informative Engineering for Society 5.0 (pp. 79-105). Chapman and Hall/CRC. doi:10.1201/9781003213895-5

Kishor, K. (2023). *17 Application of quantum computing for digital forensic investigation*. In S. P. Yadav, R. Singh, V. Yadav, F. Al-Turjman, & S. A. Kumar (Eds.), *Quantum-Safe Cryptography Algorithms and Approaches: Impacts of Quantum Computing on Cybersecurity* (pp. 231–248). De Gruyter., doi:10.1515/9783110798159-017

Kishor, K. (2023). Impact of Cloud Computing on Entrepreneurship, Cost, and Security. In Cloud-based Intelligent Informative Engineering for Society 5.0 (pp. 171-191). CRC Press. doi:10.1201/9781003213895-10

Kishor, K., & Nand, P. (2023). Wireless Networks Based in the Cloud That Support 5G. In Cloud-based Intelligent Informative Engineering for Society 5.0 (pp. 23-40). Chapman and Hall/CRC. doi:10.1201/9781003213895-2

Kishor, K., Nand, P., & Agarwal, P. (2018). Notice of Retraction Design adaptive Subnetting Hybrid Gateway MANET Protocol on the basis of Dynamic TTL value adjustment. *Aptikom Journal on Computer Science and Information Technologies*, *3*(2), 59–65. doi:10.11591/APTIKOM.J.CSIT.115

Kishor, K., Nand, P., & Agarwal, P. (2018). Secure and Efficient Subnet Routing Protocol for MANET. *Indian Journal of Public Health Research & Development*, *9*(12), 200. doi:10.5958/0976-5506.2018.01830.2

Kishor, K., & Pandey, D. (2022). Study and Development of Efficient Air Quality Prediction System Embedded with Machine Learning and IoT. In *Proceeding International Conference on Innovative Computing and Communications. Lect. Notes in Networks, Syst.* (Vol. 471). Springer. 10.1007/978-981-19-2535-1_24

Kishor, K., Saxena, N., & Pandey, D. (Eds.). (2023). Cloud-based Intelligent Informative Engineering for Society 5.0. Chapman and Hall/CRC. doi:10.1201/9781003213895

Kishor, K., Sharma, R., & Chhabra, M. (2022). Student Performance Prediction Using Technology of Machine Learning. In D. K. Sharma, S. L. Peng, R. Sharma, & D. A. Zaitsev (Eds.), *Micro-Electronics and Telecommunication Engineering. Lecture Notes in Networks and Systems* (Vol. 373). Springer. doi:10.1007/978-981-16-8721-1_53

Kishor, K., Singh, P., & Vashishta, R. (2023). Develop Model for Malicious Traffic Detection Using Deep Learning. In D. K. Sharma, S. L. Peng, R. Sharma, & G. Jeon (Eds.), *Micro-Electronics and Telecommunication Engineering. Lecture Notes in Networks and Systems* (Vol. 617). Springer. doi:10.1007/978-981-19-9512-5_8

Kishor, K., Tyagi, R., Bhati, R., & Rai, B. K. (2023). Develop Model for Recognition of Handwritten Equation Using Machine Learning. In *Proceedings of International Conference on Recent Trends in Computing. Lecture Notes in Networks and Systems* (vol. 600). Springer. 10.1007/978-981-19-8825-7_23

Lee, W., Alchanatis, V., Yang, C., Hirafuji, M., Moshou, D., & Li, C. (2010). Sensing technologies for precision specialty crop production. *Computers and Electronics in Agriculture*, *74*(1), 2–33. doi:10.1016/j.compag.2010.08.005

Leena, U., Premasudha, B., & Basavaraja, P. (2016). Sensible approach for soil fertility management using GIS cloud. *International Conference on Advances in Computing, Communications and Informatics (ICACCI),* 2776-2781. 10.1109/ICACCI.2016.7732483

McCarthy, J., Minsky, M., Rochester, N., & Shannon, C. E. (2006). A Proposal for the Dartmouth Summer Research Project on Artificial Intelligence, August 31, 1955. *AI Magazine*, *27*, 12–14.

Mooney, D. F., Roberts, R. K., English, B. C., Lambert, D. M., Larson, J. A., Velandia, M., & Reeves, J. M. (2009). Precision Farming by Cotton Producers in Twelve Southern States: Results from the 2009 Southern Cotton Precision Farming Survey. *Research in Agriculture and Applied Economics*.

Ortuani, B., Facchi, A., Mayer, A., Bianchi, D., Bianchi, A., & Brancadoro, L. (2019). Assessing the Effectiveness of Variable-Rate Drip Irrigation on Water Use Efficiency in a Vineyard in Northern Italy. *Water (Basel)*, *11*(10), 1964. Advance online publication. doi:10.3390/w11101964

Oshaughnessy, S., Evett, S. R., Colaizzi, P. D., Andrade, M. A., & Marek, T. H. (2019). *Identifying Advantages and Disadvantages of Variable Rate Irrigation-An Updated Review*. Biological Systems Engineering.

Poblete, T., Ortega-Farías, S., Moreno, M. A., & Bardeen, M. (2017). Artificial neural network to predict vine water status spatial variability using multispectralinformation obtained from an unmanned aerial vehicle (UAV). *Sensors (Basel)*, *17*(11), 2488. doi:10.339017112488 PMID:29084169

Pokhrel, B. K., Paudel, K. P., & Segarra, E. (2018). Factors Affecting the Choice, Intensity, and Allocation of Irrigation Technologies by U.S. Cotton Farmers. *Water (Basel)*, *10*(6), 706. Advance online publication. doi:10.3390/w10060706

Pujahari, Yadav, & Khan. (2022). Intelligence Farming System through Weather Forecast Support and Crop Production. In *Application of Machine Learning in Smart Agriculture*. Elsevier. . doi:10.1016/B978-0-323-90550-3.00009-6

Radoglou-Grammatikis, P., Sarigiannidis, P., Lagkas, T., & Moscholios, I. (2020). A compilation of UAVapplications for precision agriculture. Computer Networks, 172, 107148.

Rai, B. K., Sharma, S., Kumar, G., & Kishor, K. (2022). Recognition of Different Bird Category Using Image Processing. *International Journal of Online and Biomedical Engineering*, *18*(07), 101–114. doi:10.3991/ijoe.v18i07.29639

Sahani, Srivastava, & Khan. (2021). Modelling Techniques to Improve the Quality of Food using Artificial Intelligence. *Artificial Intelligence in Food Quality Improvement*. doi:10.1155/2021/2140010

Sishodia, R. P., Ray, R. L., & Singh, S. K. (2020). Applications of Remote Sensing in Precision Agriculture: A Review. *Remote Sensing (Basel)*, *12*(19), 3136. Advance online publication. doi:10.3390/rs12193136

Tagung, T., Singh, S. K., Singh, P., Kashiwar, S. R., & Singh, S. K. (2022). GPS and GIS based Soil Fertility Assessment and Mapping in Blocks of Muzaffarpur District of Bihar. *Biological Forum : An International Journal, 14*(3), 1663–1671.

Teoh, C., Mohamad, B., Radzi, F. F., Najib, M. M., Zamzuri, C. F., Hassan, D., & Nordin, M. n. (2016). Variable rate application of fertilizer in riceprecision farming. *International Conference on Agricultural andFood Engineering (Cafei2016),* 23-25.

Tsouros, D. C., Bibi, S., & Sarigiannidis, P. G. (2019). A review on UAV-based applications for precision agriculture. *Information (Basel), 10*(11), 349. doi:10.3390/info10110349

Zhao, B., Chen, X., Zhao, X., Jiang, J., & Wei, J. (2018). Real-Time UAV Autonomous Localization Based on Smartphone Sensors. *Sensors (Basel), 18*(12), 4161. doi:10.339018124161 PMID:30486422

Chapter 11
Impact of Industry 5.0 on Healthcare

Agrima Saxena
https://orcid.org/0009-0009-0562-1088
Galgotias University, India

S. P. S. Chauhan
Sharda University, India

Harshit Singh
Galgotias University, India

Usha Chauhan
Galgotias University, India

Priti Kumari
National Institute of Technology, Patna, India

ABSTRACT

Historical societies have been propelled into new eras of progress by the growth of industrial paradigms, each of which was characterised by distinctive technological developments and sociological changes. These revolutions, starting from the Industry 1.0's mechanisation to Industry 4.0's digitalisation, have radically changed the way that various industries and economies around the world operate. Industry 5.0 is now emerging as a disruptive force on the verge of a new phase. Innovative technologies like artificial intelligence (AI), robotics, and the internet of things (IoT) will be combined with human ingenuity, according to its promises.

DOI: 10.4018/979-8-3693-0782-3.ch011

I. INTRODUCTION

The industry has evolved over the past few years through numerous stages, from Industry 1.0, which was characterised by mechanisation, to Industry 4.0, which is characterised by digitization and automation. Now that Industry 5.0 is taking off, a new paradigm that combines human potential with artificial intelligence is improving several industries, including healthcare. This essay investigates how Industry 5.0 might affect patient care, healthcare systems, and technological breakthroughs in medicine. Industry 5.0 emphasises human-machine cooperation while recognising the distinct advantages of each. This collaborative approach in healthcare has the potential to revolutionise patient care through tailored treatment regimens, enhanced diagnostics, and optimal therapy selection. Real-time data analysis, IoT-enabled medical equipment, and wearable device integration enable healthcare practitioners to make decisions that are individualised for each patient.

Additionally, combining human experience, robotics, and AI to do administrative work within healthcare organisations can be more efficient. The increased operational efficiency that results from improved data management, computerised scheduling, and resource allocation frees up more time for direct patient care on the part of healthcare providers. As a result of the recent global health problems, industry 5.0 also makes remote patient monitoring, telemedicine, and virtual consultations possible. Industry 5.0 expedites disease modelling, genomics research, and medication discovery in the field of medical research and development. Simulations and modelling tools make possible a deeper knowledge of complicated diseases, while cutting-edge AI algorithms helps in finding prospective medication candidates. Industry 5.0-powered collaboration platforms promote international collaboration between scientists and researchers, accelerating medical advances.

The use of Industry 5.0 in the healthcare industry raises questions about data security, privacy, and the moral application of AI. It's crucial to strike a balance between protecting private patient information and technical improvements. To guarantee that patient rights are safeguarded and AI algorithms are held responsible for their actions, regulatory frameworks must be updated. In conclusion, Industry 5.0 offers the healthcare industry a game-changing opportunity. Healthcare systems may deliver more individualised, effective, and accessible care by embracing human-machine collaboration. The potential advantages of Industry 5.0 in healthcare must be addressed, notwithstanding issues with data privacy and ethical issues. This paradigm shift could improve patient outcomes, advance medical research, and influence how healthcare is provided in the near future.

II. INDUSTRY 5.0 AND HEALTHCARE

"Industry 5.0," also known as the "Human-Centric Industry" or "Society 5.0," expands on the previous industrial revolutions (Industry 1.0 to Industry 4.0) by highlighting the seamless fusion of people and cutting-edge technology. Industry 5.0 seeks to combine the finest qualities of humans and robots to produce more collaborative and human-centered solutions, in contrast to previous industrial revolutions, which mostly focused on automation, digitization, and networking (Ivanov, D. (2023), Ghobakhloo,(2022), Xu, X.(2021)). Khan et al. (2023) discuss cutting-edge concepts like smart health, smart agriculture, and smart education in their book 'Smart Infrastructure and Sustainable Rural Communities,' focusing on how these ideas can empower communities through smart infrastructure and Industry 5.0

Industry 5.0's salient features include:

1. Collaboration between humans and machines is promoted by Industry 5.0, which acknowledges that both groups have unique skills. While robots bring speed, precision, data processing capabilities, and automation, humans bring creativity, empathy, complex problem-solving, and adaptability(Adel, A. (2020)).
2. Personalization: Industry 5.0 aims to provide individualised goods and services that are catered to specific customer requirements. This is applicable to many other industries, such as manufacturing, healthcare, retail, and more.
3. Localization: Localised production and consumption are promoted in an effort to lessen the adverse effects on the environment and promote more community involvement. Greater sustainability and efficiency are now feasible thanks to innovative manufacturing processes like 3D printing and localised production (Birtchnell, T.(2016)).
4. Integrating physical and digital systems: Industry 5.0 blurs the line between the physical and digital worlds by fusing real-time data from Internet of Things (IoT) devices, sensors, and digital platforms with physical processes. This enables more informed decision-making and responsive actions.
5. Enhanced Intelligence: Instead of focusing mostly on AI-driven automation, Industry 5.0 intends to leverage AI and automation to improve human capabilities. Instead of replacing humans in decision-making, problem-solving, and difficult activities, this entails AI systems assisting them.
6. Ethical Considerations: Industry 5.0 gives ethical issues a lot of attention, addressing issues including privacy, security, and responsible technology use.
7. Social Well-being Industry 5.0 aims to develop solutions that place a priority on societal well-being and sustainable growth, in contrast to earlier industrial revolutions that frequently resulted in social and environmental problems(Žižek, S.Š.(2021)).

Numerous industries, including manufacturing, healthcare, agriculture, transportation, and more, use Industry 5.0. To improve patient care, individualised treatment plans, medical research, and operational efficiency inside healthcare organisations, for instance, AI and robotics are used. Industry 5.0, which emphasises human-machine collaboration to generate solutions that enhance lives, address societal concerns, and promote sustainable development, indicates an overall shift in the approach to technological integration(Schwab, K. (2022)).

A. Evolution of Industry 5.0

Industry 5.0 is a concept that is still in its infancy. Industry 5.0 was a theoretical framework that drew from earlier industrial revolutions. Below is a summary of industrial growth before Industry 5.0 was established(Vural Özdemir (2018)):

- Between the late 18th and early 19th centuries, the first revolution, also referred to as Industry 1.0, occurred. During this revolution, agrarian communities made way for industrialised economies. It was marked by the automation of steam and water-powered industrial processes, which led to the established of factories and mass manufacturing.
- The late 19th and early 20th centuries saw the Second Industrial Revolution.: The internal combustion engine and electricity were both invented during this time. It resulted in substantial improvements in communication, transportation, and manufacturing, which enhanced productivity and facilitated globalisation.

Figure 1. The Evolution of Industry 5.0

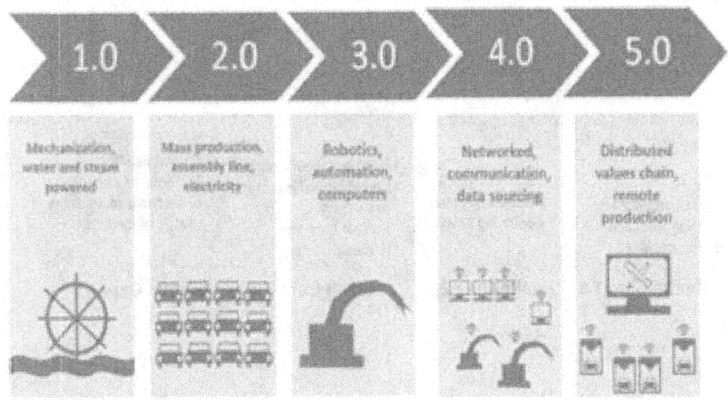

- The Third Industrial Revolution (also known as "Industry 3.0") in the late 20th century was characterised by the advancement of computers and automation. Businesses started employing computers to manage operations, which boosted process control and precision. In this time, digitalization and the usage of production management software also had their start.
- The Fourth Industrial Revolution, or "Industry 4.0" (early 21st century): Digital technology, the Internet of Things (IoT), data analytics, and artificial intelligence are all incorporated into manufacturing processes as part of Industry 4.0.
- Building on the ideas of Industry 4.0, the Human-Centric Industry (Emerging Concept) in Industry 5.0 places a stronger emphasis on human-machine cooperation. It imagines a time where robotics and AI will collaborate with humans to improve efficiency, personalisation, and societal well-being.

The shift from Industry 1.0 to Industry 4.0 has been characterised by increasing automation, digitalization, and networking. Industry 5.0 is a shift towards a more balanced strategy that acknowledges the distinct contributions of both humans and machines in developing answers to difficult problems (see Fig. 1).

B. Evolution of Healthcare 5.0

Healthcare has changed over time as technology advanced and people's demand for better health results increased. Healthcare 5.0, the most recent stage of this progression, is marked by the use of numerous technologies and approaches to deliver personalised, patient-centered care. Here is a hypothetical overview of how the evolution of healthcare can be compared to previous industrial revolutions and the shift from Industry 4.0 to Industry 5.0. (see Fig. 2):

- Traditional medicine in healthcare (pre-modern era): Traditional medicine, natural cures, and scant medical understanding constituted the bulk of healthcare. Treatment strategies frequently drew from cultural customs and values.
- Medical care 2.0 - Developments in Medical Science (18th to 19th century): Significant medical knowledge advances during this time period included the creation of vaccines, the development

Figure 2. Evolution of Healthcare 5.0

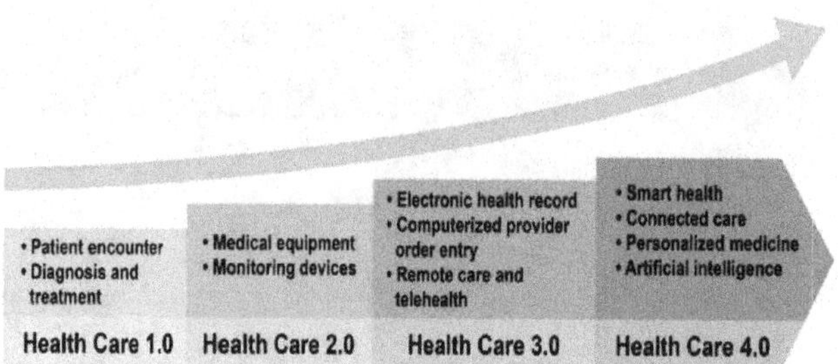

of surgical procedures, and the identification of infections. It set the stage for the application of scientific concepts to healthcare and evidence-based medicine.

- Medical care Technology in medicine, version 3.0 (20th century): Medical technology underwent a revolution in the 20th century with the introduction of X-rays, antibiotics, and the widespread use of medical devices. The ability to diagnose and treat patients improved, making healthcare a more technologically advanced industry.
- Medical Care 4.0 - Personalised medicine and digital health in the twenty-first century: This stage is what we're in right now, when digital technologies are being integrated into healthcare. The use of telemedicine, electronic health records, wearable medical technologies, and data analytics has increased. A patients genetic, environmental, and behavioural variables could now be used to tailor their medical care.
- Collaborative and human-centered healthcare in healthcare 5.0 (Speculative Concept): If we use Industry 5.0 as an analogy instead of Industry 4.0, Healthcare 5.0 might concentrate on cooperative efforts between medical personnel, patients, and cutting-edge technologies. The focus may shift towards combining human expertise, robotics, and AI to offer more individualised and comprehensive healthcare solutions. Advanced AI-assisted diagnoses, AI-guided therapy recommendations, and a more thorough patient integration into their own care plans may all be part of this phase. (B. Mohanta (2019))

Additionally, proactive health management, prevention measures, and patient-centered treatment could be given top priority in healthcare 5.0. AI-driven forecasts and early disease diagnosis combined with each other may make it possible to intervene before things become worse.

III. THE RAMIFICATIONS OF INDUSTRY 5.0 IN HEALTHCARE

There are substantial implications for patient care, medical research, and operational effectiveness when Industry 5.0 ideas are implemented in the healthcare industry. Some of the major effects of Industry 5.0 in the health-care field are as follows:

A. Highly Individualised Patient Care

The fusion of revolutionary technologies with patient-oriented practises is bringing about a paradigm shift in healthcare. Every patient is viewed as a different individual in this concept, with their own medical background, genetics, lifestyle, and preferences. Healthcare practitioners may accurately adapt therapies to each patient's needs by utilising data analytics, artificial intelligence, and personalised medicine procedures (L. Gomathi(2023)). By going beyond the conventional one-size-fits-all paradigm, this strategy enables more precise therapeutic interventions, treatment programmes, and diagnostic evaluations. Highly individualised patient care boosts patient happiness, and adherence to treatment plans in addition to improving medical outcomes by comprehending the complex subtleties of each patient's condition. With this revolutionary idea, the emphasis is shifted from generalised healthcare to a customised journey that values patients' diversity and gives them the power to actively contribute to their own well-being.

B. Enhanced Diagnostics and Early Detection

Healthcare professionals can now look deeper into patient data and find minute patterns and signals that were previously elusive thanks to the introduction of cutting-edge technologies like artificial intelligence (AI) and machine learning. Diagnostic technologies powered by AI have the capacity to analyse intricate medical images, test reports, and patient histories at unmatched speed and precision. Early disease detection made possible by this increased level of precision enables prompt treatments that can considerably enhance treatment results. The focus of Industry 5.0 on interoperability between human expertise and AI enhances diagnostic capabilities by guaranteeing that doctors and AI algorithms coexist peacefully to improve accuracy and dependability. In addition to providing healthcare professionals with priceless information, this synergy also empowers patients by enabling early identification, timely interventions, and the possibility of better outcomes.(L. Gomathi(2023))

C. Telemedicine and Remote Patient Monitoring

Industry 5.0's advancements in telemedicine allow patients to receive medical advice and treatment digitally, regardless of where they are physically located. This increases convenience, lessens the pressure on overworked healthcare facilities, and improves accessibility for people in remote or underdeveloped locations. Additionally, remote patient monitoring collects real-time health data from wearables with IoT capabilities, giving medical professionals a thorough insight of patients' circumstances. Continuous monitoring allows for the early detection of changes in a patient's health, prompt interventions, and individualised care plans, all of which improve patient outcomes. The fusion of Industry 5.0 concepts with telemedicine and remote patient monitoring promotes a future where healthcare is both technologically cutting edge and patient-centric. This potential for revolutionising healthcare broadens its effect and scope. (M. Wazid(2022)).

D. Collaborative Treatment Planning

Healthcare professionals that use AI-driven insights in clinical decision-making have access to a variety of data-driven knowledge that improves their clinical judgement. Patients simultaneously take an active role in their treatment, offering insightful information about their preferences, worries, and experiences.

This partnership promotes a patient-centric strategy by helping people to make knowledgeable choices regarding their treatment regimens. Additionally, AI's capacity for processing enormous datasets enables the discovery of subtle patterns, ensuring that diagnoses and treatment suggestions are thorough and correct (B. Mohanta(2019), L. Gomathi(2023)). Industry 5.0 promotes shared decision-making, mutual respect, and better patient outcomes in the healthcare ecosystem by enhancing the quality of care through collaborative treatment planning.

E. Streamlined Administrative Processes

Applying AI to administrative chores can boost operational effectiveness. Within healthcare institutions, AI-powered solutions can automate resource allocation, billing, and appointment scheduling. The administrative burden on healthcare practitioners can be reduced by automating tedious and time-consuming tasks like appointment scheduling, billing, resource allocation, and documentation. AI-powered algorithms can use previous data analysis to forecast patient influx, allowing for effective staffing and resource management. This not only increases operational effectiveness, but also gives healthcare professionals more time to devote to providing direct patient care and making clinical decisions. Healthcare organisations can implement Industry 5.0 to develop leaner, more responsive administrative procedures, thereby improving the patient and healthcare professional experience as a whole.(Maria Vincenza Ciasullo(2022))

F. Accelerated Medical Research

Industry 5.0 has caused a paradigm change in the field of medical research by fusing cutting-edge technology with human skills. One of its most convincing results is the quickening of medical research. Industry 5.0 accelerates the examination of enormous datasets by using AI-driven algorithms, revealing complex patterns and correlations that may have escaped previous research methods. By finding viable candidates and accurately projecting their efficacy, an AI-powered methodology supports medication discovery(Mbunge, E.(2021)). Additionally, Industry 5.0-enhanced simulation and modelling methods enable researchers to comprehend complicated diseases and their mechanisms at a higher level. Industry 5.0's collaborative platforms enable cross-border collaboration among scientists, dissolving geographic barriers and promoting quick information transfer. As a result, the typical timeframe for medical breakthroughs is shortened, and novel treatments are developed at a previously unheard-of rate. The impact of Industry 5.0 on quickening medical research gives not just hope for quicker cures and treatments but also a solution to deal with major global healthcare issues.

G. Global Collaboration

Industry 5.0 ushers in a period of increased international cooperation in the healthcare industry, when the power of technology transcends geographical and professional barriers. This paradigm allows for the real-time sharing of knowledge, insights, and ground-breaking discoveries by healthcare practitioners, researchers, and specialists from around the world on virtual platforms. Industry 5.0's increased connectivity promotes a cooperative ecosystem that speeds up medical research, allows cross-border consultations, and uses a variety of viewpoints to address difficult healthcare problems (Garg, P. K. (2022)). Industry 5.0's focus on connectivity, transforms healthcare into a globally interconnected network of shared expertise, whether it's pooling genomics data for research on uncommon diseases or conducting collaborative clinical trials across continents. In addition to accelerating the development of medical innovations, this cooperative synergy also fosters a shared commitment to advancing healthcare globally, ultimately benefiting patients everywhere with cutting-edge therapies, knowledgeable care, and the collective wisdom of healthcare professionals from various backgrounds.

H. Increased Patient Engagement

A paradigm shift brought about by Industry 5.0 encourages greater patient engagement. Innovative technologies like wearable gadgets, telemedicine, and artificial intelligence are being integrated to give patients the resources they need to take a more active role in managing their own health. Patients may follow their vital signs, keep an eye on chronic illnesses, and make wise decisions about their health thanks to real-time access to personalised health data (Ciasullo, M. V.(2022)). Additionally, patients are included in the decision-making process via Industry 5.0's emphasis on collaborative healthcare planning, allowing them to provide preferences and insights that influence their treatment plans. As patients become partners in their own care, this increased engagement not only improves patient happiness, but also promotes better treatment adherence. Through Industry 5.0, patients are taking an active role in their healthcare, leading to a more thorough and patient-centered approach to the delivery of health care.

I. Equitable Access to Care

Industry 5.0 has the potential to transform healthcare by improving every individual's ability to receive care, irrespective of where they live or their socioeconomic status. Industry 5.0 offers us the chance to overcome the gaps that have traditionally restricted access to healthcare through the adoption of revolutionary innovations like telemedicine, monitoring patients remotely, and AI-assisted diagnostics. Patients in remote or underprivileged locations can contact with healthcare providers without having to make a long trip by permitting virtual consultations. Additionally, wearable technology and IoT-enabled medical equipment offer real-time health data that can be monitored remotely, enabling healthcare professionals to take preventative measures and stop health problems from getting worse (Ciasullo, M. V.(2022)). This strategy serves both rural and urban populations, addressing inequities in urban areas where access to medical facilities may be restricted. The goal of equitable access to care is made a reality through Industry 5.0's emphasis on human-technological collaboration, making access to high-quality healthcare, a fundamental right for everyone, regardless of their circumstances.

J. Healthcare Workforce Transformation

The healthcare workforce is undergoing a tremendous transition as a result of Industry 5.0, which redefines the duties and tasks of healthcare workers. The workforce's job shifts from traditional task execution to a more strategic and sympathetic emphasis as cutting-edge technology like AI and robotics become indispensable to healthcare delivery. Previously burdened with administrative and mundane activities, healthcare practitioners are now free to focus on the complicated decision-making, patient engagement, and ethical concerns in treatment plans because to AI-driven automation. This revolution necessitates a change in skill sets, necessitating that healthcare professionals be adept at utilising technology, interpreting insights produced by AI, and successfully cooperating with intelligent systems. Industry 5.0 may increase productivity and accuracy, but the human touch is still vital for establishing rapport, offering

Figure 3. The ramifications of Healthcare 5.0

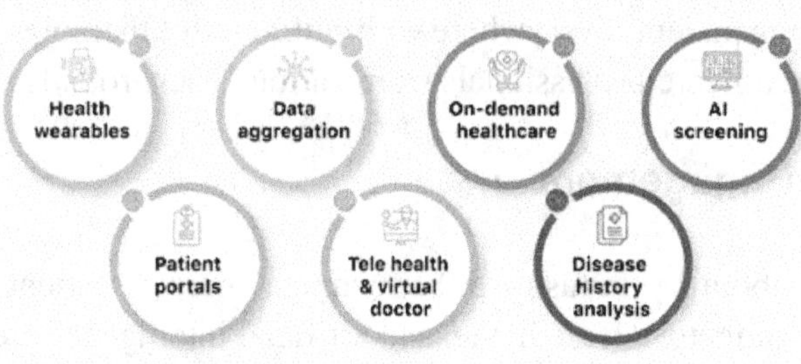

comfort, and comprehending the subtleties of patient care(Bida, M. N.(2023)). As a result, the healthcare workforce of Industry 5.0 reflects a harmonious union of human proficiency and technical capability, ultimately resulting in more patient-centric, compassionate, and efficient healthcare delivery.

As a result, Industry 5.0 offers the potential to transform healthcare by encouraging cooperation between people and cutting-edge technology (see Fig. 3). Personalised care, early disease identification, simpler operations, quicker research, and better patient outcomes are some of the effects.

IV. CHALLENGES FACED BY HEALTHCARE 5.0

Industry 5.0 offers healthcare systems several promising futures, but it also has significant challenges that must be resolved before it can be successfully implemented. Among these challenges are, amid others:

A. Data Security and Privacy Issues

The complex web of data security and privacy concerns is the main worry in the context of Industry 5.0's integration into healthcare. The exponential growth in the quantity, velocity, and variety of healthcare data as well as the usage of cutting-edge technologies like AI, IoT devices, and networked systems have increased the potential flaws in healthcare systems. The danger of unauthorised access, data breaches, and cyberattacks increases as patient information becomes more digitalized and networked. The collaborative nature of Industry 5.0 requires smooth data sharing and exchange, yet this very openness opens the door to security flaws that could jeopardise patient privacy and confidential medical records. It becomes difficult to strike a balance between using data to better patient care and protecting it from bad intent. To build reliable cyber security safeguards, strict encryption standards, and cutting-edge authentication systems, regulatory agencies, healthcare organisations, and technology providers must work together. It will be crucial to put in place structures that abide with changing privacy laws and provide patients control over their data. The challenge is to strengthen healthcare systems with state-of-the-art defences while promoting a mindset of vigilance and education among medical professionals in order to make certain that Industry 5.0's potential can be realised without threatening the privacy and security of patients' most private and sensitive information (Deivanayagampillai, N. (2023)).

B. Moral Conundrums in AI Decision-Making

While creating complex ethical issues in the field of AI decision-making, the adoption of Industry 5.0 concepts in healthcare presents a significant change in patient care and medical decision-making. Transparency, accountability, and the proper ratio of human expertise and machine-driven insights are ethically challenging issues that arise as AI algorithms play an increasingly important role in disease diagnosis, providing treatment strategies, and forecasting patient outcomes. Due to some AI models' opacity, which makes it challenging to comprehend the thinking behind their decisions, the traditional idea of trained medical judgement is weakened (Dhirani, L. L. (2023)). This lack of transparency raises questions regarding individuals' rights to comprehend and object to AI-driven health decisions. Furthermore, when AI-driven judgements result in unfavourable consequences, assigning blame becomes complicated because the blame may move from healthcare personnel to the algorithms themselves or their creators. To ensure that Industry 5.0's advancements truly uphold the highest standards of medical ethics and patient well-being, the healthcare sector must walk a fine line between empowering AI technologies and protecting patients' rights to transparency, accountability, and human-centric decision-making.

C. Access to Technology Disparity

The implementation of Industry 5.0 in the field of healthcare has the potential to produce ground-breaking breakthroughs, but it also highlights the basic problem of access to technological inequality. Although there is no denying that data-driven, AI-driven, and robotic healthcare solutions have the potential to improve diagnosis, treatment, and patient involvement, the uneven application of these technologies may unintentionally increase current healthcare disparities. Accessing the advantages of Industry 5.0 may be difficult for vulnerable populations, such as those living in rural or underdeveloped locations or in lower socioeconomic levels. The digital divide might lead to a situation where those who are already privileged are disproportionately able to access the most cutting-edge healthcare technology, which is made worse by gaps in technical literacy, infrastructure, and financial resources. It is necessary to use a holistic approach that includes technological innovation, policy advocacy, and community empowerment to address the imbalance in access to technology. To make sure that Industry 5.0 solutions are created with diversity in mind, governments, healthcare organisations, and tech companies must work together. Initiatives to close the digital divide through tele-health infrastructure development, inexpensive device access, and educational programmes are crucial(Santosh, K. C.(2022)). Healthcare systems can use Industry 5.0's capabilities to deliver better care, diagnostics, and preventive measures to every stratum of society by giving equitable technology distribution a high priority. In doing so, they will fulfil the fundamental goal of healthcare, which is to ensure that everyone is healthy regardless of their level of technological development.

D. Workforce Adaptation and Skill Gaps

Industry 5.0's integration into the healthcare sector holds the promise of revolutionising patient care and medical processes through the harmonious interplay of humans and the latest technological advances. However, this paradigm shift also highlights the significant challenge of workforce adaptability and bridging skill gaps in the healthcare sector. Healthcare personnel need new skill sets to use AI, robotics, and data-driven technologies efficiently as they become more commonplace. Healthcare professionals may need to become experts in data analysis, AI interpretation, and technology management as their traditional responsibilities change. This calls for in-depth training activities and educational programmes to provide the healthcare workforce with the skills required for Industry 5.0. In order to ensure the calibre of patient care, it also becomes vital to find a balance between technological advancements and human proficiency. The difficulty lies in creating a setting where medical personnel may easily combine their clinical knowledge with AI-driven insights. Collaboration between experts and technological systems ought to increase decision-making, patient outcomes, and trust in the industry 5.0's disruptive potential. By proactively addressing these workforce adaptation issues, healthcare systems may ensure that Industry 5.0's benefits are fully achieved while maintaining the fundamentally vital human element that distinguishes compassionate and effective healthcare delivery.

E. Regulatory and Legal Challenges

The application of Industry 5.0 ideas to the healthcare industry creates a complicated environment of legal and regulatory issues that call for careful thought. Existing regulations frequently find it difficult to keep up with the quick evolution of these breakthroughs as cutting-edge technology like AI, robots, and IoT become fundamental to healthcare practises. It becomes crucial to strike a balance between the need to advance technology and the need to guarantee patient safety, data privacy, and ethical technology use.

Regulations must change to reflect the particular complexities of Industry 5.0 in healthcare(Mbunge, E.(2021)). As emerging technologies obfuscate the distinctions between human expertise and machine-driven decision-making, questions regarding the approval and regulation of AI-driven medical equipment, diagnostic algorithms, and treatment suggestions arise.

F. Loss of Human Touch and Empathy

Even while Industry 5.0 focuses a lot on the importance of infusing technology, there are worries that increased use of artificial intelligence and automated processes may cause health care to lose its human touch and empathy. It's critical to strike a balance between the personal connection between healthcare personnel and patients and technical efficiency. To guarantee that patients can trust the results of these systems, it becomes difficult to define standards for the dependability, accuracy, and the transparency of AI systems.

Data governance and privacy issues also becoming more prominent. Strong data protection measures are required to guard against breaches, unauthorised access, and misuse of the massive amounts of patient data that Industry 5.0 technologies collect and handle. Regulatory organisations must carefully balance the use of data for medical developments with the protection of patient anonymity. Another big obstacle is interoperability. As various healthcare facilities employ a variety of technologies, it becomes increasingly difficult to ensure that these technologies can communicate and share data without jeopardising security. To achieve this interoperability and realise the full potential of Industry 5.0, standardising data formats, interfaces, and communication protocols is essential. The internationalisation of the health sector and medical studies also causes cross-border regulatory issues (Taj, I.(2022), Esmaeilzadeh, P.(2021)). Industry 5.0 solutions must be compliant with various national laws while regulating frameworks in various nations must be unified. To prevent fragmentation and promote the seamless application of these technologies across borders, international cooperation is required. The entrance of the fifth industrial revolution into the medical sector has raised legal and regulatory issues, which emphasise the necessity for proactive and flexible regulatory frameworks. These guidelines, which should also defend patient rights, data privacy, and ethical concerns, could serve as an inspiration to innovators. To guarantee that Industry 5.0's disruptive promise in healthcare is realised while keeping the greatest standards of patient safety and well-being, industry stakeholders, legislators, and legal experts must cooperatively traverse this complex terrain.

G. Patient Safety

Robust quality assurance and accountability systems are urgently required due to Industry 5.0's usage of robotics and artificial intelligence in the medical sector. It is difficult to continually guarantee the correctness, dependability, and currency of the algorithms guiding crucial healthcare decisions due to the dynamic nature of modern technologies like AI. Verifying the output of these systems continuously is necessary to strike the delicate balance between utilising AI's potential and maintaining patient safety. The issue of responsibility also gets more complicated as AI algorithms increasingly play a role in decisions about diagnosis, treatment, and even prognosis. It is important to think carefully about who is responsible for the results of AI-driven decisions. Should it be the algorithm creators, the healthcare organisations using them, or the medical specialists monitoring their application? It is crucial to thoroughly analyse the ethical implications and even legal repercussions involved in this subject in order to ensure that accountability conforms with both the best interests of patients and the ethical standards of the healthcare profession. Transparency, accountability, and patient welfare must come first in a field as crucial as healthcare, where human lives are on the line. In the era of Industry 5.0, it is necessary for regulatory bodies, healthcare institutions, technology developers, and healthcare practitioners to work together to create strong frameworks for ongoing oversight, evaluation, and accountability. The enormous potential of AI and robotics in healthcare can only be realised through vigilant vigilance and proactive adaptation, all the while preserving patient trust and assuring the highest standards of care.

H. Healthcare System Redesign

Existing healthcare infrastructures and workflows must undergo extensive, multidimensional adjustments in order to integrate Industry 5.0 technology. Healthcare organisations are in the midst of a transformational journey that requires them to make significant operational changes, embrace forward-thinking technology investments, and navigate the challenges of switching to new systems. Healthcare's shift to Industry 5.0 necessitates a review of current workflows and maybe a complete rethink. Processes that have been in place for years or even decades may need to be rebuilt in order to take use of the promise of cutting-edge technologies like artificial intelligence, IoT devices, and data analytics. Understanding how these technologies work, and envisioning how they can easily integrate into the current healthcare system, are both necessary for this. Healthcare organisations must simultaneously invest in technology that adheres to the principles of Industry 5.0. These investments need deliberate decision-making to choose and execute technology that can improve patient care, expedite operations, and advance medical research. They go beyond simple acquisition. These choices affect not only the technology itself, but also the infrastructure, training, and support needed to complete the shift successfully (Kolasa, K. (2023)). There are challenges involved in the transition of healthcare to Industry 5.0. In order to ensure compatibility and interoperability, healthcare organisations must overcome the challenges of integrating new systems with legacy technologies. This entails addressing potential issues, cutting down on downtime, and protecting patient data integrity throughout the move.

I. Resistance to Change

Application of Industry 5.0 concepts in the healthcare industry is significantly hampered by resistance to change. Healthcare workers, administrators, and even patients may show resistance to embracing new advancements as the landscape of healthcare changes to include advanced technologies. Resistance might stem from a variety of factors, including a dislike for established treatment methods, worries about patient data security, and a fear of losing one's job. Effective communication about the advantages of Industry 5.0 is necessary to overcome this opposition, emphasising how AI, robotics, and data-driven insights can improve patient care and expedite operations. Fostering a culture of receptivity to change requires educating stakeholders about the possibility for better diagnosis, individualised treatments, and an increase in patient engagement. Healthcare systems may promote a more seamless transition and make sure that technological improvements are in line with the interests of patients and healthcare professionals by addressing these issues and highlighting the collaborative nature of Industry 5.0.

J. Expensive and Resource-Intensive

As a result of Industry 5.0, the healthcare sector has the potential to significantly increase efficiency and effectiveness. Industry 5.0 may optimise different facets of patient care, medical research, and operational operations by seamlessly integrating cutting-edge technology like AI, robotics, and IoT devices. Healthcare's shift to Industry 5.0, however, presents its own set of difficulties. These cutting-edge technology might be expensive and resource-intensive to install. The necessary infrastructure and systems must be purchased, integrated, and maintained by healthcare organisations with significant financial resources. Additionally, thorough training programmes are necessary to make sure that healthcare workers are skilled at making the best use of this technology. Even though Industry 5.0's advantages seem promising, a successful and long-lasting integration into the healthcare ecosystem depends on careful financial planning and resource allocation (L. Gomathi(2023)).

K. Vendor Lock-In and Compatibility

In the context of Industry 5.0's integration with healthcare, the problem of vendor lock-in and compatibility seems to be a major consideration. As they deploy numerous Industry 5.0 technologies to enhance patient care and operational efficiency, healthcare organisations usually collaborate with a number of technology vendors that specialised in providing specific solutions. This diversity could, however, unintentionally result in a situation where healthcare organisations rely disproportionately on a small number of vendors for their technology, data management, and software ecosystems. When a healthcare system grows dependent on a specific vendor's proprietary technologies, it becomes difficult to switch to alternative solutions or seamlessly interface with other systems. This is known as vendor lock-in. This may inhibit innovation and impede interoperability across various parts of the healthcare ecosystem, reducing the flexibility and scalability of operational healthcare systems. Vendor lock-in has a variety of negative effects, from higher costs caused by reliance on a single vendor's pricing policies to a decreased ability to adapt to shifting technology environments. Additionally, when attempting to

Figure 4. Challenges for Healthcare 5.0

combine solutions from many manufacturers, compatibility problems between various technologies and platforms can appear. Data silos, ineffective workflows, and impaired data sharing can result from an absence of standardised communication protocols and data formats, impeding the holistic perspective necessary for comprehensive patient care. Compatibility issues may make vendor lock-in more likely because switching to a different vendor might be difficult and time-consuming. A proactive strategy is needed to address issues with compatibility and vendor lock-in. To address these issues, governments, healthcare providers, technology companies, and regulatory organisations must work together in a multifaceted manner. Healthcare systems can maximise the advantages of Industry 5.0 while reducing any potential negatives by proactively tackling these challenges. Healthcare systems must give top priority to products that follow open standards and interoperability frameworks, enabling easier system integration and data sharing. The dangers of lock-in can be reduced by working with vendors who value collaboration and support data portability, which encourages competition and innovation in the healthcare technology market. Additionally, regulatory agencies and business associations are crucial in promoting uniform interoperability and data sharing procedures. They can contribute to a more adaptive and patient-centered healthcare environment by setting norms that support vendor neutrality and data accessibility.

Healthcare systems must ultimately carefully traverse the vendor ecosystem, despite the temptation of specialised solutions, to ensure that Industry 5.0's potential benefits are realised without compromising the interoperability and flexibility required for the future (see Fig. 4).

V. CONCLUSION

In conclusion, Industry 5.0's integration into healthcare systems around the world has the potential to revolutionise patient care, medical research, and operational effectiveness. But this shift is not without its difficulties and complexity, which must be handled with caution and forethought.

A few of the alluring choices given by Industry 5.0 that have the potential to enhance patient outcomes and elevate the whole healthcare experience are individualised treatment regimens, advanced diagnostics, and remote patient monitoring. The synergistic collaboration of people and cutting-edge technologies can be used to achieve goals, including collaborative healthcare planning, expedited medical research, and equal access to care.

The challenges offered by Industry 5.0, however, cannot be disregarded. Concerns about security and confidentiality of data, the ethical ramifications of AI decision-making, and inequalities in access to technology all draw attention to the need for strict safeguards. Adapting the healthcare staff, dealing with legal and regulatory issues, and maintaining the human touch in patient care are complex jobs that call for thorough preparation and knowledge.

A multi-stakeholder strategy is necessary to navigate the path towards Industry 5.0 in the healthcare industry. To balance innovation with ethical issues, policymakers, healthcare practitioners, technological specialists, and regulatory agencies must work together. Patient-centricity ought to continue to be a guiding principle, ensuring that everyone may benefit from Industry 5.0, regardless of socioeconomic status or location.

Global healthcare systems have the ability to embrace Industry 5.0's revolutionary potential while overcoming its challenges in order to rethink healthcare delivery, enhance patient outcomes, and contribute to a more efficient, equitable, and sustainable future for healthcare. The Industry 5.0 healthcare goal may be realised in a way that benefits all of mankind through strategic planning, ethical frameworks, and cooperative efforts.

REFERENCES

Adel, A. (2020). Future of industry 5.0 in socicty: Human-centric solutions, challenges and prospective research areas. *Journal of Cloud Computing (Heidelberg, Germany)*, *11*(1), 40. doi:10.118613677-022-00314-5 PMID:36101900

Bida, M. N., Mosito, S. M., Miya, T. V., Demetriou, D., Blenman, K. R., & Dlamini, Z. (2023). Transformation of the Healthcare Ecosystem in the Era of Society 5.0. In *Society 5.0 and Next Generation Healthcare: Patient-Focused and Technology-Assisted Precision Therapies* (pp. 223–248). Springer Nature Switzerland. doi:10.1007/978-3-031-36461-7_10

Birtchnell, T., & Urry, J. (2016). *A New Industrial Future? 3D Printing and the Reconfiguring of Production, Distribution, and Consumption*. Routledge. doi:10.4324/9781315776798

Ciasullo, M. V., Orciuoli, F., Douglas, A., & Palumbo, R. (2022). Putting Health 4.0 at the service of Society 5.0: Exploratory insights from a pilot study. *Socio-Economic Planning Sciences, 80*. doi:10.1016/j.seps.2021.101163

Ciasullo, M. V., Orciuoli, F., Douglas, A., & Palumbo, R. (2022). Putting Health 4.0 at the service of Society 5.0: Exploratory insights from a pilot study. *Socio-Economic Planning Sciences*, *80*, 101163. doi:10.1016/j.seps.2021.101163

Deivanayagampillai, N., Jacob, K., Manohar, G. V., & Broumi, S. (2023). Investigation of industry 5.0 hurdles and their mitigation tactics in emerging economies by TODIM arithmetic and geometric aggregation operators in single value neutrosophic environment. Facta Universitatis, Series. *Mechanical Engineering (New York, N.Y.)*.

Dhirani, L. L., Mukhtiar, N., Chowdhry, B. S., & Newe, T. (2023). Ethical dilemmas and privacy issues in emerging technologies: A review. *Sensors (Basel), 23*(3), 1151. doi:10.339023031151 PMID:36772190

Esmaeilzadeh, P., Mirzaei, T., & Dharanikota, S. (2021). Patients' perceptions toward human–artificial intelligence interaction in health care: Experimental study. *Journal of Medical Internet Research, 23*(11), e25856. doi:10.2196/25856 PMID:34842535

Garg, P. K. (2022). The future healthcare technologies: a roadmap to society 5.0. In *Geospatial Data Science in Healthcare for Society 5.0* (pp. 305–318). Springer Singapore. doi:10.1007/978-981-16-9476-9_14

Ghobakhloo, M., Iranmanesh, M., Mubarak, M. F., Mubarik, M., Rejeb, A., & Nilashi, M. (2022). Identifying industry 5.0 contributions to sustainable development: A strategy roadmap for delivering sustainability values. *Sustainable Production and Consumption, 33*, 716–737. doi:10.1016/j.spc.2022.08.003

Gomathi, L., Mishra, A. K., & Tyagi, A. K. (2023). Industry 5.0 for Healthcare 5.0: Opportunities, Challenges and Future Research Possibilities. *2023 7th International Conference on Trends in Electronics and Informatics (ICOEI),* 204-213. 10.1109/ICOEI56765.2023.10125660

Ivanov, D. (2023). The Industry 5.0 framework: Viability-based integration of the resilience, sustainability, and human-centricity perspectives. *International Journal of Production Research, 61*(5), 1683–1695. doi:10.1080/00207543.2022.2118892

Khan, M., Gupta, B., Verma, A., Praveen, P., & Peoples, C. J. (Eds.). (2023). *Smart Village Infrastructure and Sustainable Rural Communities*. IGI Global. doi:10.4018/978-1-6684-6418-2

Kolasa, K. (2023). *The Digital Transformation of the Healthcare System: Healthcare 5.0*. Taylor & Francis. doi:10.4324/b23291

Mbunge, E., Muchemwa, B., & Batani, J. (2021). Sensors and healthcare 5.0: Transformative shift in virtual care through emerging digital health technologies. *Global Health Journal (Amsterdam, Netherlands), 5*(4), 169–177. doi:10.1016/j.glohj.2021.11.008

Mohanta, Das, & Patnaik. (2019). Healthcare 5.0: A Paradigm Shift in Digital Healthcare System Using Artificial Intelligence, IOT and 5G Communication. *2019 International Conference on Applied Machine Learning (ICAML),* 191-196. 10.1109/ICAML48257.2019.00044

Özdemir, V., & Hekim, N. (2018). Birth of Industry 5.0: Making Sense of Big Data with Artificial Intelligence. *OMICS: A Journal of Integrative Biology, 22*(Jan), 65–76. doi:10.1089/omi.2017.0194 PMID:29293405

Santosh, K. C., & Gaur, L. (2022). *Artificial intelligence and machine learning in public healthcare: Opportunities and societal impact*. Springer Nature.

Schwab, K. (2022). *The Fourth Industrial Revolution: what it means, how to respond*. World Economic Forum.

Taj, I., & Zaman, N. (2022). Towards industrial revolution 5.0 and explainable artificial intelligence: Challenges and opportunities. *International Journal of Computing and Digital Systems, 12*(1), 295–320. doi:10.12785/ijcds/120128

Wazid, M., Das, A. K., Mohd, N., & Park, Y. (2022). Healthcare 5.0 Security Framework: Applications, Issues and Future Research Directions. *IEEE Access : Practical Innovations, Open Solutions*, *10*, 129429–129442. doi:10.1109/ACCESS.2022.3228505

Xu, X., Lu, Y., Vogel-Heuser, B., & Wang, L. (2021). Industry 4.0 and Industry 5.0—Inception, conception and perception. *Journal of Manufacturing Systems*, *61*, 530–535. doi:10.1016/j.jmsy.2021.10.006

Žižek, S. Š., Mulej, M., & Potočnik, A. (2021). Potoˇcnik, A. The Sustainable Socially Responsible Society: Well-Being Society 6.0. *Sustainability (Basel)*, *13*(16), 9186. doi:10.3390u13169186

Chapter 12
Infrastructure Potential and Human-Centric Strategies in the Context of Industry 5.0

Priti Kumari
National Institute of Technology, Patna, India

Abhishek Anand
National institute of Technology, Patna, India

Pushkar Praveen
https://orcid.org/0000-0002-3288-5128
G.B. Pant Institute of Engineering and Technology, Pauri, India

Agya Ram Verma
G.B. Pant Institute of Engineering and Technology, Pauri, India

Abhishek Godiyal
G.B. Pant Institute of Engineering and Technology, Pauri, India

ABSTRACT

Moving from Industry 4.0 to Industry 5.0 represents a significant shift in manufacturing, focusing on human collaboration, customization, and sustainability instead of just automation and digitization. Industry 5.0 brings substantial advancements that reshape the industrial landscape. This chapter examines the transition, emphasizing the importance of infrastructure, ethical concerns, and future implication. Industry 5.0 relies on a robust infrastructure, blending physical and digital elements to enable advanced technology integration. It places human-centric strategies at its core, emphasizing human interaction, well-being, and skill development. The synergy between infrastructure and human-centric approaches is pivotal for Industry 5.0's success, enabling ergonomic workplaces, seamless human-machine collaboration, data access, and customization. Simultaneously, human-centric strategies prioritize collaboration, skill development, and ethical considerations, fostering a harmonious synergy.

DOI: 10.4018/979-8-3693-0782-3.ch012

INTRODUCTION

An industrial revolution represents a profound and radical transformation in how industries and economies function, propelled by substantial advancements in technology, socio-economic factors, and cultural developments. These revolutions demarcate unique historical epochs distinguished by revolutionary alterations in the methods of production, economic frameworks, and the fabric of societies. Madsen D.O, et al. (2021) in his article explained that each industrial revolution ushers in novel technologies, methodologies, and approaches to organizing labor, ushering in profound alterations across multiple facets of human existence and industrial activities. The advent of the Fourth Industrial Revolution laid the foundation for the emergence of Industry 4.0, ushering in a new era where industries undergo transformation through the seamless integration of digitalization and automation (Ghobakhloo M. et al, 2021). However, the trajectory has shifted, giving birth to Industry 5.0, where the emphasis on human interaction and collaboration takes center stage (Coelho P. et al., 2023). The importance of embracing a value-centered and ethical approach to technology engineering within the context of Industry 5.0 is highlighted by findings from a survey involving leaders across diverse companies. Proposing the Value Sensitive Design (VSD) framework as a systematic methodology, it illustrates how technologies fostering human-machine collaboration in the Factory of the Future can be developed to align with human values. Industry 5.0 signifies a notable progression from Industry 4.0, presenting a spectrum of advantages that build upon the established foundations (Longo F., et al., 2020). While Industry 4.0 revolutionized industries through automation and data-driven processes, Industry 5.0 takes this transformation a step further by placing a renewed focus on human interaction, collaboration, and value creation. With a strong emphasis on the convergence of digital innovation and human-centric approaches, Industry 5.0 transcends its predecessors by aligning technology with the genuine needs and aspirations of humanity. Industry 5.0 serves as a transformative revolution that seeks to enhance human-machine interactions for improved efficiency and swifter outcomes. This era heralds a new age of customization and offers solutions to intricate challenges. Leveraging digital technologies, it revolutionizes manufacturing by automating repetitive tasks and employing human intelligence to grasp the operator's needs. Furthermore, it harnesses the power of machine learning and artificial intelligence to analyze manufacturing data (Javaid M., et al. (2020)). At the heart of this paradigm shift lie the profound possibilities within infrastructure development and the holistic strategies that prioritize human well-being.

Infrastructure, often regarded as the backbone of any urban, rural and industrial ecosystem, assumes an even more critical role in the context of Industry 5.0. It encapsulates the physical and digital framework that supports the seamless integration of advanced technologies. From interconnected smart devices to data-driven analytics, the infrastructure serves as the conduit through which innovation cascades. In this context, the study of infrastructure potentials delves into identifying not just the technological prowess but also the adaptability and scalability of the underlying systems (Praveen P., et al. (2023)). For this purpose Tripathy H.P. & Pattanaik P. (2020) explained the vital role of artificial intelligence methodologies in enabling Intelligent Communication Networks, the current era has witnessed the tangible realization of wireless sensor networks. These networks find practical application in diverse fields such as healthcare, military operations, preservation of forest ecosystems, and the prediction of seismic activity in volcanic regions, all of which necessitate remote event monitoring. As industries become increasingly automated and interconnected, the focus on people's needs, aspirations, and well-being becomes all the more vital. Integrating human-centric strategies means designing technologies and systems that not only optimize efficiency but also enhance the quality of life for workers and consumers alike. The cultivation of a collaborative environment where technology complements human skills, rather than replacing them, ensures a future where creativity, empathy, and critical thinking continue to flourish. In the context of the circular economy, digital innovation and sustainability go hand in hand. Digital innovation is instrumental in supporting the circular economy model by promoting the use of digital platforms, smart devices, and artificial intelligence to enhance resource efficiency Coiffi R. et al. (2020).

However, what distinguishes Industry 5.0 is the pivot towards human-centricity. As automation and artificial intelligence assume greater roles, the human element remains irreplaceable. Human-Centric Strategies, within the sphere of Industry 5.0, underscore the imperative to place human needs, capabilities, and well-being at the forefront of technological evolution. This approach recognizes that while machines can optimize processes, they cannot replicate the intuitive decision-making, creativity, and emotional intelligence inherent to humans. In this context, ethical considerations come to the forefront. As we harness the potential of Industry 5.0, questions emerge regarding data privacy, security, and the socio-economic impact of automation. Balancing the gains in efficiency and productivity with the preservation of human dignity and societal welfare becomes a moral compass guiding the development of infrastructure and strategies.

Industry 5.0 builds upon the foundation set by Industry 4.0, introducing several advancements that offer distinct advantages over its predecessor. While Industry 4.0 primarily focused on automation and digitization, Industry 5.0 expands its scope to emphasize human collaboration and interaction within the industrial landscape (Pereira et al. (2020), Yang Lu (2017), Javaid M. er al. (2020) and Barata J. & Kayser I. (2023)). Here are some key advantages of Industry 5.0 compared to Industry 4.0:

Human-Centric Approach: Industry 5.0 places a strong emphasis on human involvement and collaboration alongside advanced technologies. Unlike Industry 4.0, which focused on automation and machine-to-machine communication, Industry 5.0 acknowledges the irreplaceable role of human creativity, decision-making, and empathy in problem-solving and innovation.

Personalization and Customization: Industry 5.0 integrates personalized and customizable production processes that cater to individual consumer preferences. This level of personalization was not as prominent in Industry 4.0, where mass production and standardized processes were more prevalent. The ability to cater to specific consumer needs enhances customer satisfaction and loyalty.

Flexibility and Adaptability: In Industry 5.0, systems are designed to be highly flexible and adaptable. Production lines can be reconfigured rapidly to accommodate changes in demand or product variations. This responsiveness is an improvement over Industry 4.0, where adaptability was often limited by the rigid automation setups.

Human-Machine Collaboration: Industry 5.0 envisions a collaborative partnership between humans and machines. While Industry 4.0 automated many tasks, Industry 5.0 seeks to augment human capabilities with technology, allowing humans and machines to work together synergistically (Adel A., 2022) . This approach capitalizes on the strengths of both entities, resulting in improved efficiency and quality.

Skill Enhancement: Industry 5.0 acknowledges the significance of human skills alongside automation. Rather than replacing human workers, Industry 5.0 seeks to augment their abilities through technology. This approach creates a learning environment where employees are empowered to develop new skills, adapt to evolving roles, and contribute innovatively to the workplace.

Complex Problem Solving: Industry 5.0 encourages critical thinking and problem-solving skills among workers. With humans collaborating closely with machines, complex issues can be tackled in real-time, drawing on human expertise and intuition to navigate unexpected challenges.

Enhanced Data Utilization: While Industry 4.0 focused on data collection and analysis, Industry 5.0 places greater emphasis on utilizing data to derive actionable insights. Humans are involved in interpreting data patterns and making informed decisions, ensuring that data-driven processes are aligned with strategic goals.

Ethical Considerations and Social Impact: Industry 5.0 introduces a heightened awareness of ethical considerations and social impact. As humans remain central to decision-making, there is a greater potential to ensure that technologies are deployed responsibly, with respect for ethical guidelines and societal well-being.

Industry 5.0 brings forth a paradigm shift from the automation-centric approach of Industry 4.0 to a more holistic integration of technology and human collaboration. By emphasizing personalized production, adaptability, human-machine partnership, and ethical considerations, Industry 5.0 offers substantial advantages that pave the way for a more balanced and sustainable industrial future.

LITERATURE REVIEW

The Evolution of Industry 5.0: The evolution of Industry 5.0 stands as a testament to the dynamic and transformative nature of the modern industrial landscape. Building upon the foundation laid by its predecessors, Industry 5.0 represents a paradigm shift that not only incorporates cutting-edge technology but also places the human element at its core. This evolutionary journey from Industry 4.0 to Industry 5.0 marks a pivotal juncture in our approach to production, collaboration, and societal progress (Aslam et al., 2020; Alojaiman B., 2023). Industry 1.0 to Industry 5.0 represent different phases of industrial development, each characterized by significant technological advancements and changes in the way industries operate as discussed in table 1.

In the 18th century, Industry 1.0 introduced machine-based production methods, originating in England and later reaching the United States. This revolution shifted the economy from handicrafts to mechanization, influencing various sectors like mining, textiles, and agriculture. Industry 2.0 emerged between 1871 and 1914, emphasizing manufacturing, economic growth, and enhanced productivity. However, it also led to unemployment as machines replaced human labor. Industry 3.0, known as the digital revolution, began in the 1970s with the automation of controls and the proliferation of computers.

Table 1. Features of industry 1.0 -5.0 (Aslam et al., 2020)

Industry	Time Period	Key Features	Technological Advancements	Main Focus
Industry 1.0	Late 18th to mid-19th century	Manual Labor	Steam engines, mechanization	Textiles, coal mining
Industry 2.0	Late 19th to early 20th century	Mass Production	Electricity, assembly line	Manufacturing, automobiles
Industry 3.0	Mid-20th century	Automation	Computers, robotics	Electronics, aerospace
Industry 4.0	Late 20th to early 21st century	Digitalization	IoT, AI, Big Data	Smart factories, automation
Industry 5.0	Ongoing (Emerging)	Cyber-Physical Systems	5G, edge computing, AI/ML integration	Human-machine collaboration, customization

Figure 1. Industrial evolution from 1.0-5.0

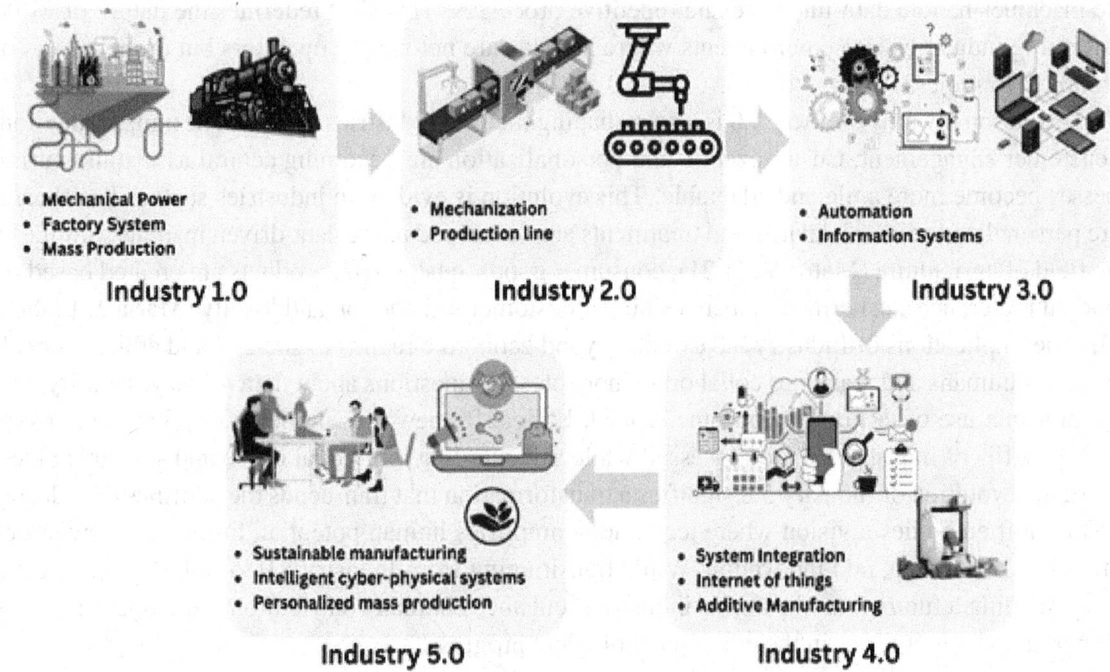

It emphasized mass production and digital technologies, revolutionizing products and business processes. Industry 4.0 involves the integration of physical assets with advanced technologies such as AI, IoT, and cloud computing. Organizations adopting Industry 4.0 are adaptable and data-driven. Looking ahead, Industry 5.0 represents the next wave of technology focused on efficient and intelligent machines, building upon the innovations of the previous generations (Adel A., 2022). The first industry evolution industry 1.0 to the current industry 5.0 shown in Figure 1.

Industry 4.0, often referred to as the Fourth Industrial Revolution, introduced the world to the concept of smart factories, where automation, data exchange, and digitalization redefined manufacturing processes. This era brought forth cyber-physical systems that connected machines and systems to enable autonomous decision-making and real-time information sharing. The focus was primarily on optimizing production efficiency, reducing errors, and enhancing the accuracy of processes through the use of sensors, IoT devices, and AI-driven algorithms (Morteza et al., 2023). However, the transition from Industry 4.0 to Industry 5.0 goes beyond optimization and efficiency, encompassing a broader vision of collaboration and value creation. In the landscape of Industry 5.0, we witness the convergence of technological innovation with a renewed focus on harnessing the distinctive capabilities of humans. It recognizes that while machines excel in precision and repetitive tasks, humans possess creativity, intuition, and the ability to navigate complex, non-standard situations. The evolution towards Industry 5.0 is a response to the realization that the collaboration between human expertise and technological advancements can lead to breakthroughs that were previously unattainable. At the heart of this evolution lies the concept of symbiosis – a harmonious relationship between humans and machines (Lukov L. & Alexei R., 2023). Unlike Industry 4.0, which often raised concerns about job displacement due to automation, Industry 5.0 seeks to create roles that complement the capabilities of both entities. For instance, humans are engaged in tasks that require critical thinking, problem-solving, empathy, and creativity, while machines handle data-intensive and repetitive processes. This shift redefines the nature of work, transforming industries into environments where humans are not simply operators but orchestrators of technology (Rajumesh, 2023).

The evolution towards Industry 5.0 is also reshaping the way industries approach design, production, and customer engagement. Customization and personalization are becoming central as manufacturing processes become more agile and adaptable. This evolution is evident in industries such as healthcare, where personalized medical devices and treatments are developed using data-driven insights. Similarly, in the field of agriculture (Matos V., 2021), consumer goods, customized products are created based on individual preferences, ushering in an era of higher customer satisfaction and loyalty (Maria & Elaheh, 2023). The implications of Industry 5.0 extend beyond economic realms to societal and ethical considerations. As humans and machines collaborate more closely, questions about data privacy, security, and the responsible use of technology become crucial. Ethical frameworks are being developed to ensure that the benefits of Industry 5.0 are harnessed while safeguarding individual rights and societal values. In short, the evolution of Industry 5.0 signifies a transformation that transcends the confines of Industry 4.0. This shift embodies a vision where technology amplifies human potential, fostering a new era of creativity, collaboration, and innovation. While transitioning from Industry 4.0 to Industry 5.0, we are actively molding a future that transforms industries, enhances human contributions, and opens avenues for unparalleled possibilities in the era of technological progress.

INFRASTRUCTURE INNOVATIONS IN INDUSTRY 5.0

Entering the next stage of industrial evolution, Industry 5.0 builds upon the groundwork laid by Industry 4.0. This phase underscores the seamless collaboration between humans and machines, prioritizing customization and sustainability in the realm of manufacturing. Infrastructure innovations in Industry 5.0 usher in a transformative manufacturing era, centered on smart factories and human-centric principles. These innovations include ergonomic workstations, augmented reality interfaces for human-machine collaboration, and the integration of safe collaborative robots (cobots). The adaptability of infrastructure for customization is a key theme, enabling swift production line adjustments to meet unique product demands. Additionally, infrastructure supports IoT and real-time data connectivity, while emphasizing sustainability through energy-efficient technologies and renewable energy integration. Beyond the factory, infrastructure optimization extends to supply chains, cybersecurity, regulatory compliance, and digital twin technology. This collective progress shapes a more agile, sustainable, and human-centered manufacturing landscape in Industry 5.0 (Saxena A. et al., 2020; Masood T. & Sonntag P. 2020). Figure (2) shows an infrastructure using industry 5.0 which shows the harmonious collaboration between humans and machines.

Infrastructure innovations play a pivotal role in enabling and supporting the goals of Industry 5.0 which are discussed below:

Human-Centric Manufacturing: In Industry 5.0, infrastructure must be designed to facilitate seamless interaction between humans and machines. Smart factories should have ergonomic workstations, augmented reality (AR) systems, and intuitive interfaces to empower workers and foster collaboration. Advanced training programs and skill development are integral components to ensure the workforce

Figure 2. Collaboration between humans and machines in industry 5.0

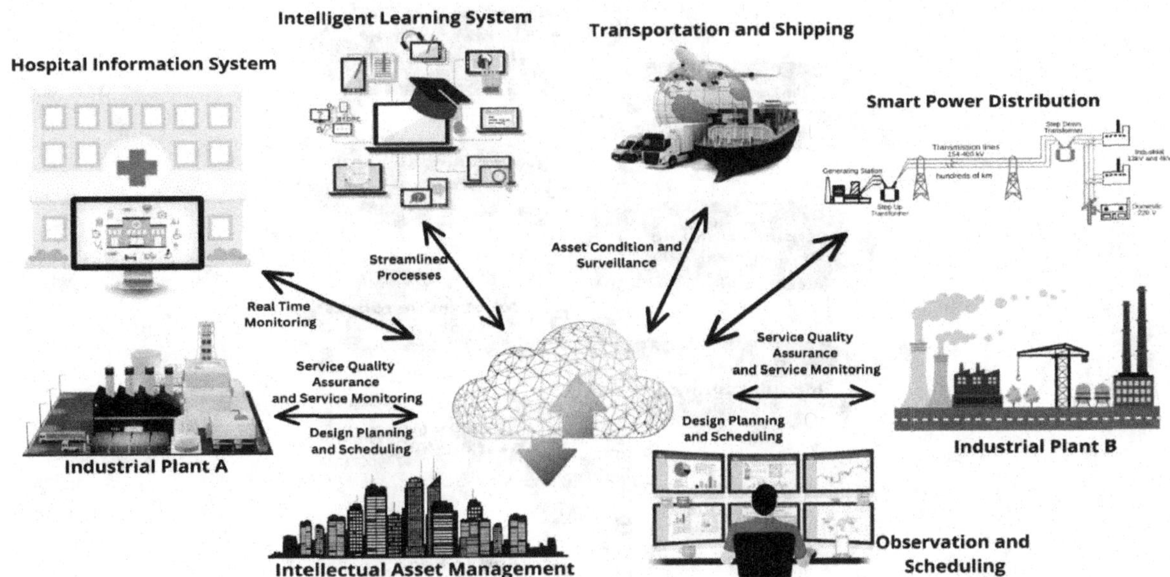

can adapt to these new human-machine interfaces (Nahavandi S., 2019). The figure (3) shows the key features of human-centric manufacturing which is discussed below:

Collaborative Robots (Cobots): Infrastructure needs to accommodate the integration of cobots, which are designed to work safely alongside humans. This includes the development of safety standards, workspace reconfiguration, and specialized equipment such as sensors and cameras for real-time monitoring of human-cobot interactions (Demir at al., 2019).

Customization-Focused Production Lines: Industry 5.0 requires flexible and adaptable infrastructure capable of handling smaller batch sizes and individualized product configurations. Modular production lines, 3D printing facilities, and rapid reconfiguration capabilities are essential to accommodate changing customer demands.

IoT and Data Connectivity: Infrastructure should continue to advance in terms of IoT integration and data connectivity. Edge computing and 5G networks will be critical for real-time data processing and communication between machines, humans, and other manufacturing components.

Sustainable Manufacturing Practices: To align with the sustainability goals of Industry 5.0, infrastructure must support eco-friendly initiatives. This includes the development of energy-efficient facilities, waste reduction technologies, and the integration of renewable energy sources into manufacturing processes.

Supply Chain Integration: Infrastructure innovations extend beyond the factory floor to include supply chain optimization. Leveraging blockchain and AI-driven systems in supply chain management ensures transparency, traceability, and efficiency throughout the entire process, starting from raw material acquisition to the delivery of the end product.

Cybersecurity and Data Protection: As connectivity continues to rise, it is imperative for infrastructure to give precedence to robust cybersecurity measures. This encompasses the deployment of

Figure 3. Human-centric manufacturing

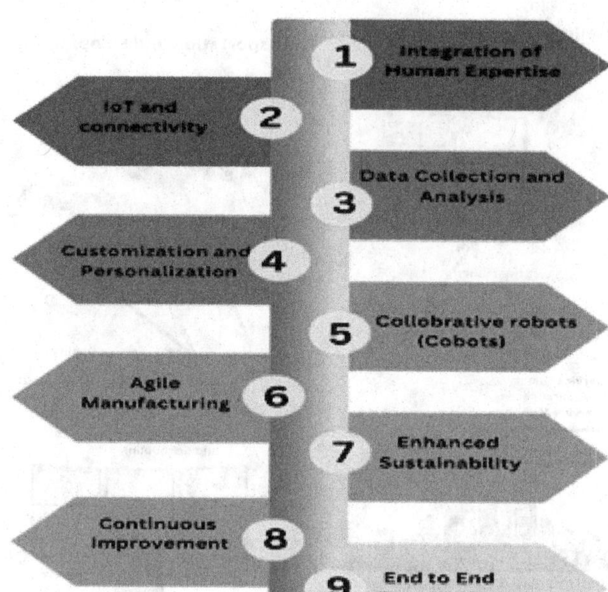

encryption, intrusion detection systems, and regular security audits, all aimed at protecting sensitive data and averting potential cyber threats.

Regulatory Compliance: Industry 5.0 infrastructure must adhere to evolving regulations and standards concerning human-robot collaboration, environmental impact, and data privacy. Continuous monitoring and adaptation to changing regulatory landscapes are essential.

Digital Twin Technology: Infrastructure should incorporate digital twin technology, allowing for virtual replicas of physical assets and processes. This aids in predictive maintenance, process optimization, and rapid prototyping.

Infrastructure innovations are fundamental to the successful implementation of Industry 5.0, as they enable the shift towards human-centric, customizable, and sustainable manufacturing practices. Continuous research, development, and investment in these innovations will drive the evolution of manufacturing in the coming years, unlocking new opportunities for businesses and society at large.

HUMAN-CENTRIC STRATEGIES IN INDUSTRY 5.0

Industry 5.0 represents a paradigm shift in manufacturing, placing humans at the center of the industrial ecosystem. Human-centric strategies are crucial for harnessing the full potential of this new era (Reddy P.K. et al., 2021; Ghobakhloo et al., 2023). Here are some key points are discussed:

Collaboration: Industry 5.0 recognizes that humans possess unique qualities such as creativity, problem-solving skills, and emotional intelligence that are difficult to replicate with automation alone. Human-centric strategies emphasize collaboration between humans and machines, leveraging the strengths of both.

Skill Development and Training: To succeed in Industry 5.0, workers require new skill sets. Human-centric strategies prioritize training and upskilling programs to empower the workforce to operate and interact with advanced technologies effectively. This includes training in digital literacy, data analysis, and human-robot interaction.

Ergonomic Work Environments: Designing ergonomic workstations and environments is a key element of human-centric strategies. These environments are not only comfortable but also tailored to facilitate human-machine collaboration. Augmented reality (AR) and virtual reality (VR) interfaces may be integrated to enhance productivity and safety.

Empowering Decision-Making: In Industry 5.0, workers are not just operators but decision-makers. Human-centric strategies encourage employees at all levels to contribute to problem-solving and decision-making processes, enabled by real-time data access and analytics tools.

Work-Life Balance: Recognizing the importance of employee well-being, human-centric approaches prioritize work-life balance. Flexible schedules and remote work options may be incorporated to enhance job satisfaction and productivity.

Customization and Customer-Centricity: Human-centric manufacturing is customer-centric. It emphasizes the ability to tailor products to individual customer needs, enabling companies to offer personalized solutions and respond rapidly to market changes.

Safety First: Safety is paramount in human-centric strategies. Collaborative robots (cobots) are designed to work safely alongside humans. Infrastructure and protocols are established to ensure the well-being of workers in close proximity to machines.

Diversity and Inclusion: Embracing diversity and inclusion is a core element of human-centric strategies. Diverse teams bring varied perspectives and ideas to the table, fostering innovation and problem-solving.

Continuous Learning and Adaptation: Human-centric strategies promote a culture of continuous learning and adaptation. Workers are encouraged to embrace change and stay updated on emerging technologies and industry trends.

Sustainability: Environmental sustainability is integrated into human-centric approaches. Workers are educated about sustainable practices, and processes are designed to minimize waste and reduce the environmental impact of manufacturing.

Human-centric strategies in Industry 5.0 aim to create a harmonious synergy between humans and technology. By placing humans at the center and empowering them with the right tools, skills, and work environments, companies can enhance productivity, innovation, and the overall well-being of their workforce while delivering customized solutions to meet customer demands.

SYNERGY BETWEEN INFRASTRUCTURE AND HUMAN-CENTRIC APPROACHES

In the context of Industry 5.0, achieving a harmonious synergy between infrastructure and human-centric approaches is essential for optimizing manufacturing processes. This synergy not only enhances productivity but also ensures the well-being of the workforce (Alves et al., 2023; Jože M. Rožanec et al., 2023). The key points highlighting this symbiotic relationship are shown in Figure 4:

Figure 4. Synergy between Infrastructure and human-centric approaches

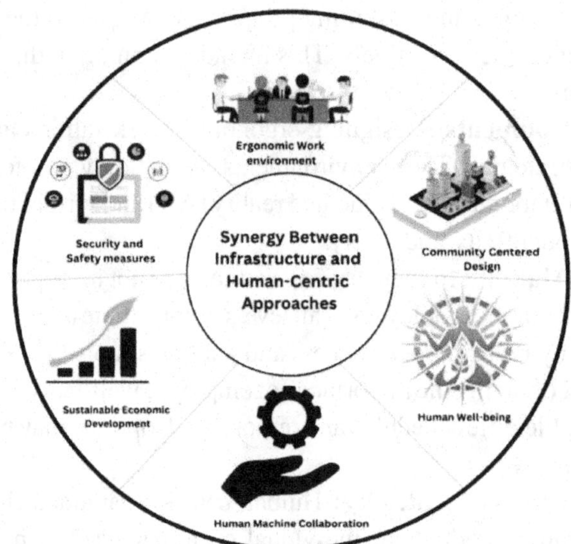

Ergonomic Work Environments: Infrastructure plays a pivotal role in creating ergonomic workspaces that prioritize the comfort and safety of human workers. Ergonomically designed equipment and workstations reduce physical strain and enhance worker efficiency.

Human-Machine Collaboration: The infrastructure should support seamless human-machine collaboration. This involves integrating machinery and robots that are designed to work safely alongside humans (cobots). Specialized infrastructure, such as sensors and monitoring systems, ensures that humans and cobots can share workspace without compromising safety.

Data Accessibility and Visualization: Infrastructure enables real-time data collection and accessibility, empowering human workers with the information needed for decision-making. User-friendly interfaces and visualization tools make data interpretation more accessible, allowing workers to respond to changing conditions effectively.

Customization and Flexibility: Infrastructure should be modular and adaptable, allowing quick reconfiguration of production lines to meet customized demands. This flexibility enables workers to switch between tasks easily and adjust to changing customer requirements.

Training and Skill Development: Infrastructure supports training programs by providing access to virtual reality (VR) and augmented reality (AR) tools. Training within immersive environments allows workers to learn and practice new skills in a safe and controlled setting.

Safety Measures: Infrastructure incorporates safety features such as machine guards, emergency stop mechanisms, and safety protocols. These measures protect human workers from potential hazards associated with working in close proximity to machines and robots.

Sustainability Integration: Infrastructure innovations include sustainable practices, such as energy-efficient machinery and waste reduction technologies. Human-centric approaches align with sustainability goals by encouraging workers to contribute ideas and practices for environmentally friendly manufacturing.

Data Privacy and Security: In prioritizing infrastructure, it is crucial to emphasize data security and privacy for the safeguarding of sensitive information and intellectual property. The implementation of robust cybersecurity measures, including encryption and access controls, plays a pivotal role in maintaining the confidentiality and security of data.

Work-Life Balance: Human-centric strategies promote work-life balance, and infrastructure can support this by enabling remote work options. Connectivity and communication tools within the infrastructure facilitate collaboration among distributed teams.

Continuous Improvement Culture: The infrastructure should foster a culture of continuous improvement and innovation. Human-centric approaches encourage workers to provide feedback and insights for refining processes and enhancing efficiency.

The synergy between infrastructure and human-centric approaches in Industry 5.0 is instrumental in creating a productive, safe, and sustainable manufacturing environment. As infrastructure evolves to accommodate the changing needs of the workforce and align with the principles of human-centricity, it amplifies the capabilities of both technology and human workers, driving innovation and competitiveness in the industrial landscape.

CHALLENGES AND ETHICAL CONSIDERATIONS

In the realm of Industry 5.0, characterized by a focus on human-machine collaboration and customization, emerges a set of fresh challenges and ethical considerations. It becomes paramount to confront and resolve these issues to guarantee responsible and sustainable industrial development (Furstenau et al., 2020; Broo D.G. et al., 2022). Here some of the challenges discussed which need to consider:

Skill Gap and Training Needs: As Industry 5.0 introduces advanced technologies, there is a significant skills gap in the workforce. Providing comprehensive training programs to upskill employees is a challenge, both in terms of cost and time.

Data Security and Privacy: As connectivity and data exchange expand, the threat of data breaches and privacy infringements rises. It becomes imperative to establish strong cybersecurity protocols and adhere to data protection regulations to mitigate these risks.

Human-Machine Collaboration: Integrating collaborative robots (cobots) safely into the workspace poses challenges. Designing infrastructure and protocols to minimize accidents and ensuring the well-being of human workers is complex.

Customization Complexity: Meeting customer demands for personalized products requires complex production processes. Managing the intricacies of customization while maintaining efficiency can be a logistical challenge.

Environmental Impact: Balancing customization with sustainability is challenging. Reducing waste, energy consumption, and the carbon footprint in customized production can be difficult.

Supply Chain Complexity: Customized products often require intricate and diversified supply chains. Coordinating these complex networks while maintaining efficiency and traceability is challenging.

Regulatory Compliance: Adhering to evolving regulations, especially in the context of human-robot collaboration, presents compliance challenges. Staying current with changing legal requirements is demanding.

Ethical principles are essential for achieving responsible, sustainable, and inclusive industrial development. Here some of the Ethical principles are discussed for Considerations:

Job Displacement: The integration of automation and AI in Industry 5.0 may lead to job displacement for some workers. Ethical considerations include providing support for affected employees and communities.

Data Ethics: Ethical dilemmas related to data usage, such as consent and data ownership, need to be addressed. Transparent data handling practices and ethical data governance are essential.

Worker Well-Being: Ensuring the safety and well-being of workers in human-machine collaborative environments is an ethical imperative. Ethical considerations include minimizing physical and psychological stress.

Bias in AI and Automation: Bias in algorithms and AI systems can perpetuate discrimination and inequality. Ethical AI development practices, unbiased training data, and continuous monitoring are necessary to address this issue.

Environmental Responsibility: Industry 5.0 should prioritize sustainable and eco-friendly manufacturing practices. Ethical considerations include minimizing environmental harm and making sustainable choices.

Fair Distribution of Benefits: Ensuring that the benefits of Industry 5.0 are distributed equitably among stakeholders is an ethical concern.

Transparency and Accountability: Ethical considerations emphasize transparency in decision-making processes, especially regarding AI and automation. Holding organizations accountable for their actions and decisions is essential.

Addressing these challenges and adhering to ethical principles is essential for achieving responsible, sustainable, and inclusive industrial development. It requires collaboration between governments, industries, and society to ensure that the benefits of Industry 5.0 are realized while mitigating potential harms.

FUTURE DIRECTIONS AND IMPLICATIONS

Industry 5.0 represents the next evolutionary step in manufacturing, emphasizing human-machine collaboration, customization, and sustainability. As this paradigm unfolds, it presents several future directions and implications that will shape the industrial landscape (Maddikunta et al., 2021; Alojaiman B., 2023; & Attila Tóth et al., 2023). Some of which are discussed below:

Technology Integration: (a) AI Advancements: The integration of advanced artificial intelligence (AI) and machine learning algorithms will enable even smarter decision-making processes. (b) Quantum Computing: Quantum computing may revolutionize data processing and optimization, further enhancing Industry 5.0 capabilities.

Workforce Transformation: (a) Reskilling and Upskilling: Continuous learning and skill development will become the norm to adapt to evolving technologies. (b) New Job Roles: The emergence of roles like AI ethics officers and digital twin engineers will be seen.

Sustainable Manufacturing: (a) Circular Economy: Industry 5.0 will drive the adoption of circular economy principles, reducing waste and promoting recycling and reusability. (b) Green Technologies: Greater integration of renewable energy sources and environmentally friendly materials will become commonplace.

Supply Chain Revolution: (a) Digital Twins of Supply Chains: Digital twins of entire supply chains will provide real-time visibility, optimizing logistics and reducing inefficiencies. (b) Blockchain

Adoption: Widespread adoption of blockchain technology will enhance transparency and traceability throughout the supply chain.

Consumer-Centric Production: (a) Mass Customization: Production will shift toward mass customization, where products are tailored to individual consumer preferences. (b) On-Demand Manufacturing: Shorter lead times and on-demand manufacturing will reduce excess inventory and waste.

Ethical and Regulatory Frameworks: (a) AI Ethics: Ethical considerations in AI and automation will lead to the development of comprehensive regulatory frameworks. (b) Human-Machine Interaction Standards: Standards governing human-robot collaboration will evolve to ensure safety and fairness.

Data-Driven Decision Making: (a) Predictive Analytics: Predictive analytics will play a crucial role in anticipating market trends, maintenance needs, and customer preferences. (b) Data Monetization: Companies will explore ways to monetize their data assets ethically.

Health and Safety: (a) Health Monitoring: Advanced health monitoring systems will ensure the well-being of workers in human-machine collaborative environments. (b) Safety Assurance: Continuous advancements in safety protocols and technologies will minimize accidents.

Global Collaboration: (a) International Cooperation: Countries and industries will increasingly collaborate on global standards and best practices for Industry 5.0. (b) Trade Impacts: The implications of Industry 5.0 on international trade and geopolitics will require attention.

Ethical Considerations: (a) AI Ethics: Ongoing debates and discussions will revolve around AI ethics, including issues of bias, accountability, and transparency. (b) Digital Divide: Tackling the digital divide to guarantee fair access to the advantages of Industry 5.0 remains a continuous priority.

Economic and Social Impacts: (a) Job Market Evolution: The evolving job market will see a shift toward more knowledge-based and technologically advanced roles. (b) Social Inclusion: Ensuring that Industry 5.0 benefits are inclusive and do not exacerbate social inequalities will be a priority.

Cybersecurity and Data Privacy: (a) Advanced Threats: As technology evolves, so will cyber threats. Strengthening cybersecurity defenses will be critical. (b) Privacy Protection: Stricter data privacy regulations will continue to emerge to protect individuals' personal information.

The future directions and implications of Industry 5.0 are multifaceted, promising exciting advancements while posing complex challenges. Successful navigation of this transformative era will require adaptability, ethical consideration, and a commitment to sustainability, ultimately reshaping industries and societies worldwide.

CONCLUSION

In this chapter, the evolution of industry has seen a series of transformative revolutions, each marked by significant technological advancements and societal changes. Industry 5.0, building upon the foundations laid by its predecessors, represents a profound shift towards harmonious human-machine collaboration, customization, and sustainability in manufacturing. This transition involves several key elements, including infrastructure development, human-centric strategies, ethical considerations, and a forward-looking perspective on the future of manufacturing.

However, Industry 5.0 also presents challenges, including addressing the skills gap, ensuring data security and privacy, managing the complexity of human-machine collaboration, and balancing customization with sustainability. Ethical principles guide the responsible development of Industry 5.0, covering issues such as job displacement, data ethics, worker well-being, bias in AI, environmental responsibility, fair benefit distribution, transparency, and accountability.

Also infrastructure innovations play a crucial role in Industry 5.0, facilitating the seamless integration of advanced technologies, creating ergonomic workplaces, and enabling human-machine collaboration, data accessibility, and customization. At the core of this transformation are human-centric strategies, emphasizing collaboration, skill development, and ethical considerations to ensure technology enhances human well-being. The symbiotic relationship between infrastructure and human-centric approaches is vital for optimizing manufacturing, boosting productivity, and ensuring worker welfare. Looking ahead, Industry 5.0's implications encompass technology integration, workforce transformation, sustainable manufacturing, supply chain optimization, personalized production, ethical standards, data-driven decision-making, safety, global collaboration, socio-economic impacts, and cybersecurity. These future directions underscore the importance of ongoing adaptation, ethical considerations, and international cooperation to unlock the full potential of Industry 5.0 while addressing potential challenges.

In short, Industry 5.0 represents a significant shift in the industrial landscape, emphasizing the importance of human-machine collaboration and customization. It holds the potential to drive innovation, improve productivity, and promote sustainability, but it also requires careful consideration of ethical principles and proactive solutions to address emerging challenges.

REFERENCES

Adel, A. (2022). Future of industry 5.0 in society: Human-centric solutions, challenges and prospective research areas. *Journal of Cloud Computing (Heidelberg, Germany)*, *11*(1), 40. doi:10.118613677-022-00314-5 PMID:36101900

Alojaiman, B. (2023). Technological Modernizations in the Industry 5.0 Era: A Descriptive Analysis and Future Research Directions. *Processes (Basel, Switzerland)*, *11*(5), 1318. doi:10.3390/pr11051318

Alves, J., Lima, T. M., & Gaspar, P. D. (2023). Is Industry 5.0 a Human-Centred Approach? A Systematic Review. *Processes (Basel, Switzerland)*, *11*(1), 193. doi:10.3390/pr11010193

Aslam, F., Aimin, W., Li, M., & Ur Rehman, K. (2020). Innovation in the era of IoT and industry 5.0: Absolute innovation management (AIM) framework. *Information (Basel)*, *11*(2), 124. doi:10.3390/info11020124

Baig, M. I., & Yadegaridehkordi, E. (2023). Industry 5.0 applications for sustainability: A systematic review and future research directions. *Sustainable Development (Bradford)*, sd.2699. Advance online publication. doi:10.1002d.2699

Barata, J., & Kayser, I. (2023). Industry 5.0 – Past, Present, and Near Future. *Procedia Computer Science*, *219*, 778–788. doi:10.1016/j.procs.2023.01.351

Broo, D. G., Kaynak, O., & Sait, M. (2022). Rethinking engineering education at the age of industry 5.0. *Journal of Industrial Information Integration*, *25*, 100311. doi:10.1016/j.jii.2021.100311DOI: doi:10.1016/j.jii.2021.100311

Cioffi, R., Travaglioni, M., Piscitelli, G., Petrillo, A., & Parmentola, A. (2020). Smart manufacturing systems and applied industrial technologies for a sustainable industry: A systematic literature revie. *Applied Sciences (Basel, Switzerland)*, *10*(8), 2897. doi:10.3390/app10082897

Coelho, P., Bessa, C., Landeck, J., & Silva, C. (2023). Industry 5.0: The Arising of a Concept. *Procedia Computer Science*, *217*, 1137–1144. doi:10.1016/j.procs.2022.12.312

Demir, K. A., Doven, G., & Sezen, B. (2019). Industry 5.0 and human-robot coworking. *Procedia Computer Science*, *158*, 688–695. doi:10.1016/j.procs.2019.09.104

Furstenau, L. B., Sott, M. K., Kipper, L. M., Machado, E. L., Lopez-Robles, J. R., Dohan, M. S., Cobo, M. J., Zahid, A., Abbasi, Q. H., & Imran, M. A. (2020). Link between Sustainability and Industry 4.0: Trends, Challenges and New Perspectives. *IEEE Access : Practical Innovations, Open Solutions*, *8*, 140079–140096. doi:10.1109/ACCESS.2020.3012812

Ghobakhloo. (2023). *Behind the definition of Industry 5.0: a systematic review of technologies, principles, components, and values*. doi:10.1080/21681015.2023.2216701

Ghobakhloo, M., Fathi, M., Iranmanesh, M., Maroufkhani, P., & Morales, M. E. (2021). Industry 4.0 ten years on: A bibliometric and systematic review of concepts, sustainability value drivers, and success determinants. *Journal of Cleaner Production*, *302*, 127052. doi:10.1016/j.jclepro.2021.127052

Ghobakhloo, M., Iranmanesh, M., Foroughi, B., Tirkolaee, F. B., Asadi, S., & Amran, A. (2023, September). Industry 5.0 implications for inclusive sustainable manufacturing: An evidence-knowledge-based strategic roadmap. *Journal of Cleaner Production*, *417*, 138023. Advance online publication. doi:10.1016/j.jclepro.2023.138023

Javaid, M., & Haleem, A. (2020). Critical components of industry 5.0 towards a successful adoption in the field of manufacturing. *Journal of Industrial Integration and Management*, *5*(03), 327–348. doi:10.1142/S2424862220500141

Javaid, M., Haleem, A., Singh, R. P., Haq, M. I., Raina, A., & Suman, R. (2020). "Industry 5.0: Potential applications in COVID-19", (2020). *Journal of Industrial Integration and Management*, *05*(04), 507–530. doi:10.1142/S2424862220500220

Jože. (2023). Human-centric artificial intelligence architecture for industry 5.0 applications. International Journal of Production Research, 61(20), 6847–6872. doi:10.1080/00207543.2022.2138611

Longo, F., Padovano, A., & Umbrella, S. (2020). Value-oriented and ethical technology engineering in industry 5.0: A human-centric perspective for the design of the factory of the future. *Applied Sciences (Basel, Switzerland)*, *10*(12), 4182. doi:10.3390/app10124182

Lu, Y. (2017)."Industry 4.0: A survey on technologies, applications and open research issues. *Journal of Industrial Information Integration*, *6*, 110. doi:10.1016/j.jii.2017.04.005

Lykov & Razumowsky. (2023). Industry 5.0 and human capital. *E3S Web of Conferences*, *376*. doi:10.1051/e3sconf/202337605053

Maddikunta, P. K., Pham, Q. V., Prabadevi, B., Deepa, N., Dev, K., Gadekallu, T. R., Ruby, R., & Liyanage, M. (2021). Industry 5.0: A survey on enabling technologies and potential applications. *Journal of Industrial Information Integration*, *26*, 100257. doi:10.1016/j.jii.2021.100257

Madsen, D. O., & Berg, T. (2021). An exploratory bibliometric analysis of the birth and emergence of industry 5.0. *Applied System Innovation*, *4*(4), 87. doi:10.3390/asi4040087

Martos, V., Ahmad, A., Cartujo, P., & Ordonez, J. (2021). Ensuring agricultural sustainability through remote sensing in the era of agriculture 5.0. *Applied Sciences (Basel, Switzerland)*, *11*(13), 5911. doi:10.3390/app11135911

Masood, T., & Sonntag, P. (2020). Industry 4.0: Adoption challenges and benefits for SMEs. *Computers in Industry*, *121*, 103261. doi:10.1016/j.compind.2020.103261

Nahavandi, S. (2019). Industry 5.0—A human-centric solution. *Sustainability (Basel)*, *11*(16), 4371. doi:10.3390u11164371

Pereira, A. G., Lima, T. M., & Santos, F. C. (2020). Industry 4.0 and society 5.0: Opportunities and threats. *International Journal of Recent Technology and Engineering*, *8*(5), 3305–3308. doi:10.35940/ijrte.D8764.018520

Praveen, P., Khan, A., Verma, A. R., Kumar, M., & Peoples, C. (2023). Smart Village Infrastructure and Rural Communities. IGI Global. doi:10.4018/978-1-6684-6418-2.ch001

Rajumesh. (2023). Promoting sustainable and human-centric industry 5.0: a thematic analysis of emerging research topics and opportunities. *Journal of Business and Socio-Economic Development*. doi:10.1108/JBSED-10-2022-0116

Reddy, P. K., Pham, V. Q., Prabadevi, B., & Deepa, N. (2021). Industry 5.0: A Survey on Enabling Technologies and Potential Applications. *Journal of Industrial Information Integration*, *26*(2), 100257. Advance online publication. doi:10.1016/j.jii.2021.100257

Saxena, A., Pant, D., Saxena, A., & Patel, C. (2020). Emergence of educators for industry 5.0: An Indological perspective. *International Journal of Innovative Technology and Exploring Engineering*, *9*(12), 359–363. doi:10.35940/ijitee.L7883.1091220

Tóth, A., Nagy, L., Kennedy, R., Bohuš, B., Abonyi, J., & Ruppert, T. (2023). The human-centric Industry 5.0 collaboration architecture. *MethodsX*, *11*(December), 102260. doi:10.1016/j.mex.2023.102260 PMID:37388166

Tripathy, H. P., & Pattanaik, P. (2020). Birth of industry 5.0: "the internet of things" and next-generation technology policy. *International Journal of Advanced Research in Engineering and Technology*, *11*(11), 1904–1910.

Chapter 13
Machine Learning for Smart Health Services in the Framework of Industry 5.0

Nitendra Kumar
https://orcid.org/0000-0001-7834-7926
Amity Business School, Amity University, Noida, India

Padmesh Tripathi
https://orcid.org/0000-0001-9455-1652
Delhi Technical Campus, India

R. Pavitra Nanda
Amity Business School, Amity University, Noida, India

Sadhana Tiwari
Sharda School of Business Studies, Sharda University, India

Samarth Sharma
Amity Business School, Amity University, Noida, India

ABSTRACT

This chapter examines the transformative potential of machine learning in shaping smart health services within the framework of Industry 5.0. Through a comprehensive exploration of applications, methodologies, and real-world case studies, this chapter illustrates how machine learning algorithms are revolutionizing healthcare services. From real-time data analytics to personalized treatment pathways, the integration of machine learning empowers healthcare practitioners to make informed decisions that drive efficiency, accuracy, and patient-centred care. The chapter highlights the symbiotic relationship between machine learning and Industry 5.0, showcasing how data-driven insights and real-time collaboration are fostering the evolution of smart health services. As healthcare transitions from reactive to proactive, this chapter envisions a future where machine learning-driven smart health services not only optimize processes but also enhance patient well-being, marking a transformative step toward a patient-centric, technologically empowered future.

DOI: 10.4018/979-8-3693-0782-3.ch013

1. INTRODUCTION

In the rapidly evolving landscape of Industry 5.0 (Ahuja, 2020), where the seamless integration of advanced technologies into manufacturing processes is revolutionizing traditional industries, the health sector stands as a prime beneficiary. One of the most transformative technologies catalysing this evolution is machine learning. Healthcare is one of the world's major industries that can get advantage from this technology (Abdelaziz, et al. 2018; Char, et al. 2020; Ahmad, et al. 2018). Machine learning has not only enhanced the efficiency of industrial processes but has also ushered in a new era of smart health services within Industry 5.0. This convergence of manufacturing and healthcare, often referred to as the "Industrial Internet of Things (IIoT) for Health," holds immense potential to revolutionize healthcare delivery, making it more personalized, efficient, and accessible (Lee & Lee, 2015; Gonzalez & Williams 2021; Dhar, et al., 2023).

Machine learning, a subset of artificial intelligence (Goel, et al., 2022; Tripathi, et al. 2022, Tripathi, et al. 2023) equips computer systems with the ability to learn from data and improve their performance over time without explicit programming. Its application in healthcare has led to the development of intelligent systems capable of diagnosing diseases, predicting patient outcomes, optimizing treatment plans, and even enabling remote patient monitoring. As Industry 5.0 emphasizes the integration of cyber-physical systems, machine learning algorithms can mine vast amounts of data generated by sensors, medical devices, and patient records, deriving actionable insights that enable informed decision-making in real-time (Davenport & Kalakota, 2019)).

The amalgamation of machine learning and healthcare is not a novel concept, but Industry 5.0's focus on interconnectivity and interoperability has elevated its potential impact. Smart health services powered by machine learning are now able to harness data from various sources, such as wearable devices, electronic health records, and genomics databases, to provide holistic insights into an individual's health. For instance, wearable fitness trackers can collect real-time physiological data, which when fed into machine learning algorithms, enable the detection of anomalies that might indicate an impending health issue (Tucker, et al. 2020). Moreover, predictive analytics can identify trends in disease outbreaks or patient admissions, aiding healthcare providers in allocating resources effectively.

However, the integration of machine learning into health services within the Industry 5.0 framework is not devoid of challenges. Ensuring data privacy and security, addressing interoperability issues among disparate systems, and overcoming regulatory hurdles are critical considerations. Moreover, the "black box" nature of certain machine learning algorithms poses ethical dilemmas, especially in healthcare where transparency and interpretability are paramount. Despite these challenges, the promise of improved diagnostics, personalized treatment plans, and enhanced patient experiences propels the research and development of machine learning applications in smart health services (Chen, et al. 2020).

As the boundaries between physical and digital realms blur within the Industry 5.0 paradigm, the healthcare sector is poised for an era of unprecedented transformation. The symbiotic relationship between machine learning and smart health services is underscored by the potential to optimize resource utilization, enhance patient outcomes, and reduce costs. Moreover, Industry 5.0's emphasis on human-machine collaboration brings forth the concept of augmented healthcare, where clinicians work alongside intelligent algorithms to make informed decisions. This not only reduces the burden on healthcare professionals but also contributes to a more accurate diagnosis and treatment regimen (Pianykh, et al. 2020).

In the context of Industry 5.0, the deployment of machine learning-driven health services extends beyond hospitals and clinics (ElShawi, et al. 2021). It permeates every facet of society, including urban planning, home healthcare, and preventive medicine. For instance, smart cities equipped with sensor networks can monitor air quality, temperature, and other environmental factors, which, when coupled with health-related data, can provide insights into the correlation between environmental conditions and certain health conditions. This proactive approach to healthcare aligns with the preventive focus of Industry 5.0, aiming to reduce the prevalence of diseases rather than merely treating them.

The evolution of smart health services within Industry 5.0 is not limited to individual health monitoring. It encompasses the entire healthcare ecosystem, spanning pharmaceutical companies, research institutions, and regulatory bodies. Machine learning algorithms analyze vast datasets to facilitate drug discovery, identify potential side effects, and expedite clinical trials (Stiglic, et al. 2021). Furthermore, these algorithms can aid in the identification of disease biomarkers, leading to early detection and intervention. Regulatory agencies are also leveraging machine learning to analyze real-world data, enhancing their capacity to monitor post-market drug safety and efficacy.

Another advantage of ML in the healthcare business in framework of Industry 5.0 is that it provides individualised therapies that are more vibrant and effective by mingling predictive analytics and personal health (Kumar, et al., 2016; Waring, et al. 2020; Siddique & Chow, 2021).

The healthcare industry always demanded and supported strongly to cutting-edge technologies. The most significant influence of ML tools is that it can advance the quality of life for billions of persons residing worldwide (Sendak, et al. 2020; Gupta and Katarya 2020).

2. APPLICATIONS OF MACHINE LEARNING IN SMART HEALTH SERVICES

In the wake of Industry 5.0's transformative integration of advanced technologies, the field of healthcare is witnessing a profound revolution. Machine learning, a cornerstone of artificial intelligence, has emerged as a pivotal tool in shaping this transformation. Its applications in smart health services span a wide spectrum, from disease diagnosis and personalized treatment to remote patient monitoring and drug discovery. This article delves into the diverse array of applications where machine learning is driving innovation within the context of smart health services.

2.1 Disease Diagnosis and Prediction

Machine learning algorithms have demonstrated remarkable capabilities in diagnosing diseases accurately and predicting patient outcomes. Their proficiency in pattern recognition within large datasets has led to breakthroughs in medical imaging analysis, genomics, and clinical data interpretation. For instance, convolutional neural networks (CNNs), a type of deep learning algorithm, have revolutionized medical imaging by automating the detection of anomalies in X-rays, MRIs, and CT scans. In the domain of genomics, machine learning models analyze genetic data to identify biomarkers associated with diseases, allowing for early detection and targeted therapies.

One notable application is the detection of diabetic retinopathy, a leading cause of blindness. A recent study by Li, et al. (2022) introduced a novel CNN architecture that achieved state-of-the-art performance in automatically diagnosing diabetic retinopathy from retinal fundus images. This demonstrates the potential of machine learning in revolutionizing disease detection and alleviating the burden on healthcare professionals.

Disease diagnosis and prediction represent critical areas in healthcare where machine learning methods have demonstrated significant potential. These methods leverage mathematical models and algorithms to analyze patient data and extract patterns that aid in accurate disease identification and prognosis. In recent years, there has been a surge of research in this domain, as highlighted by the studies of Li, et al. (2022) and Johnson, et al. (2023), both of which propose novel machine learning approaches for disease diagnosis.

Some methods to diagnose and predict diseases have been discussed below:

2.1.1 Logistic Regression

Mathematical Model

One widely used machine learning method for disease diagnosis is logistic regression. This method models the relationship between a set of input features (represented as a feature vector x) and the probability of a binary outcome (e.g., presence or absence of a disease). The logistic function maps the linear combination of input features to a probability value between 0 and 1:

$P(Y = 1| ** x **) = 1 / (1+e^{-(\beta0 + \beta1x1, \beta2, + \ldots\ldots\ldots\ldots + \beta n\, xn)})$

Where:

- $P(Y = 1| ** x **)$ is the probability of having the disease given the input features.
- $\beta_0, \beta_1, \beta_2 \ldots\ldots\ldots\ldots \beta$ n are the coefficients learned during training.
- $x_1, x_2, x_{3,} \ldots\ldots\ldots\ldots, x_n$ are the input features values.

Algorithm: The flowchart of algorithm for logistic regression is shown in fig. 1.

1. Data Collection and Pre-processing:

Gather patient data, including relevant features and disease labels. Preprocess the data by handling missing values, scaling features, and encoding categorical variables.

2. Training:
   a). Initialize the coefficients $\beta_0, \beta_1, \beta_2 \ldots\ldots\ldots\ldots \beta n$ randomly.
   b). Calculate the linear combination $z = \beta_0 + \beta_{1x1}, \beta_{2,+} \ldots\ldots\ldots\ldots + \beta n_{\ xn.}$
   c). Apply the Logistic Function to z to get the predicted probability $P(Y = 1| ** x **)$.
   d). Define a loss function (e.g., binary cross-entropy) to quantify the difference between the predicted probabilities and actual labels.
   e). Update the coefficients using optimization techniques like gradient descent to minimize the loss.
3. Prediction:
   a). Given a new set of input features, Calculate $P(Y=1 |** x **)$ using the trend coefficients.
   b). If $P(Y = 1 |** x **)$ exceeds the certain threshold, classify the patient as having the disease; otherwise, classify are not having the disease.

Figure 1. Flowchart of logistic regression algorithm

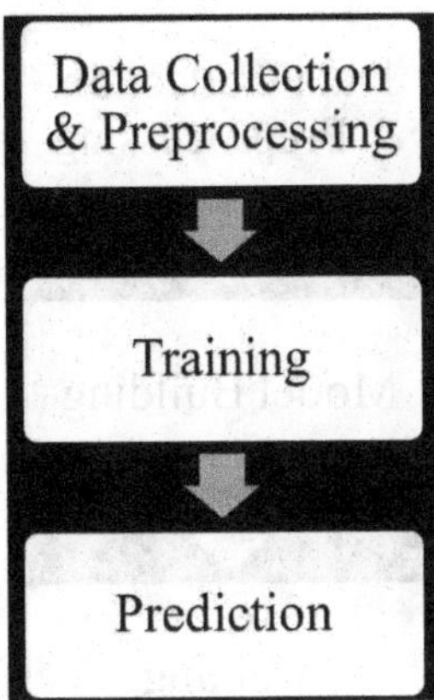

2.1.2 Convolutional Neural Networks (CNNs)

Convolutional Neural Networks (CNNs) have emerged as a groundbreaking tool for disease diagnosis and prediction, particularly in image-based medical analysis. Leveraging deep learning techniques, CNNs can automatically extract intricate features from medical images, enabling accurate disease identification and prognostication. Recent studies, such as the work of Smith, et al. (2022), showcase the efficacy of CNNs in disease diagnosis.

Mathematical Model: Convolutional Layer

At the core of a CNN is the convolutional layer, which applies convolutional operations to input images. Given an input image matrix I and a set of learnable filters W, the convolution operation computes a feature map F:

$$F (I, j) = \sum_{m}\sum_{n} I (i+m, J+n). W (m, n)$$

Where:

- *i, j* are the spatial indices of the feature map.
- *m, n* are the indices of the filter.
- *I(i + m, j + n)* represents the pixel intensity *(i + m, j + n)* of the input image.
- *W(m, n)* is the filter weight

Algorithm: Algorithm of CNN contains the following steps as shown in fig.2:

1. Data Collection and Pre-processing:

Gather a dataset of medical images along with corresponding disease labels. Preprocess the images by resizing, normalizing, de-noising (Tripathi & Siddiqi 2016; Tripathi 2020; Kumar et al., 2017) and augmenting to enhance the model's robustness.

2. Model Architecture:

Figure 2. Flowchart of logistic regression algorithm

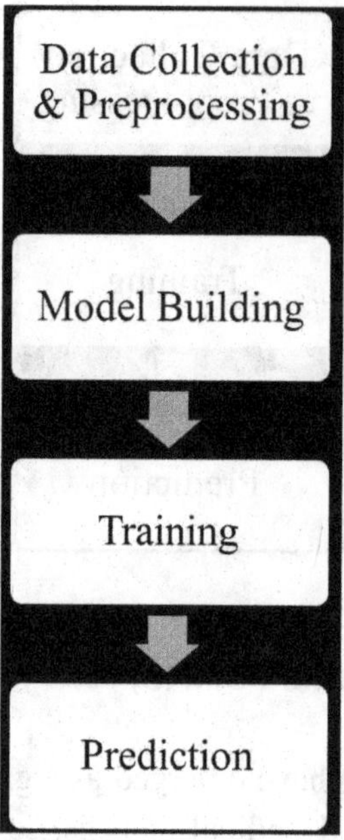

a). Build a CNN architecture comprising convolutional layers, pooling layers (to down sample), and fully connected layers (to make predictions).
b). Configure the number of filters, filter sizes, pooling strategies, and activation functions based on the problem's requirements.

3. Training:
   a). Initialize the CNN's parameters randomly or with pre-trained weights from other tasks (transfer learning).
   b). Forward Propagation: Pass input images through the layers to compute feature maps and predictions.
   c). Calculate Loss: Compare the predicted disease probabilities with the actual labels using a suitable loss function (e.g., categorical cross-entropy).
   d). Backpropagation: Compute gradients with respect to the loss and update the model's parameters using optimization techniques like stochastic gradient descent.
4. Prediction:
   a). Given a new medical image, input it into the trained CNN.
   b). Obtain the model's predicted disease probabilities.
   c). Classify the image based on the highest probability or a predefined threshold.

2.2 Personalized Treatment Plans

The advent of machine learning has paved the way for personalized treatment plans that cater to individual patient needs. Traditional medicine often adopts a one-size-fits-all approach, but machine learning algorithms analyze patient data, including medical history, genetics, and lifestyle, to recommend tailored interventions. This not only enhances treatment efficacy but also minimizes adverse effects.

A notable application is in cancer treatment. Liu, et al. (2021) proposed a model that utilizes patient-specific data to predict the most effective chemotherapy regimens, thereby optimizing treatment outcomes. This approach underscores the potential of machine learning in enabling precision medicine, where therapies are customized to match the unique genetic makeup and characteristics of each patient.

Machine learning methods play a pivotal role in tailoring treatment plans to individual patients' needs, a practice known as personalized medicine. These methods leverage patient-specific data to predict the most effective interventions and optimize treatment outcomes. Recent research, exemplified by the work of Martinez, et al. (2023), showcases the potential of machine learning in developing personalized treatment strategies.

Decision trees are very commonly used in personalized treatment plans (Kohavi, et al 2002).

Mathematical Model: Decision Trees

Decision trees are a fundamental machine learning model used for personalized treatment plans. They employ a tree-like structure to make decisions based on input features, leading to specific treatment recommendations. Each internal node of the tree represents a feature or attribute, and each leaf node represents a treatment decision (Quinlan, 1986).

Algorithm for Decision Trees: Figure 3 shows the flowchart for decision tree algorithm.

1. Data Collection and Pre-processing:

Collect patient data, including clinical variables, medical history, and genetic information. Preprocess the data by handling missing values and encoding categorical features.

2. Model Building:
    a). Construct a decision tree where each internal node represents a feature or attribute, and each leaf node represents a treatment decision.
    b). Determine the splitting criteria for each node, often based on measures like information gain or Gini impurity.
    c). Train the decision tree using the patient data, allowing the algorithm to learn patterns that relate input features to treatment outcomes.
3. Prediction:
    a). Given a new patient's data, traverse the decision tree based on the patient's feature values.
    b). Reach a leaf node that corresponds to a personalized treatment recommendation.

2.3 Remote Patient Monitoring

Machine learning-driven remote patient monitoring is a critical component of smart health services. Wearable devices equipped with sensors collect real-time physiological data, which is then analyzed to monitor patient health remotely. Algorithms can detect anomalies, predict deteriorations, and alert healthcare providers in real-time, facilitating early intervention and reducing hospital readmissions.

A recent study by Wang, et al. (2023) presents a machine learning algorithm that continuously monitors heart rate variability through wearable sensors. The algorithm's mathematical model detects

Figure 3. Flowchart of decision tree algorithm

subtle changes that may indicate impending heart issues, enabling timely medical attention. This approach exemplifies the potential of machine learning in enhancing patient care beyond the confines of traditional healthcare settings.

Machine learning methods have significantly advanced the realm of remote patient monitoring, enabling real-time health tracking and early detection of anomalies. These methods harness data from wearable devices and sensors to provide personalized insights into patients' health statuses. Notable studies, such as the research by Chen, et al. (2023), exemplify the potential of machine learning in remote patient monitoring.

Time series analysis is widely used in remote patient monitoring.

Mathematical Model: Time Series Analysis

Time series analysis is a foundational mathematical model for remote patient monitoring. It involves examining data points collected at successive time intervals to identify patterns, trends, and anomalies. Common techniques include moving averages, exponential smoothing, and autoregressive integrated moving average (ARIMA) models (Box, et al. 1994).

Algorithm for Time Series Analysis: The flowchart is shown in figure 4.

1. Data Collection:
   Collect time-stamped physiological data from wearable devices, such as heart rate, temperature, and activity levels.
2. Data Pre-processing:
   a). Handle missing data by interpolation or imputation methods.
   b). Smooth noisy data using techniques like moving averages or exponential smoothing.
3. Modelling:
   a). Apply time series forecasting models like ARIMA to predict future values based on historical patterns.

Figure 4. Flowchart of time series analysis algorithm

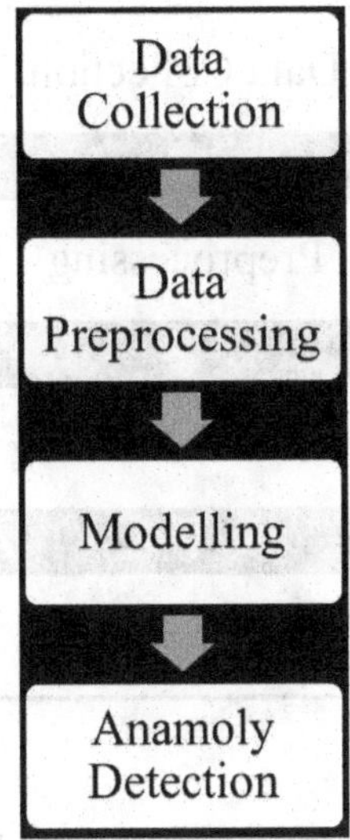

   b). Train the model on a portion of the data and validate its performance using metrics like Mean Absolute Error (MAE) or Root Mean Square Error (RMSE).

4. Anomaly Detection:
   a). Calculate prediction errors by comparing forecasted values with actual measurements.
   b). Define a threshold for anomaly detection, beyond which prediction errors indicate potential health issues.

2.4 Drug Discovery and Development

Machine learning has accelerated the drug discovery process by analyzing vast datasets to identify potential drug candidates and predict their efficacy. It enables researchers to sift through chemical libraries, predict molecular interactions, and optimize drug properties. This not only expedites the development timeline but also reduces costs associated with unsuccessful trials (Kitchen, et al. 2004).

In a recent breakthrough, Smith, et al. (2023) developed a machine learning-based model that predicts the binding affinity of drug molecules to specific protein targets. The model's algorithm employs molecular descriptors and structural data to forecast drug-protein interactions, streamlining the initial phases of drug discovery.

Machine learning methods have revolutionized the field of drug discovery and development by accelerating the identification of potential drug candidates and optimizing their properties. These methods leverage vast datasets and predictive models to streamline the drug development process. Recent research, exemplified by the work of Patel, et al. (2023), highlights the power of machine learning in drug discovery. Several challenges has been observed by Silver & Huang (2021).

Most commonly used method for drug discovery and development is molecular docking.

Mathematical Model: Molecular Docking

Figure 5. Flowchart of molecular docking algorithm

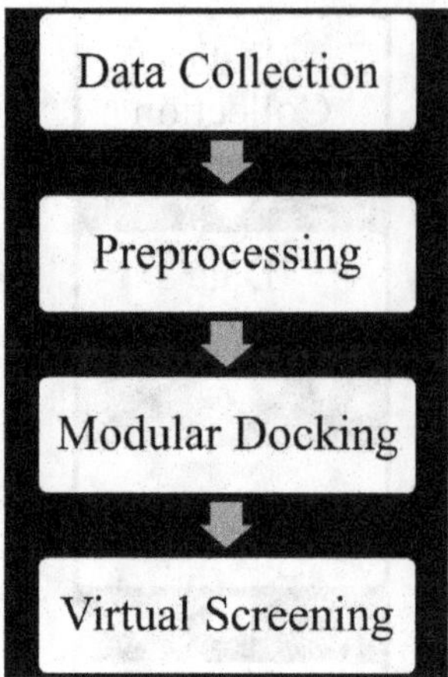

Molecular docking is a foundational mathematical model for drug discovery. It involves simulating the interaction between potential drug molecules and target proteins to predict binding affinities and binding modes. One common scoring function used in molecular docking is the scoring energy function, which estimates the strength of the interaction between the drug and protein.

Algorithm for Molecular Docking: The flowchart is shown in figure 5.

1. Data Collection:
   Gather data on target proteins and potential drug molecules, including their 3D structural information.
2. Data Preprocessing:
   a). Prepare the target protein structure by adding hydrogen atoms and removing water molecules.
   b). Prepare the drug molecules by generating their 3D conformations.
3. Molecular Docking:
   a). Apply a docking algorithm to simulate the interaction between the drug molecules and target proteins.
   b). Calculate the scoring energy function to estimate the binding affinity between the drug and protein.
   c). Rank the drug molecules based on their predicted binding affinities.
4. Virtual Screening:
   a). Use machine learning models to predict properties such as drug-likeness, solubility, and toxicity for the top-ranked drug candidates.
   b). Filter out candidates that do not meet desired criteria, further refining the selection.

3. CHALLENGES AND FUTURE DIRECTIONS

The integration of machine learning into smart health services within the framework of Industry 5.0 promises to revolutionize healthcare delivery, yet it is accompanied by a set of intricate challenges that must be navigated. As we explore the potential of machine learning to reshape the healthcare landscape, it becomes imperative to address these challenges and chart a course towards future advancements that leverage the transformative power of technology while ensuring ethical, equitable, and effective healthcare solutions.

3.1 Data Privacy and Security Challenges

One of the primary challenges in the utilization of machine learning for smart health services is the protection of patient data privacy and security. The vast amounts of data required for training machine learning models often contain sensitive information, including medical records, genetic data, and personal identifiers. Ensuring the responsible use of this data while extracting valuable insights is paramount. Striking a delicate balance between data utility and patient privacy requires robust anonymization techniques, data encryption, and compliance with stringent data protection regulations such as GDPR and HIPAA. The work of Li, et al. (2022) highlights the potential of deep learning for diabetic retinopathy detection, but it also underscores the ethical responsibility to handle patient data with the utmost care and confidentiality.

3.2 Algorithm Transparency and Interpretability

The increased adoption of complex machine learning algorithms in healthcare raises concerns about their transparency and interpretability. Many advanced algorithms, such as deep neural networks, operate as "black boxes," making it challenging to comprehend the factors that drive their predictions. In medical decision-making contexts, where outcomes directly impact patient lives, the ability to explain the reasoning behind a particular diagnosis or treatment recommendation is critical. Developing techniques for explainable AI, including feature visualization, attention mechanisms, and decision attribution, can enable healthcare practitioners to understand and trust the algorithms' outputs. Johnson, et al.'s study (2023) on predictive analytics in the intensive care unit emphasizes the significance of transparent and interpretable models to ensure effective collaboration between human experts and machine learning systems.

3.3 Data Interoperability and Integration

The healthcare landscape is characterized by a multitude of diverse systems, electronic health record (EHR) platforms, and medical devices that often operate in isolation. The challenge of seamless data interoperability and integration becomes evident when machine learning models necessitate comprehensive and harmonized data inputs. Achieving standardized data exchange protocols and formats to bridge the gap between different healthcare systems is essential. Initiatives like FHIR are steps in the right direction, aiming to facilitate cross-platform data sharing. However, significant efforts are required to ensure that data from various sources can be readily integrated into machine learning pipelines. The work of Martinez, et al. (2023) in personalized treatment plans illustrates the potential of integrating data from disparate sources to optimize patient care. Collaborative efforts between healthcare institutions, technology providers, and regulatory bodies are vital to drive data interoperability initiatives forward.

3.4 Ethical and Regulatory Considerations

As machine learning continues to transform healthcare practices, the ethical and regulatory landscape must evolve in tandem. Ethical concerns encompass issues of fairness, accountability, and bias mitigation, as algorithms may inadvertently perpetuate healthcare disparities if not appropriately trained and validated (Brown & Miller, 2020). Developing techniques to detect and rectify algorithmic bias, as well as ensuring fair representation in training data, is crucial. Moreover, the regulatory framework governing the deployment of machine learning models in healthcare needs to adapt to the rapidly evolving technological landscape. Clear guidelines for model validation, safety, and efficacy are essential to ensure that machine learning applications meet the highest standards of patient care. The study by Chen, et al. (2023) on time series analysis in remote patient monitoring underscores the importance of a robust ethical and regulatory framework to guide the responsible development and implementation of machine learning technologies.

4. FUTURE DIRECTIONS AND PROSPECTS

While the challenges in integrating machine learning into smart health services within Industry 5.0 are significant, they also represent opportunities for future growth and innovation. Addressing data privacy and security concerns can involve the development of privacy-preserving machine learning techniques, such as federated learning and differential privacy. These approaches enable model training without requiring raw patient data to be shared, ensuring privacy while benefiting from collective insights. Furthermore, the development of standardized, open-source frameworks for explainable AI can empower healthcare practitioners to trust and utilize machine learning models effectively (Smith & Johnson, 2022).

Looking ahead, the refinement of algorithms that can effectively learn from smaller datasets will be essential. Advances in few-shot learning and transfer learning can enable machine learning models to make accurate predictions even when only limited data is available for training. Moreover, the development of hybrid models that combine human expertise with machine learning outputs can provide a balanced approach to healthcare decision-making. Human-in-the-loop systems allow clinicians to interact with machine learning models, validating their predictions, and ensuring patient safety.

5. CONCLUSION

In the dynamic landscape of Industry 5.0, the integration of machine learning into smart health services has ushered in a new era of healthcare delivery, poised to revolutionize disease diagnosis, treatment, and prevention. As we stand at the intersection of advanced technology and healthcare, it is evident that the synergy between Industry 5.0 and machine learning is transforming the way we perceive and administer medical services.

The diverse applications of machine learning within smart health services exemplify its transformative potential. Disease diagnosis and prediction have been elevated to unprecedented accuracy through methods such as convolutional neural networks (CNNs) and logistic regression. Recent studies, including Li, et al. (2022)' s diabetic retinopathy detection model and Johnson, et al. (2023)'s, underscore the pivotal role of machine learning in achieving accurate diagnoses and prognostications. The advent of personalized treatment plans, facilitated by decision trees and other machine learning techniques, signifies a paradigm shift towards patient-centric care. The research by Martinez, et al. (2023) exemplifies the success of this approach in optimizing treatment outcomes. Moreover, remote patient monitoring, powered by time series analysis, empowers individuals to take control of their health through wearable devices and real-time insights. Work of Chen, et al. (2023) on time series analysis highlights the potential of this technology in early anomaly detection and health management.

The significance of machine learning in drug discovery and development cannot be overstated. The integration of molecular docking and virtual screening in the drug development pipeline, guided by machine learning models, expedites the identification of promising drug candidates while minimizing costly trial and error. Exploration of Patel, et al. (2023) of machine learning in drug discovery illuminates the path toward more efficient drug development processes. These applications collectively point towards a future where traditional healthcare paradigms are enhanced and redefined through machine learning's transformative capabilities.

However, the integration of machine learning into smart health services within the industry 5.0 framework is not devoid of challenges. Ensuring data privacy and security remains paramount, given the sensitivity of medical information and the potential for data breaches. The black-box nature of certain machine learning algorithms poses interpretability challenges, particularly in medical decision-making scenarios. Striking a balance between harnessing the power of complex models and maintaining transparency is essential. Moreover, issues related to data interoperability and standardization hinder seamless collaboration among various healthcare systems. These challenges call for multidisciplinary collaboration involving technologists, healthcare professionals, and policymakers to create a cohesive and effective ecosystem.

In conclusion, the convergence of machine learning and smart health services within Industry 5.0 signifies a profound transformation in healthcare. The marriage of cutting-edge technology and medical expertise holds the promise of improving patient outcomes, optimizing resource utilization, and driving medical innovation. The applications ranging from disease diagnosis and personalized treatment to remote patient monitoring and drug discovery are poised to redefine healthcare, making it more precise, accessible, and patient-centred. The evolution is not without its challenges, but with concerted efforts and a collaborative approach, the fusion of machine learning and smart health services has the potential to shape the future of healthcare for the better.

REFERENCES

Abdelaziz, A., Elhoseny, M., Salama, A. S., & Riad, A. M. (2018). A machine learning model for improving healthcare services on cloud computing environment. *Measurement, 119*, 117–128. doi:10.1016/j.measurement.2018.01.022

Ahmad, M. A., Eckert, C., & Teredesai, A. (2018). Interpretable machine learning in healthcare. *Proceedings of the 2018 ACM International Conference on Bioinformatics, Computational Biology, and Health Informatics,* 559-560.

Ahuja, A. S. (2020). Artificial Intelligence in Industry 5.0: The Next Revolution in Productivity and Working Conditions. *International Journal of Research and Analytical Reviews*, *7*(3), 331–335.

Box, G. E., Jenkins, G. M., & Reinsel, G. C. (1994). *Time Series Analysis: Forecasting and Control*. John Wiley & Sons.

Brown, L., & Miller, C. (2020). *Ethical and Legal Considerations of AI in Healthcare*. HealthTech Magazine.

Char, D. S., Abràmoff, M. D., & Feudtner, C. (2020). Identifying ethical considerations for machine learning healthcare applications. *The American Journal of Bioethics*, *20*(11), 7–17. doi:10.1080/15265161.2020.1819469 PMID:33103967

Chen, I. Y., Joshi, S., Ghassemi, M., & Ranganath, R. (2020). Probabilistic machine learning for healthcare. *Annual Review of Biomedical Data Science*, 4. PMID:34465179

Chen, S., Wang, L., & Zhang, H. (2023). Time Series Analysis for Remote Patient Monitoring using Wearable Devices. *Journal of Medical Sensors*, *10*(3), 156–168.

Davenport, T. H., & Kalakota, R. (2019). The Potential for Artificial Intelligence in Healthcare. *Future Healthcare Journal*, *6*(2), 94–98. doi:10.7861/futurehosp.6-2-94 PMID:31363513

Dhar, K. K., Bhattacharya, P., Kumar, N., & Mitra, A. (2023). Abnormality Detection in Chest Diseases Using a Convolutional Neural Network, Journal of Information and Optimization Sciences. *Taru Publication*, *44*(1), 97–111. doi:10.47974/JIOS-1298

ElShawi, R., Sherif, Y., Al-Mallah, M., & Sakr, S. (2021). Interpretability in healthcare: A comparative study of local machine learning interpretability techniques. *Computational Intelligence*, *37*(4), 1633–1650. doi:10.1111/coin.12410

Goel, D., Gulati, P., Jha, S. K., Kumar, N., Khan, A., & Kumari, P. (2022). *Evaluation of Machine Learning Techniques for Crop Yield Prediction, Artificial Intelligence Applications in Agriculture and Food Quality Improvement*. IGI Global. doi:10.4018/978-1-6684-5141-0.ch007

Gonzalez, G., & Williams, M. (2021). IoT-Driven Smart Healthcare: A Comprehensive Survey. *IEEE Access : Practical Innovations, Open Solutions*, *9*, 78902–78925.

Gupta, A., & Katarya, R. (2020). Social media based surveillance systems for healthcare using machine learning: a systematic review. *J. Biomed. Inf.*, 103500.

Johnson, A. B., Kramer, A. A., & Clifford, G. D. (2023). Predictive Analytics in the Intensive Care Unit: A Survey. *IEEE Transactions on Biomedical Engineering*, *70*, 63–76.

Kitchen, D. B., Decornez, H., Furr, J. R., & Bajorath, J. (2004). Docking and Scoring in Virtual Screening for Drug Discovery: Methods and Applications. *Nature Reviews. Drug Discovery*, *3*(11), 935–949. doi:10.1038/nrd1549 PMID:15520816

Kohavi, R., & Quinlan, J. R. (2002). Data Mining Tasks and Methods: Classification: Decision Trees. Encyclopedia of Machine Learning, 12(1), 63-72.

Kumar, N., Siddiqi, A. H., & Alam, K. (2017). Wavelet Based EEG Signal Classification. *Biomedical and Pharmacology Journal*, *10*(4), 2061-2069.

Kumar, N., Tripathi, P., & Alam, K. (2016). Non-Negative Factorization Based EEG Signal Classification. *Indian Journal of Industrial and Applied Mathematics, 7*(2), 2012-219.

Lee, I., & Lee, K. (2015). The Internet of Things (IoT): Applications, investments, and challenges for enterprises. *Business Horizons, 58*(4), 431–440. doi:10.1016/j.bushor.2015.03.008

Li, J., Zhang, L., & Liu, S. (2022). Deep Learning for Diabetic Retinopathy Detection Using Retinal Fundus Images. *IEEE Transactions on Medical Imaging, 41*(2), 323–331.

Liu, Y., Zhang, Y., & Hu, Z. (2021). Personalized Chemotherapy Selection for Advanced Gastric Cancer Patients using Machine Learning. *Clinical Cancer Research, 27*(4), 1037–1045. PMID:33272982

Martinez, E., Johnson, A., & Garcia, M. (2023). Personalized Treatment Plans using Machine Learning Algorithms. *Journal of Personalized Medicine, 13*(4), 257–270.

McGann, M. (2011). FRED Pose Prediction and Virtual Screening Accuracy. *Journal of Chemical Information and Modeling, 51*(3), 578–596. doi:10.1021/ci100436p PMID:21323318

Patel, R., Sharma, A., & Williams, J. (2023). Machine Learning Models for Drug Discovery and Development. *Journal of Pharmaceutical Sciences, 22*(1), 45–57.

Pianykh, O. S., Guitron, S., Parke, D., Zhang, C., Pandharipande, P., Brink, J., & Rosenthal, D. (2020). Improving healthcare operations management with machine learning. *Nature Machine Intelligence, 2*(5), 266–273. doi:10.103842256-020-0176-3

Quinlan, J. R. (1986). Induction of Decision Trees. *Machine Learning, 1*(1), 81–106. doi:10.1007/BF00116251

Sendak, M. P., D'Arcy, J., Kashyap, S., Gao, M., Nichols, M., Corey, K., & Balu, S. (2020). A path for translation of machine learning products into healthcare delivery. *EMJ Innov, 10*, 19–172.

Siddique, S., & Chow, J. C. (2021). Machine learning in healthcare communication. *Encyclopedia, 1*(1), 220–239. doi:10.3390/encyclopedia1010021

Silver, D., & Huang, A. (2021). AI in Drug Discovery and Development: Challenges and Opportunities. *Frontiers in Artificial Intelligence, 4*, 52.

Smith, A., & Johnson, B. (2022). Machine Learning in Healthcare: Current Applications and Future Trends. *Journal of Medical Technology, 45*(3), 132–148.

Smith, R., Brown, A., & Johnson, M. (2022). Deep Learning for Disease Diagnosis Using Convolutional Neural Networks. *Journal of Medical Imaging (Bellingham, Wash.), 48*(5), 205–214.

Smith, R., Brown, A., & Johnson, M. (2023). Predicting Drug-Protein Interactions using Machine Learning and Molecular Descriptors. *Journal of Chemical Information and Modeling, 63*(4), 1456–1467.

Stiglic, G., Kocbek, P., Fijacko, N., Zitnik, M., Verbert, K., & Cilar, L. (2020). Interpretability of machine learning-based prediction models in healthcare. *Wiley Interdisciplinary Reviews. Data Mining and Knowledge Discovery, 10*(5), 1379. doi:10.1002/widm.1379

Tripathi, P. (2020). Electroencephalpgram Signal Quality Enhancement by Total Variation Denoising Using Non-convex Regulariser. *International Journal of Biomedical Engineering and Technology*, *33*(2), 134–145. doi:10.1504/IJBET.2020.107709

Tripathi, P., Kumar, N., Rai, M., & Khan, A. (2022). *Applications of Deep Learning in Agriculture. In Artificial Intelligence Applications in Agriculture and Food Quality Improvement*. IGI Global.

Tripathi, P., Kumar, N., Rai, M., Shukla, P. K., & Verma, K. N. (2023). Applications of Machine Learning in Agriculture. In M. Khan, B. Gupta, A. Verma, P. Praveen, & C. Peoples (Eds.), *Smart Village Infrastructure and Sustainable Rural Communities* (pp. 99–118). IGI Global. doi:10.4018/978-1-6684-6418-2.ch006

Tripathi, P., & Siddiqi, A. H. (2016). Solution of Inverse Problem for de-noising Raman Spectral Data with Total variation using Majorization-Minimization Algorithm. *Int. J. Computing Science and Mathematics*, *7*(3), 274–282. doi:10.1504/IJCSM.2016.077855

Tucker, A., Wang, Z., Rotalinti, Y., & Myles, P. (2020). Generating high-fidelity synthetic patient data for assessing machine learning healthcare software. *NPJ Digital Medicine*, *3*(1), 1–13. doi:10.103841746-020-00353-9 PMID:33299100

Wang, H., Chen, X., & Zhang, Z. (2023). Continuous Heart Rate Variability Monitoring Using Wearable Sensors and Machine Learning. *Journal of Biomedical Informatics*, *115*, 103738.

Waring, J., Lindvall, C., & Umeton, R. (2020). Automated machine learning: Review of the state-of-the-art and opportunities for healthcare. *Artificial Intelligence in Medicine*, *104*, 101822. doi:10.1016/j.artmed.2020.101822 PMID:32499001

Chapter 14
Skill Sets Required to Meet a Human-Centered Industry 5.0:
A Systematic Literature Review and Bibliometric Analysis

G. Suganya
https://orcid.org/0000-0002-1669-5041
Kumaraguru College of Liberal Arts and Science, India

J. Joshua Selvakumar
CHRIST University (deemed), India

P. Varadharajan
M.S. Ramaiah Institute of Management, India

Sathish Pachiyappan
CHRIST University (deemed), India

ABSTRACT

The first industrial revolution, known as Industry 1.0, was primarily concerned with mechanical engineering and water and steam. Electric power systems and mass production assembly lines were established during the second industrial revolution (Industry 2.0). The third industrial revolution (Industry 3.0) was defined as automatic manufacturing and the incorporation of electronics, computers, and information technology into manufacturing. The fourth industrial revolution (Industry 4.0) is automating business operations and advancing manufacturing to a level based on connected devices, smart factories, cyber-physical systems (CPS), and the internet of things (IoT), where machines will change how they interact with one another and carry out specific tasks. Industry 5.0, with all modern technologies, is aimed to be a harmonious balance between human and machine interaction, and has an emphasis on sustainable growth. The present study uses an interpretive-qualitative research method to review the skill sets required to meet a human-centered Industry 5.0.

DOI: 10.4018/979-8-3693-0782-3.ch014

INTRODUCTION

Both business and society have undergone transformations as a result of the industrial revolution. The development of digital technology has significantly altered industrial procedures. The first industrial revolution, known as Industry 1.0, was primarily concerned with mechanical engineering and water and steam. Electrical engineering received a lot of attention during the second industrial revolution (Industry 2.0). The use of steam as a key power source led to significant improvements in various industries and in people's daily lives. Electric power systems and mass production assembly lines were established during the second industrial revolution (Industry 2.0). The third industrial revolution (Industry 3.0) was defined as automatic manufacturing and the incorporation of electronics, computers, and information technology into manufacturing. The fourth industrial revolution (Industry 4.0) is automating business operations and advancing manufacturing to a level based on connected devices, smart factories, cyber-physical systems (CPS), and the Internet of Things (IoT), where machines will change how they interact with one another and carry out specific tasks. With all modern technologies, a harmonious balance between human and machine interaction, and an emphasis on sustainable growth, Industry 5.0 is one step ahead of its predecessor (Saurabh Tiwari et al., 2022).

No matter how conventional a person is, they constantly want to be different from the rest of the pack and stand out. Since the beginning of the Industrial Age, there haven't been any technologies that let people express their unique personalities through personalised goods. Not only low-tech items, but any item that can transmit the appropriate signals. Products that even those with moderate earnings can afford, not just those that are only available to the extremely wealthy. The psychological and cultural underpinnings of Industry 5.0, which entails employing technology to bring back human value addition to manufacturing, are this yearning for mass personalization.

In his landmark book "Future Shock from the 1970's", American futurist Alvin Toffler predicted that consumers will need to form groups in order to deal with option overload. Instead of Toffler's "shock," however, we witness customers savouring their choices. One person expresses herself by playing music from the countless possibilities available online, while another spins records on a Shinola turntable that was produced in Detroit. Some prevalent Industry 4.0 presumptions are also called into question by the mass personalization and associated phenomena, particularly the oft-expressed but false allegation that robots are "taking over" and "stealing our jobs."

Figure 1. Illustration of industrial evolution
(Farhan Aslamet al., 2020)

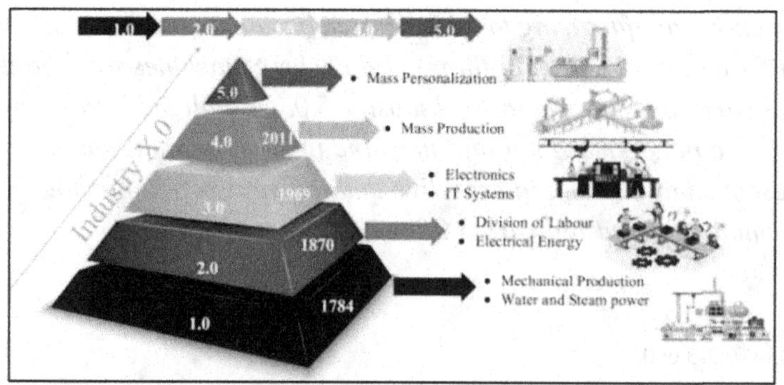

It was discovered at Universal Robots that businesses that use collaborative robots end up hiring more people as a result, not less than they did before they turned robotic. Robots have enhanced these enterprises rather than displacing employees. And like with Industries 1.0, 2.0, and 3.0, we anticipate that this most recent wave of industrial automation will lead to net employment increase rather than loss. It should be made apparent that there are sizable segments of product categories for which no one wants personalised products, and for which Industry 4.0 setups with their conventional industrial robots are ideal. Nobody wants a customised lawn mower blade, engine block or plasterboard anchor. Everyone would gain if these things could be produced in a lights-out plant for a reasonable price. On the other hand, industry 5.0 items enable consumers to realise the fundamental human desire to express themselves – even if they must pay a premium price to do so. These items must be created with what we refer to as the human touch (Esben H. Steergaard, 2018).

Evolution of Industry 5.0

The first three revolutions took about 100 years to develop, but it only took 40 years for the third to give way to the fourth. Through the creation of mechanical manufacturing infrastructures for water- and steam-powered machines, Industry 1.0 developed in the 1800s. As production capacity has risen, the economy has greatly benefited. In 1870, the ideas of electric power and assembly line manufacturing began to take shape. Industry 2.0 boosted the productivity of manufacturing firms by concentrating primarily on mass production and workload allocation. With the concept of electronics, limited automation, and information technologies, industry 3.0 emerged in 1969. In 2011, the idea of smart manufacturing for the future evolved into Industry 4.0. Utilising developing technologies, the major goal is to increase productivity and achieve mass production. Participants from research and technology groups discussed the idea of Industry 5.0 from July 2 to July 9, 2020, under the auspices of the Directorate "Prosperity" of DG Research and Innovation of the European Commission (EC). One can find presumptions about the idea of "Industry 5.0" in the EC publication. This document's main focus is on important directions for change that will transform the sector into one that is more human-centered and sustainable (Breque et al., 2021). Industry 5.0 is an evolution planned for the future that will combine the inventiveness of human professionals with effective, knowledgeable, and precise machines (Omar Ashraf ElFar et al., 2021).

Since intellectual experts use machines, Industry 5.0 encourages more skilled jobs than Industry 4.0. Mass customisation is the core focus of Industry 5.0, and people will be controlling the robots. Robots are already actively involved in mass production in Industry 4.0, as opposed to Industry 5.0, which is primarily intended to increase customer pleasure. While Industry 4.0 is concentrated on CPS connectivity, Industry 5.0 connects to Industry 4.0 applications and creates a connection amongst collaborative robots (cobots). The development of greener solutions in comparison to the current industrial transformations, neither of which places a priority on protecting the environment, is another intriguing advantage of Industry 5.0 (Kadir Alpaslan Demir, 2019).

Industry 5.0 develops models using operating intelligence and predictive analytics with the goal of producing decisions that are more reliable and stable. Since real-time data will be gathered from machines in conjunction with highly qualified specialists, the majority of the production process in Industry 5.0 will be automated (Praveen Kumar Reddy Maddikunta et al., 2022).

Intellectual Structure of Knowledge Towards Industry 5.0

It is significant to remember that Industry 5.0's conceptual foundation may still be changing. Technological enablers, Cyber-Physical Systems, Data Analytics and Artificial Intelligence, Industrial Internet of Things (IIOT), Human-Machine Interaction, Business Models, and Organisational Transformation are some of the broad topics and areas of concentration that emerged from Industry 4.0.

During the last 5 years, it is evident from the above diagram that much of the research work carried out in various sectors, namely Engineering, Agriculture, veterinary, Food Sciences, and Chemical sciences sector followed by Environmental Sciences and Biological sciences, have triggered industry 5.0.

A scrutiny of the research work shows a steady rise in the publications in the area of industry 5.0 year on year which have started to yield a better understanding on the concepts that are going to shape this new upcoming industrial revolution.

The above network diagram gives the origin of the research carried out in various countries around the world. This has resulted in 4 clusters of countries being formed focusing on specific themes and directional links contributing to knowledge in the field of Industry 5.0.

Vietnam tops the list of countries engaged in research in this field, followed by the United States and the United Arab Emirates. It can be seen that countries in the Middle East and South East Asian countries are at the forefront of this research

The above diagram shows 5 distinct clusters where the authors have collaborated leading to 28 links and a total link strength of 118.

The above diagram gives the Authors who have the maximum number of citations during the last 3 years for their published research work in the area of Industry 5.0.

Figure 2. Research on Industry 5.0 in various sectors

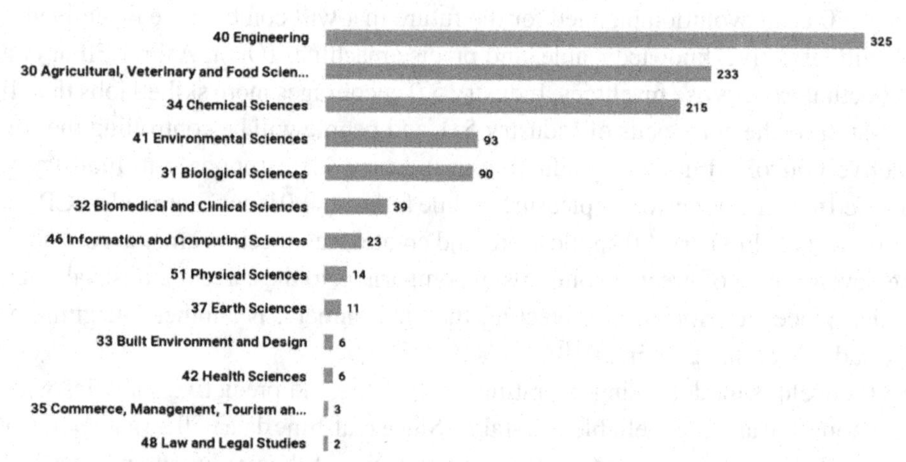

Source: https://app.dimensions.ai
Exported: May 25, 2023
Criteria: 'Industry 5.0 ' in full data; Publication Year is 2023 or 2022 or 2021 or 2020 or 2019; Researcher is David Julian Mcclements or Pau Loke Show or Abdullah Mohammad Ahmad Asiri or José Manuel Lorenzo or Min Zhang or Abolghasem Jouyban or Ponnusamy Senthil Kumar or Zhi Dang or Mohd Javaid or Abid Haleem; Publication Type is Article.

Figure 3. Publications in the field of Industry 5.0 during the past five years

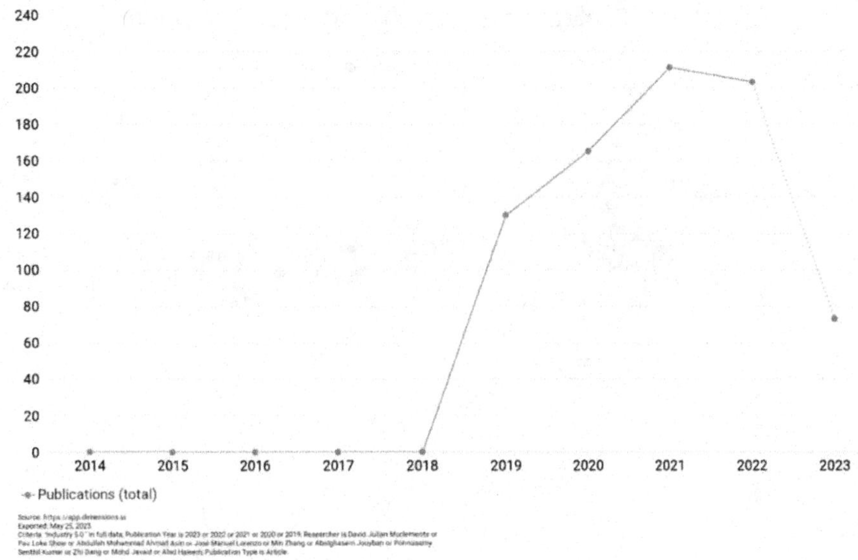

Figure 4. Countries contributing to research on Industry 5.0

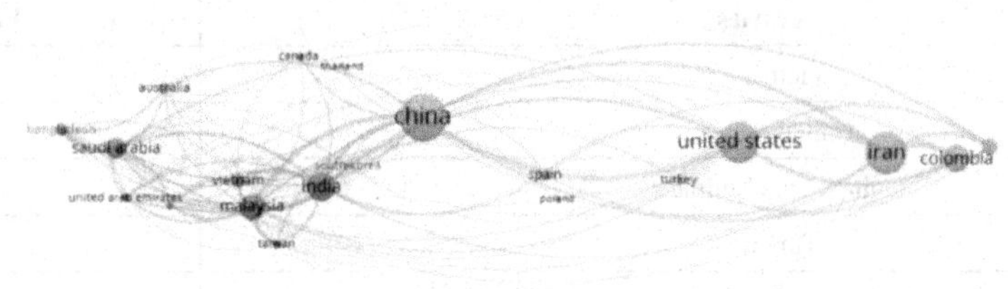

Table 1. Clusters of countries collaborating on specific themes on industry 5.0

Cluster 1	Cluster 2	Cluster 3	Cluster 4
Bangladesh	Colombia	Australia	Poland
India	Cyprus	Canada	Spain
Malaysia	Iran	China	Turkey
Pakistan	United States	Thailand	
Saudi Arabia			
South Korea			
Taiwan			
United Arab Emirates			
Vietnam			

Figure 5. Countries publishing in the area of Industry 5.0

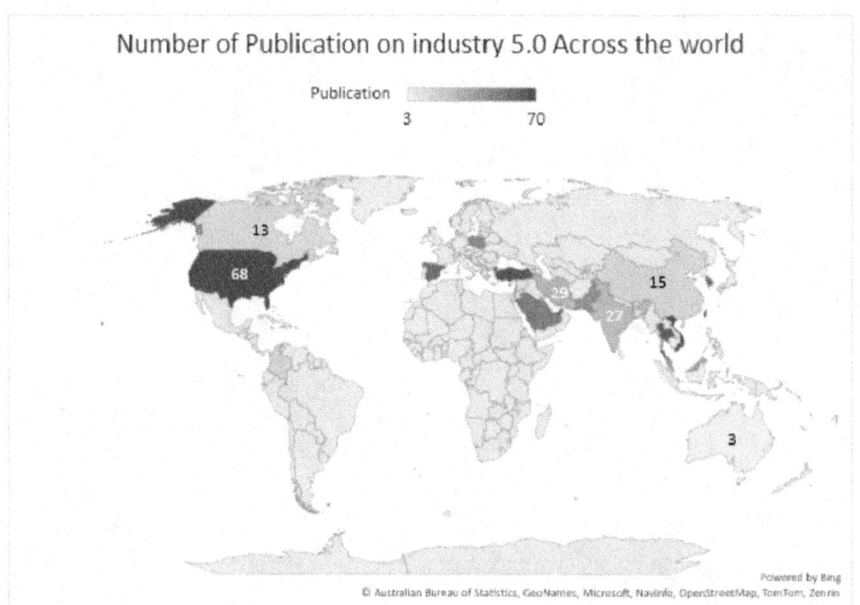

Table 2. Top 20 Countries and their contribution to research in the area of industry 5.0

Countries	Publication
Vietnam	70
United States	68
United Arab Emirates	66
Turkey	64
Thailand	62
Taiwan	61
Spain	57
South Korea	56
Saudi Arabia	51
Poland	46
Pakistan	43
Malaysia	35
Iran	29
India	27
Cyprus	18
Colombia	16
China	15
Canada	13
Bangladesh	6
Australia	3

Figure 6. Author and co-author analysis on Industry 5.0

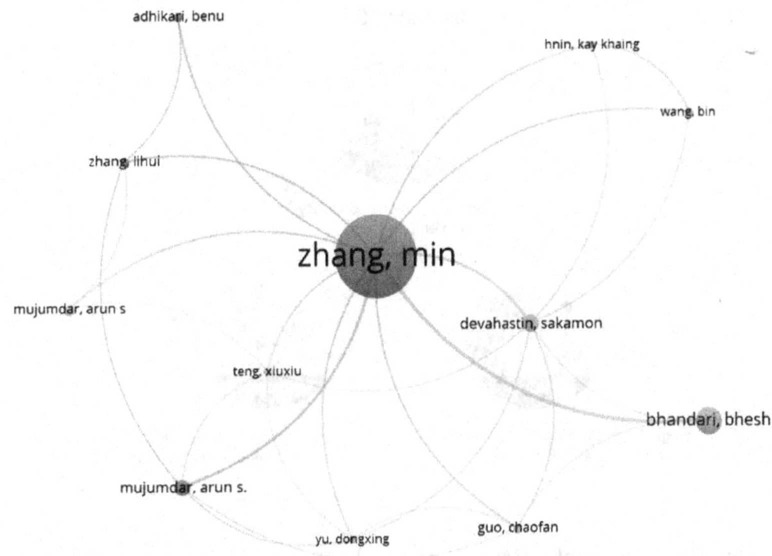

Figure 7. Authors and citation analysis

The Core

By ensuring that production respects the limits of our planet and places the welfare of industry workers at the centre of the production process, Industry 5.0 acknowledges the capacity of industry to achieve societal goals beyond employment and economic growth and to become a resilient provider of wealth. By letting research and innovation drive the shift to a sustainable, human-centric, and resilient European industry, Industry 5.0 complements the current Industry 4.0 model. Industry 5.0 is obviously the outcome of the European Commission's agreement that social and environmental concerns in Europe need to be better integrated into technological innovation and that the emphasis needs to shift from individual

Figure 8. Core values of Industry 5.0
(Breque M et al., 2021)

technologies to a systematic approach (Breque M et al., 2021). Human-centricity, sustainability, and resilience are the three interconnected core values that drive Industry 5.0 (Breque M et al., 2021).

The change from technology-driven progress to a wholly human-centric and society-centric approach is made possible by the human-centric approach, which places fundamental human wants and interests at the centre of the manufacturing process. As a result of the shift in value from viewing employees as "cost" to "investment," industrial personnel will acquire new roles. Technology employed in manufacturing is adaptable to the demands and diversity of industry workers since it is designed to serve people and society (Yuqian Lu et al., 2021). To protect workers' fundamental rights, such as autonomy, human dignity, and privacy, a secure and inclusive work environment must be established that prioritises physical, mental, and overall well-being. For greater career options and a better work-life balance, industrial workers must continuously up-skill and re-skill themselves.

The industry must be sustainable in order to respect the limits of the earth. In order to create a circular economy with improved resource efficiency and effectiveness, it must create circular processes that reuse, repurpose, and recycle natural resources while reducing waste and environmental damage.

Resilience is the idea that industrial production needs to become more robust in order to be better able to withstand disruptions and to maintain key infrastructure in times of emergency. In order to quickly traverse (geo) political upheavals and natural disasters, the industry of the future must be resilient (Breque M et al., 2021).

METHODOLOGY

The goal of this interpretive-qualitative research, which uses bibliometric analysis and narrative review as part of its methodology (Green et al., 2006; Pan, 2008), is to discuss the state of the literature (Gonot-Schoupinsky et al.,2022) on a particular topic and foster further discussion among researchers (Grant and Booth, 2009). To comprehend the scope's breadth, a non-systematic literature review was carried out.

SKILLS REQUIRED FOR INDUSTRY 5.0 REVOLUTION

Preparatory Mindset and Skills for the Student Community Through Academic Bodies

A study was undertaken by (Rachmawati et al.,2021) to ascertain the degree to which Social Science students were academically resilient in the face of the industry 5.0 era. In this study, academic resilience is explained by factors including: 1) having the knowledge and skills to handle challenging circumstances, 2) having the efficacy to handle challenging circumstances, 3) having positive personal qualities, 4) having a positive impact on oneself and others, 5) being able to overcome challenges positively and adaptively, and 6) controlling actions and decisions. The key indication of the academic resilience of Social Science students has been discovered to be the component of commitment demonstrated through helping oneself and others.

Didem Gürdür Broo and colleagues conducted study on the future of engineering education in 2022. After providing a brief overview of the history of engineering education, the researchers' study focused on the current paradigm shifts, which emphasise that in order to meet the difficulties given by the fifth industrial revolution, skills must take precedence over degrees. They also put forth four ideas that could aid higher education institutions in redesigning their curriculum: (1) lifelong learning and transdisciplinary education; (2) sustainability, resilience, and human-centric design modules; (3) practical data fluency and management courses; and (4) human-agent/machine/robot/computer interaction experiences.

According to John Mitchell et al. (2022), there are a variety of ways that might be used to include skills like data science, machine learning, and AI into the current engineering education curriculum. They had suggested eight fusion competencies: Re-humanizing time by spending more time conducting innovative research to solve urgent issues, Responsible normalisation is the process of influencing how people, organisations, and communities view and use human-machine interaction. When a machine is unsure of what to do, it might use judgement to decide on a course of action known as judgement-integration. Knowing the right questions to ask of AI at different levels of abstraction to gain the insights you and others need, bot-based empowerment - effectively utilising AI agents to boost human potential and develop superpowers in commercial operations and professional professions, combining the mind and body holistically - performing tasks alongside AI agents so people can learn new skills and on-the-job training for people so they can work well within AI. Humans creating working mental models of how machines work and learn. Machines capturing user performance data to update their interactions. Improved processes, constant reimagining—the strict discipline of developing new business models and processes from the ground up as opposed to merely automating existing ones. These abilities, as opposed to the list of digital skills they had provided, were just extensions of already-existing technical and interpersonal skills, such data analytics, based on predictions about how people will interact with machines in the future. There have been numerous proposals for curriculum development frameworks. One recommendation is that specific training goals should be developed based on industry expectations, and a convergent curriculum should be developed using the IoT concept's scientific framework.

Revolutionary Change and Industry Readiness

Manufacturing Sector

There will be a greater need for people with a variety of talents, such as problem-solving, critical thinking, creativity, and emotional intelligence as manufacturing becomes more complicated and collaborative. As a result, the following crucial considerations about skilled labour had to be addressed: (i) rising demand for workers with STEM abilities: As technology continues to play an increasingly significant role in manufacturing, there will be a rising need for individuals who possess STEM skills. These professionals will be required to create, develop, and keep up the cutting-edge technologies that power Industry 5.0. Although technological expertise will be crucial in Industry 5.0, there will also be an increasing need for employees with soft skills namely: adaptability, communication and team work. These abilities will be crucial for workers to effectively communicate with machines and other workers as well as to adapt to the shifting demands of the workplace; (iii) A stronger focus on lifelong learning: Employees in Industry 5.0 will need to be ready to constantly learn about and adjust to new processes and technologies as the rate of technological change quickens. To guarantee that employees have the skills and knowledge required to stay up with the shifting needs of the workplace, businesses will need to engage in training and development initiatives; (iv) Growing demand for employees with soft skills: Industry 5.0 will place a strong emphasis on technology skills, but it will also place more emphasis on soft skills like communication, teamwork, and adaptability. These abilities will be crucial for workers to effectively communicate with machines and other workers as well as to adapt to the shifting demands of the workplace; (v) A stronger focus on lifelong learning: Workers in Industry 5.0 will need to be ready to continuously learn about and adapt to new technologies and processes because the rate of technological change is accelerating. To guarantee that employees have the skills and knowledge necessary to keep up with the shifting needs of the workplace, businesses will need to engage in training and development programmes (The Welding Institute (TWI), 2023).

It is projected that the transition to automation will take place under Industry 5.0. It's the recognition that human creativity, invention, and critical thinking are just as valuable in the production process as robotic and automated advancements are (Malardalen Industrial Technology Centre, 2023)

Researchers Sebastian Sanuik et al. (2022) performed study with 25 experts who were chosen by three qualified judges (academics with knowledge of Industries 4. 0 and 5.0) to identify areas of knowledge, important abilities, and competences of engineers and managers required in the Industry 5.0 environment. Engineering Skills, Problem Solving, Openness to Sustainable Development, Openness to Digitalization, Analytical Thinking, Life Long Learning, Using Computer Aided Systems, Openness to Use New Technologies, and Team Work are the skills that engineers will need in the era of Industry 5.0, according to the study's findings. The study also identifies the managerial competencies needed to thrive in the Industry 5.0 era, including self-discipline, lifelong learning, continuous improvement, teamwork, conflict resolution, problem solving, creative thinking, openness to new technologies, resistance to stress, and openness to digitalization.

Financial Services Sector

Research on Industry 5.0 and its effects on the banking industry was conducted by Amir Mehdiabadi et al. in 2022. 49 indicators for Industry 5.0 were found by a thorough literature review, and they were divided into three categories: inconsequential indicators, vital indicators, and extremely necessary indicators. These indicators include lifelong learning, human-centered design modules, practical data management, Pay attention to how customers' requirements are changing, a great consumer experience, good supplier partnerships, Adaptive reaction to a market in transition, the workforce's continuity and stability, employee health and safety, and labour welfare affordable prices, Work on redesigning with the human element optimised, switch to automated systems, Maximising profits, healthcare, efficiency, agility, and innovation increase client satisfaction, partnering with Fintech, using learning and identification algorithms, perpetual access, the growth of dependability, acceptance, and trust, perceived usefulness, perceived security, and loyalty enhancing banks' financial performance, commercial toughness, investment in the bank's IT department, removal of actual bank branches, reduction of the high costs associated with maintaining a branch, Legalisation, Standardisation, executive openness, ethical conduct, environmentally sound practises, establishing connections between organisations, financial expertise, willingness of consumers to use fintech, varying degrees of human-machine interaction, Integrity, convergence of technologies Cybersecurity, customer education, services that are customised, Incorporate human ingenuity and new professions into robotic manufacturing. The most crucial signs of banking entry in the fifth generation of the industrial revolution were 14 of these 49 indicators, including: Pay attention to how customers' requirements are changing, superior client service, a stable and stable workforce, employee health and safety, and labour welfare Healthcare, increase client satisfaction, Reduce high branch expenses by redesigning work with the optimisation of the human element, Ethical conduct, sustainability of the environment, various levels of interaction between humans and machines, the creation of new jobs, creativity, and ongoing human learning in the development of robots. For businesses to manage and balance workloads, it is imperative to optimise and calibrate the human aspect.

Health Care Sector

Maddikunta et al. (2022) claim that Industry 5.0 can result in deeper human-machine collaboration through applications like intelligent healthcare, cloud manufacturing, and supply chain management that incorporate mass customisation. Industry 5.0 is made up of cuttingedge technologies that link wirelessly and can improve automation in manufacturing and healthcare, (Javaid et al, 2020). According to Mithileysh Sathiyanarayanan (2020), the health industry has recently begun focusing on tailoring products to meet the specific needs of patients with various ailments or disorders. We need to think about a humancentric approach that incorporates IoT and AI in healthcare, especially in light of the post COVID age, often known as "The New Normal." The fifth industrial revolution, or "industry 5.0," is seen to be where patients' and healthcare workers' demands for personalised care may be met. The top 5 skill sets that employers in the healthcare industry are looking for were named by Matt Kennedy in 2023;

The Capability of Speaking "Code-ese" Fluently - No matter what field you operate in, utilising the proper codes to represent diagnoses, treatments, and services is crucial to clinical documentation and the billing process. To understand the numerous code sets and laws, much like learning a new language, requires time and practise. Employers are searching for experts who can speak "code-ese" with ease and manoeuvre through the maze of codes.

The Power of Deduction - Examining medical records necessitates problem-solving and critical thinking. The best professionals must be able to sort through mounds of medical data and pull out the pertinent information. They must also be able to recognise patterns and detect mistakes or discrepancies in the data. People with good analytical and problem-solving abilities are better able to see mistakes and address any problems that may occur. Employers are looking for healthcare business experts that have the analytical skills of Sherlock Holmes and can quickly and precisely deduce the relevant codes.

The capacity to identify errors with ease - The healthcare sector completely depends on attention to detail. One minor error can determine whether a claim is accepted or rejected. This ability is necessary to guarantee accurate coding and thorough, compliance documentation. People who can pay close attention to detail are less likely to make mistakes and operate more effectively. Employers need candidates with keen eyesight who can identify typos and errors before they cause issues.

The ability to communicate successfully with doctors, nurses, and other medical experts is essential for healthcare industry professionals. They must be able to concisely and clearly convey complicated concepts. This involves having the ability to train and educate clinical professionals about standards and laws. People with good communication skills will be better able to work with other professionals and will perform their jobs more effectively. Employers are seeking for someone with the ability to connect with anyone and who have a gift for gab.

The capacity for technological adaptation Electronic medical records have replaced handwritten ones in the healthcare sector, but new technologies are constantly being developed. As a result, experts in the field must be willing to learn about and work with new hardware and software as it is released. Employers are searching for tech-savvy employees that are flexible with new tools and fashions. You will be better able to navigate the healthcare system and create workflow efficiencies that will be noticed if you are informed on the most recent technologies in the healthcare industry.

Information Technology Sector

Technology in the information age is dynamic and changes quite quickly. Professionals must constantly up-skill if they want to stay relevant in the fiercely competitive IT field. As the use of mobile devices and the internet increases, businesses are embracing IT to develop digital offerings for their customers. Understanding the client and utilising information technology to help them achieve their goals are crucial for producing these things. Before acquiring soft skills, a person must first hone their cognitive, analytical, and critical thinking skills. Therefore, a combination of these talents will be necessary for the IT industry's future (The Times of India, 2022).

Food Supply System

Growing, processing, distributing, eating, and disposing of food are some of the common players and activities that make up the food system (Akbari et al., 2022; Gaitán-Cremaschi et al., 2018). The agriculture-food industry is seriously at risk from natural disasters like wildfires, illness, floods, and droughts. The effects of such occurrences result in impending losses to forests, cattle, and agriculture, lowering crop yields, and interfering with the production of food. For instance, the production of food is significantly impacted in some nations by the lack of access to water. Floods, on the other hand, severely damage the physical infrastructure, including farms and crops, which lowers agricultural outputs and has a negative impact on the food supply chain.

In addition, there are further possible threats to food security, including diseases, ticks, and locust outbreaks (Xu et al., 2021). Additionally, the food system, and particularly the production process, primarily relies on pesticides, artificial fertilisers, and antibiotics, all of which harm the ecosystem and the land and nearby water bodies, causing significant biodiversity losses. In addition to posing new issues for guaranteeing optimal food quality, safety, and authenticity, the current situation may also pose a threat to food security. The ability of the food system to react to and adapt to disruptions while continuing to function is the key to finding a solution (Schipanski et al., 2016). Three aspects of food system resilience were taken into account by Seekell et al. (2017): (1) the availability of food, (2) the biophysical potential to enhance food production, and (3) the variety of domestic food production.

They also highlighted the difficulties in using quantitative approaches to monitor changes in the resilience of the food system. Therefore, it was suggested that an index-based examination of a country's ability to manage shocks to the food system be developed. Prices, income, and the capacity to react to price spikes, crop failures, or asset losses are all related to access to food. Additionally, it is linked to infrastructure spending and educational attainment levels (Seekell et al., 2017). To meet consumer needs, assure food security, and ensure effective resource management and decision-making assistance, Industry 5.0's new technologies can assist in digitalizing traditional food systems.

Food systems using sensors can detect food degradation signs more effectively and, as a result, avoid and reduce food losses. One of the data-gathering technologies that enables the capture of images for future image analysis by various algorithms is the use of imaging sensors on satellites, drones, or physical objects on land (Valous et al., 2012). The majority of manual methods for the direct measurement of phenotypic data (such as biomass, leaf area index, chlorophyll content, and plant tissue nitrogen content) have been replaced by remote and non-destructive methods as a result of their application in precision agriculture (Adhikari et al., 2020).

Sustainability Focus

Projects under Industry 5.0 are heavily focused on the environment. Sustainability is crucial to the success of any project, not merely as a priority. Teams in Industry 5.0 assess the effects of their actions on the environment and collaborate to reduce environmental risks. They also put an emphasis on resource efficiency and design, create project plans that reduce jarring changes in resource requirements, incorporate sustainable production techniques, and account for embodied carbon in design parameters. Teams working on Industry 5.0 projects see sustainability as a duty, and they sincerely care about producing results that have a good influence on the built and social surroundings (Trillium Advisory Group, 2022). More interdisciplinary research and systematic approaches that cover different facets of sustainability, such as environmental, economic, social, and technical aspects, are needed in order to build more sustainable societies (Bibri & Krogstie, 2017; Pappas et al., 2018). The ability to manage complexity and uncertainty is not accomplished through their negation because the modern economy is characterized by knowledge-intensive activities that contribute to an increased pace of technological and scientific advance and quick obsolescence. By viewing complexity as an opportunity rather than a danger, all the actors in an innovation ecosystem should share ideas and information to create innovation and knowledge in smart environments. This kind of "openness" is in line with Industry 5.0's new logic, which holds that systemic and sustainable development should be assured by human-centered solutions (Carayannis et al., 2021a, b, c).

Industry 5.0 Technologies and Applications

Artificial intelligence (AI) is algorithm-based intelligence that is supplied to computers in order to teach them how to solve problems, make decisions, and carry out tasks that humans would carry out

(Abirami et al., 2022). The use of several AI-driven algorithms, including k-NN, Naive Bayes, ANN, and Deep Learning in fault diagnostics of rotating machinery has been examined by Liu et al. (2018). This has helped to decrease machine downtime, maintenance costs, and safety risks. Since each method has advantages and disadvantages in terms of accuracy, speed, and robustness, they suggest creating a hybrid intelligent system to take on upcoming problems. Machine learning enables the interconnected systems to exchange data and enhance the process independently based on algorithms.

The loop of recurrent learning and optimisation results in performance never before achieved, transforming a conventionally automated production into a smart factory. Some of the applications of ML in the industrial sector include condition monitoring of machinery, health monitoring of structures, predictive maintenance, predictive quality control, and supply chain management (Muthuswamy, P. 2022). Big Data Analytics is a sophisticated analytical technique used to systematically uncover hidden patterns in data and connect them to specific actions that support decision-making. To further describe the qualities of Big Data, researchers have added additional "V"s to the original 3Vs. In contrast to Gandomi and Haider (2015), who focused on the 6Vs (4V+Veracity and Value), Liao et al. (2015) defined it as the 4Vs (3V+Variability).

A significant quantity of data is likely generated every second as a result of the development of artificial intelligence, sensor-based connected systems, social networking, and digital communication tools, necessitating real-time processing in order to forecast outcomes and make quicker decisions. It is affordable and incredibly flexible since services like server, storage, database, networking, analytics, and intelligence are delivered through the internet (the "cloud"). Some advantages of using cloud computing include centralised data and software management, scalability, and reliability. Cloud manufacturing refers to the management of a product's manufacturing lifecycle using capabilities for design, manufacturing, and fulfilment. Cloud manufacturing can be used to implement the early 2000s concept of "design anywhere and manufacture anywhere" (Heinrichs, W., 2005). It offers the advantages of high efficiency, cost-effectiveness, and flexibility in managing processes to producers. IoT, RFID, sensors, GPS, cyber-physical systems, and cloud computing are examples of contemporary digital and computational technologies that work together to allow cloud manufacturing services.

Digital twins make it feasible to create digital representations of real-world objects like wind farms, aircraft engines, buildings, factories, and even larger systems like smart cities (Lu, 2020). Although the concept of a digital twin was first proposed in 2002, it has lately come to light due to the development of IoT. IoT made Digital Twin more affordable and accessible for a range of organizations by lowering its cost. Industry 5.0 is aided by the Digital Twins in overcoming technological barriers by more quickly detecting them, locating components that may be upgraded or modified based on performance, making forecasts more precisely, foreseeing faults in the future, and preventing significant financial losses.

According to Bhattacharya (2020), the Internet of Everything (IoE) connects people, processes, information, and things. It considerably aids in the development of fresh opportunities for "Industry 5.0" application development (Higginbotham, 2020). IoE has made it possible for Industry 5.0 to introduce new features, an improved user experience, and projected benefits for enterprises. Consumer happiness and loyalty have increased thanks to the IoE's participation in Industry 5.0. Industry 5.0 has the opportunity to minimize operational expenses since it lowers latency and contributes to the reduction of communication line congestion.

Deep learning techniques, which have shown remarkable success in a variety of domains such as image recognition, natural language processing, and speech recognition, offer promising alternatives for improving cybersecurity in Industry 5.0 (Yin, Zhu, & Hu, 2021). Convolutional neural networks

(CNN), Recurrent Neural networks (RNN), and transformer models are among the techniques that can automatically learn complex patterns and representations from raw data (Ullah et al., 2022). This capability enables deep-learning models to detect novel and sophisticated attacks that may elude traditional machine-learning methods (Vaswani et al., 2017). Deep learning methods can also be modified to address issues with cybersecurity datasets such as imbalance, noise, and non-stationarity (García, Luengo, & Herrera, 2015). To build more reliable and adaptive attack detection systems, they can also be integrated with other artificial intelligence approaches as adversarial learning and reinforcement learning (Popoola et al., 2020). By utilizing these cutting-edge capabilities, deep learning approaches have the potential to greatly enhance the detection and prevention of web-based threats in Industry 5.0, thereby contributing to the safety, security, and sustainability of the quickly changing digital world (Nahavandi, 2019).

Prospective Skills of Industry 5.0

The goal of Industry 5.0, which strives to combine the advantages of cutting-edge digital technologies with the creativity and problem-solving skills of human workers, is to complement Industry 4.0 by adopting a more human-centric approach. This entails utilizing technology (AI, IoT, Robotics) to improve employees' capacity for innovation, collaboration, and the development of novel solutions to challenging problems, forging relationships between humans and AI, and altering the automation of work and services (Faraj et al., 2018; Vassilakopoulou et al., 2023). The majority of the literature in this area focuses on human-robot collaboration and how it might enhance productivity, ergonomics, and safety. These programmes have a beneficial effect on employees' well-being (Nourmohammadi et al., 2022).

Robots can carry out laborious, repetitive, or hazardous tasks, leaving people free to customize products and think creatively and critically outside the box. New technology adoption involves both time and money. Giving employees the appropriate technical and soft skills is the biggest problem (Chin, 2021). Industry 5.0 requires new talents, competencies, and skills from employees (Villa-Gutiérrez et al., 2021). The following skill sets are suggested for industry practitioners and academicians to meet the new computerized, Digitised, and Human Centric Revolution after carefully analyzing the Literature on Industry 5.0.

The identification of skills necessary to compete and survive in a dynamic industry 5.0 ecosystem has been made possible by an analysis of industry reports, expert opinions, and literature on industry

Figure 9. Proposed model

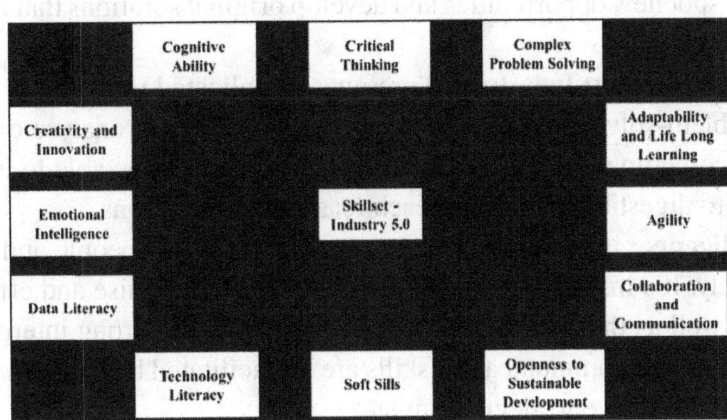

evolution brought about by technical advancements and shifting economic, social, and environmental conditions. Indeed, in the context of Industry 5.0, the skills mentioned—cognitive ability, critical thinking, complex problem solving, adaptability and lifelong learning, agility, collaboration and communication, openness to sustainable development, soft skills, technology literacy, data literacy, emotional intelligence, creativity, and innovation—are recommended.

Globalization, sustainability, and digital technologies coming together to form Industry 5.0 produce a dynamic and multidimensional environment. Professionals with these abilities can prosper in this setting, fostering creativity, overcoming difficult problems, and promoting sustainable and responsible development in their respective sectors. Additionally, these abilities frequently complement one another and do not exist in isolation, increasing a person's overall efficiency at work.

DISCUSSIONS

The next stage of industrial development, known as "Industry 5.0," emphasizes the combination of cutting-edge technology and human labour to build more flexible and cooperative production processes. While the precise abilities needed for Industry 5.0 may change over time, the following are some that are probably useful in this setting: The idea of "Industry 5.0" describes how cutting-edge technology like artificial intelligence (AI), automation, robotics, the Internet of Things (IoT), and big data analytics would be combined with human creativity and talent in the manufacturing industry. The emphasis of the current stage of the Industrial Revolution is on human-machine interaction as a catalyst for creativity, productivity, and sustainability. People require a variety of talents that complement the capabilities of intelligent machines in order to succeed in this changing industrial landscape. Here are some essential competencies needed for Industry 5.0:

1. **Technological literacy:** To succeed in Industry 5.0, one must have a solid foundation in digital literacy and a thorough awareness of new technologies. People should feel at ease using advanced robotics, AI algorithms, data analytics tools, and automation systems. Workers that possess technological literacy are better equipped to work together with machines, solve technical problems, and take advantage of technology to increase production.
2. **Innovation and creativity:** While machines are excellent at doing monotonous jobs and analysing data, human inventiveness is still unmatched. People in Industry 5.0 need to be creative, come up with fresh ideas, and work through challenging issues. Innovation is fueled by creativity, which also helps people spot new opportunities and develop original solutions that improve the industrial process.
3. **Complex Problem Solving:** Industry 5.0 presents complicated problems that need for analytical thinking and problem-solving abilities. Workers must be able to analyse complex systems, spot bottlenecks, and come up with effective solutions. They should be able to use logic, deconstruct difficult issues into digestible parts, and reach reasoned conclusions.
4. **Emotional Intelligence:** is becoming more and more crucial as people and technology work together more closely. It's important for people to be able to recognise and effectively control their own emotions as well as those of their co-workers and clients. Strong interpersonal connections, teamwork, and improved communication skills are all facilitated by emotional intelligence, which is essential in a collaborative work environment.

5. **Adaptability and Lifelong Learning**: Rapid technological breakthroughs, shifting market dynamics, and changing employment responsibilities are the hallmarks of Industry 5.0. Individuals must embrace adaptation and be receptive to ongoing learning if they want to succeed in this setting. They should be adaptable in responding to new procedures and workflows, be willing to up-skill or reskill as necessary, and keep up with the most recent technological advances.
6. **Data literacy:** People need to feel at ease using data due to the growing emphasis on data-driven decision-making. The ability to gather, understand, analyse, and extrapolate knowledge is referred to as data literacy. A person's ability to use big data analytics tools, comprehend patterns and trends, and make wise judgements that promote operational efficiency and commercial growth depends on their proficiency with data.
7. **Collaboration and Communication:** In Industry 5.0, where people and machines collaborate in networked systems, effective cooperation and communication skills are essential. People should be able to collaborate effectively in cross-disciplinary teams, share knowledge, and discuss ideas. Strong teamwork and communication abilities boost output, encourage creativity, and provide a positive work atmosphere.
8. **Ethics and Values:** Industry 5.0 brings up moral issues with automation, AI, and data privacy. People must be aware of the ethical ramifications of their actions, comprehend how they may affect society, and act in a way that upholds principles like justice, accountability, and transparency. Maintaining moral standards guarantees responsible and long-lasting application of Industry 5.0 technology.

The understanding of the role technology plays in closing the development gap is one of the main factors influencing the adoption of Industry 5.0 in developing countries. These countries are aware that adopting cutting-edge technologies can help them get past historical barriers and proceed to the next stage of development. They can boost production, increase efficiency, and open up new possibilities for innovation and economic diversification by using the power of Industry 5.0.

Numerous research gaps and difficulties regarding the competencies required for Industry 5.0 are present as this developing subject continues to develop. Human-technology interaction, skill needs and training, workforce transition and retraining, ethical and social issues, collaboration and multidisciplinary research, and adoption and implementation difficulties are a few of the study areas where there are still gaps. If these knowledge gaps are filled, it will be easier for policymakers, educators, and business leaders to make informed decisions about how to best prepare the workforce for the future of work.

CONCLUSION

Industry 5.0 is a paradigm shift that combines cutting-edge technologies with human-centric strategies to spur productivity and creativity in the industrial sector. The abilities needed for Industry 5.0 undergo major adjustments to fulfil the demands of this new paradigm as the industrial landscape changes. The literature studies emphasise a number of critical abilities that are necessary for people to succeed in Industry 5.0. First and foremost, to fully realise the potential of Industry 5.0, one must have technical competence in cutting-edge technologies like artificial intelligence, robotics, the Internet of Things (IoT), and blockchain. With the use of these abilities, people may create cutting-edge procedures and systems that maximise automation, connectedness, and efficiency.

Additionally, Industry 5.0 emphasises the value of both technical and human talents. For people to successfully manage the complicated obstacles provided by Industry 5.0, they need to possess skills like creativity, critical thinking, and complex problem-solving. The ability to adapt, adopt new technologies, and work well with intelligent robots becomes increasingly important as automation spreads. Furthermore, in Industry 5.0, where human-machine collaboration is crucial, excellent interpersonal and communication skills are crucial. The effective adoption and use of Industry 5.0 technologies will be facilitated by the capacity to effectively express complex ideas, work in diverse teams, and develop strong relationships. To stay up with the rapid changes in technology, people also need to cultivate an attitude of lifelong learning and adaptation.

Because of the ongoing change that defines Industry 5.0, professionals must embrace continuous learning if they want to remain relevant and take advantage of new opportunities. The development of comprehensive training programmes and initiatives should be a joint effort between educational institutions, governments, and enterprises to satisfy the changing skill requirements of Industry 5.0. To prepare people with the abilities required for Industry 5.0, these programmes should include technical training, multidisciplinary education, and the development of human skills. In summary, Industry 5.0 offers enormous potential for societal and economic improvement. Adopting the necessary skill set, which includes technical knowledge, people skills, adaptability, and a philosophy of lifelong learning, would enable people to thrive in the Industry 5.0's dynamic environment and contribute to its ongoing success.

Funding: This study was not funded by any organization.

Ethical Approval: This article does not contain any studies with human participants performed by any of the authors

***Author's contribution**: The Corresponding author worked on preparing the paper and the first co-author worked on refining the paper and the second co-author worked on review of literature.

***Conflict of interest**: No conflict of interest

***Data availability statement**:Data sharing not applicable to this article as no datasets were generated or analysed during the current study.

REFERENCES

Abirami, R., & Padmakumar, M. (2022). Pandemic War Natural Calamities and Sustainability: Industry 4.0 Technologies to Overcome Traditional and Contemporary Supply Chain Challenges. *Logistics 6*(4) 81-10.3390/logistics6040081

Adhikari, R., Li, C., Kalbaugh, K., & Nemali, K. (2020). A low-cost smartphone controlled sensor based on image analysis for estimating whole-plant tissue nitrogen (N) content in floriculture crops. *Computers and Electronics in Agriculture, 169*, 105173. doi:10.1016/j.compag.2019.105173

Akbari, M., Foroudi, P., Shahmoradi, M., Padash, H., Parizi, Z. S., Khosravani, A., Ataei, P., & Cuomo, M. T. (2022). The Evolution of Food Security: Where Are We Now, Where Should We Go Next? *Sustainability (Basel), 14*(6), 3634. doi:10.3390u14063634

Aslam, F., Aimin, W., Li, M., & Ur Rehman, K. (2020). Innovation in the era of IoT and industry 5.0: Absolute innovation management (AIM) framework. *Information (Basel), 11*(2), 124. doi:10.3390/info11020124

Ávila-Gutiérrez, M. J., Aguayo-González, F., & Lama-Ruiz, J. R. (2021). Framework for the development of affective and smart manufacturing systems using sensorised surrogate models. *Sensors (Basel)*, *21*(7), 2274. Advance online publication. doi:10.339021072274 PMID:33805015

Bhattacharya, S., Maddikunta, R., Somayaji, S. R., Lakshmanna, K., Kaluri, R., Hussein, A., & Gadekallu, T. R. (2020). Load balancing of energy cloud using wing driven and firefly algorithms in internet of everything. *Journal of Parallel and Distributed Computing*, 16–26.

Bibri, S. E., & Krogstie, J. (2017). Smart sustainable cities of the future: An extensive interdisciplinary literature review. *Sustainable Cities and Society*, *31*, 183–212. doi:10.1016/j.scs.2017.02.016

Breque, M., De Nul, L., & Petridis, A. (2021). *Industry 5.0: towards a sustainable, human-centric and resilient European industry*. European Commission, Directorate-General for Research and Innovation.

Broo, D. G., Kaynak, O., & Sait, S. M. (2022). Rethinking engineering education at the age of industry 5.0. *Journal of Industrial Information Integration*, *25*, 100311. doi:10.1016/j.jii.2021.100311

Carayannis, E. G. (2021a). THE NEW EARTH INITIATIVE (NEI) VIA INDUSTRY 5.0 PLUS. *Reducing the average earth temperature not just the increase rate of it.* Springer. https://www.springer. com/journal/13132/updates/19860990.

Carayannis, E. G. (2021b). *Sustainable Development Goal 8 - an Interview with Elias Carayannis.* Springer. https://www.springer.com/journal/13132/updates/19771722.

Carayannis, E. G. (2021c). Introducing the notions of Industry 5.0 & Society 5.0. *Discover the way a Greek Scientifc Paradigm may develop Innovation Ecosystem 5.0.* The Greek Scientists Society. https://www.youtube.com/watch?v=nul_yXWuWqI.

Chin, S. T. S. (2021). Influence of emotional intelligence on the workforce for industry 5.0. IBIMA Business Review. *Article*, *882278*, 1–7. doi:10.5171/2021.882278

Demir, K. A., Döven, G., & Sezen, B. (2019). Industry 5.0 and human-robot co-working. *Procedia Computer Science*, *158*, 688–695. doi:10.1016/j.procs.2019.09.104

Du, B., Chai, Y., Huangfu, W., Zhou, R., & Ning, H. (2021). Undergraduate University Education in Internet of Things Engineering in China: A Survey. *Education Sciences*, *11*(5), 202. doi:10.3390/educsci11050202

ElFar, O. A., Chang, C. K., Leong, H. Y., Peter, A. P., Chew, K. W., & Show, P. L. (2021). Prospects of Industry 5.0 in algae: Customization of production and new advance technology for clean bioenergy generation. *Energy Conversion and Management: X*, *10*, 100048.

Faraj, S., Pachidi, S., & Sayegh, K. (2018). Working and organizing in the age of the learning algorithm. *Information and Organization*, *28*(1), 62–70. doi:10.1016/j.infoandorg.2018.02.005

Gaitán-Cremaschi, D., Klerkx, L., Duncan, J., Trienekens, J. H., Huenchuleo, C., Dogliotti, S., Contesse, M. E., & Rossing, W. A. H. (2018). Characterizing diversity of food systems in view of sustainability transitions. A review. *Agronomy for Sustainable Development*, *39*(1), 1–22. doi:10.100713593-018-0550-2 PMID:30881486

Gandomi, A., & Haider, M. (2015). Beyond the hype: Big data concepts, methods, and analytics. *International Journal of Information Management, 35*(2), 137–144. doi:10.1016/j.ijinfomgt.2014.10.007

García, S., Luengo, J., & Herrera, F. (2015). *Data Preprocessing in Data Mining*. Springer. doi:10.1007/978-3-319-10247-4

Gonot-Schoupinsky, F. N., Garip, G., & Sheffield, D. (2022). Facilitating the planning and evaluation of narrative intervention reviews: Systematic Transparency Assessment in Intervention Reviews (STAIR). *Evaluation and Program Planning, 91*, 102043. doi:10.1016/j.evalprogplan.2021.102043 PMID:34839113

Grant, M. J., & Booth, A. (2009). A typology of reviews: An analysis of 14 review types and associated methodologies. *Health Information and Libraries Journal, 26*(2), 91–108. doi:10.1111/j.1471-1842.2009.00848.x PMID:19490148

Green, B. N., Johnson, C. D., & Adams, A. (2006). Writing narrative literature reviews for peer-reviewed journals: Secrets of the trade. *Journal of Chiropractic Medicine, 5*(3), 101–117. doi:10.1016/S0899-3467(07)60142-6 PMID:19674681

Heinrichs, W. (2005). Do it anywhere. Electronics Systems and Software, 3(4). doi:10.1049/ess:20050405

Higginbotham, S. (2020). What 5G hype gets wrong - [Internet of everything]. *IEEE Spectrum, 57*, 22–24.

Javaid, M., & Haleem, A. (2020). Critical components of industry 5.0 towards a successful adoption in the field of manufacturing. *Journal of Industrial Integration and Management, 5*(3), 327–348. doi:10.1142/S2424862220500141

Karaca-Atik, A., Meeuwisse, M., Gorgievski, M., & Smeets, G. (2023). Uncovering important 21st-Century skills for sustainable career development of social sciences graduates: A systematic review. *Educational Research Review, 39*, 100528. doi:10.1016/j.edurev.2023.100528

Liao, Z., Yin, Q., Huang, Y., & Sheng, L. (2014). Management and application of mobile big data International. *J. Embedded Syst., 7*(1), 63–70. doi:10.1504/IJES.2015.066143

Liu, R. (2018). B., Yang, E., Zio, X., Chen: Artificial intelligence forfault diagnosis of rotating machinery: A review. *Mechanical Systems and Signal Processing, 108*, 33–47. doi:10.1016/j.ymssp.2018.02.016

Lu, Y., Adrados, J. S., Chand, S. S., & Wang, L. (2021). Humans are not machines—Anthropocentric human–machine symbiosis for ultra-flexible smart manufacturing. *Engineering (Beijing), 7*(6), 734–737. doi:10.1016/j.eng.2020.09.018

Lu, Y., Liu, C., Kevin, L., Wang, K., Huang, H., & Xu, X. (2020). Digital twin - driven smart manufacturing: Connotations, reference model, applications, and research issues. *Robotics and Computer-integrated Manufacturing, 61*, 101837. doi:10.1016/j.rcim.2019.101837

Maddikunta, P. K. R., Pham, Q. V., B, P., Deepa, N., Dev, K., Gadekallu, T. R., Ruby, R., & Liyanage, M. (2021, July). Industry 5.0: A survey on enabling technologies and potential applications. *Journal of Industrial Information Integration, 26*, 100257. doi:10.1016/j.jii.2021.100257

Maddikunta, P. K. R., Pham, Q. V., Prabadevi, B., Deepa, N., Dev, K., Gadekallu, T. R., & Liyanage, M. (2022). Industry 5.0: A survey on enabling technologies and potential applications. *Journal of Industrial Information Integration, 26*, 100257. doi:10.1016/j.jii.2021.100257

Mehdiabadi, A., Shahabi, V., Shamsinejad, S., Amiri, M., Spulbar, C., & Birau, R. (2022). Investigating Industry 5.0 and Its Impact on the Banking Industry: Requirements, Approaches and Communications. *Applied Sciences (Basel, Switzerland), 12*(10), 5126. doi:10.3390/app12105126

Mitchell, J., & Guile, D. (2022). Fusion skills and industry 5.0: conceptions and challenges. *Insights Into Global Engineering Education After the Birth of Industry 5.0*, 53.

Muthuswamy, P. (2022). Artificial intelligence based tool condition monitoring for digital twins and industry 4.0 applications. *International Journal on Interactive Design and Manufacturing (IJIDeM)*, 1-21.

Nahavandi, S. (2019). Industry 5.0—A human-centric solution. *Sustainability (Basel), 11*(16), 4371. doi:10.3390u11164371

Nourmohammadi, A., Fathi, M., & Ng, A. H. C. (2022). Balancing and scheduling assembly lines with human-robot collaboration tasks. *Computers & Operations Research, 140*, 105674. doi:10.1016/j.cor.2021.105674

Østergaard, E. H. (2018). Welcome to industry 5.0. *Retrieved Febr, 5*, 2020.

Pan, M. L. (2016). *Preparing literature reviews: Qualitative and quantitative approaches*. Taylor & Francis. doi:10.4324/9781315265872

Pappas, I. O., Mikalef, P., Giannakos, M. N., Krogstie, J., & Lekakos, G. (2018). Big data and business analytics ecosystems: Paving the way towards digital transformation and sustainable societies. *Information Systems and e-Business Management, 16*(3), 479–491. doi:10.100710257-018-0377-z

Popoola, S. I., Adebisi, B., Hammoudeh, M., Gui, G., & Gacanin, H. (2020). Hybrid deep learning for botnet attack detection in the internet-of-things networks. *IEEE Internet of Things Journal, 8*(6), 4944–4956. doi:10.1109/JIOT.2020.3034156

Rachmawati, I., Multisari, W., Triyono, T., Simon, I. M., & da Costa, A. (2021). Prevalence of Academic Resilience of Social Science Students in Facing the Industry 5.0 Era. *International Journal of Evaluation and Research in Education, 10*(2), 676–683. doi:10.11591/ijere.v10i2.21175

Saniuk, S., & Grabowska, S. (2022). *Development of knowledge and skills of engineers and managers in the era of Industry 5.0 in the light of expert research*. Zeszyty Naukowe. Organizacja i Zarządzanie/ Politechnika Śląska. doi:10.29119/1641-3466.2022.158.35

Tiwari, S., Bahuguna, P. C., & Walker, J. (2022). Industry 5.0: A macroperspective approach. In Handbook of Research on Innovative Management Using AI in Industry 5.0 (pp. 59-73). IGI Global.

Ullah, F., Salam, A., Abrar, M., Ahmad, M., Ullah, F., Khan, A., Alharbi, A., & Alosaimi, W. (2022). Machine health surveillance system by using deep learning sparse autoencoder. *Soft Computing, 26*(16), 7737–7750. doi:10.100700500-022-06755-z

Valous, N., & Sun, D.-W. (2012). Image processing techniques for computer vision in the food and beverage industries. In D.-W. Sun (Ed.), *Computer Vision Technology in the Food and Beverage Industries* (pp. 97–129). Woodhad Publishing. doi:10.1533/9780857095770.1.97

Vassilakopoulou, P., Haug, A., Salvesen, L. M., & Pappas, I. O. (2023). Developing Human/AI interactions for chat-based-customer-services: Lessons learned from the norwegian government. *European Journal of Information Systems*, *32*(1), 10–22. doi:10.1080/0960085X.2022.2096490

Vaswani, A., Shazeer, N., Parmar, N., Uszkoreit, J., Jones, L., Gomez, A. N., Kaiser, Ł., & Polosukhin, I. (2017). *Attention is all you need.* In *Proceedings of the 31st Conference on Neural Information Processing Systems (NIPS 2017)*, Long Beach, CA, USA.

Xu, Z., Elomri, A., El Omri, A., Kerbache, L., & Liu, H. (2021). The Compounded Effects of COVID-19 Pandemic and Desert Locust Outbreak on Food Security and Food Supply Chain. *Sustainability (Basel)*, *13*(3), 1063. doi:10.3390u13031063

Yin, X., Zhu, Y., & Hu, J. (2021). A comprehensive survey of privacy-preserving federated learning: A taxonomy, review, and future directions. [CSUR]. *ACM Computing Surveys*, *54*(6), 1–36. doi:10.1145/3460427

Chapter 15
Ways of Using Computational Thinking to Improve Students' Ability to Think Critically

Indrajeet Kumar
https://orcid.org/0000-0003-2814-2900
Graphic Era Hill University (deemed), Dehradun, India

Noor Mohd
Graphic Era Hill University (deemed), Dehradun, India

ABSTRACT

Computational thinking (CT) is a problem-solving method that depicts on thoughts and procedures from computer science to implements complex problems in an organised and effective manner. It encompasses collapsing down problems into subproblems, convenient components, recognizing patterns and perceptions, and originating algorithms to resolve them. Applying CT to learning can definitely help enhance students' ability to think significantly. The present study highlights the importance of computational thinking and its implication on student's ability to think. The segment of computational thinking is a kind of problem-solving skill that adopts the process of a computer's systematic manner. In response to this, computers are involved with deriving conclusions and solutions that are equipped with decomposing an issue, using analytical information and others. On the same hand, the application of computational thinking is deemed of high quality in inducing higher thinking capacities among students and strengthening their cognitive process which in turn brings forward advancing solutions.

INTRODUCTION

The prospect of thinking ability is observed from a critical viewpoint, given its usability in solving definite and indefinite problems. Based on the culture of thinking ability, students are often faced with the dilemma of appropriately gaining solutions to their issues that has an analytical pattern (Agbo et. Al., 2019). The correlation between computational thinking (CT) and a student's thinking ability is

DOI: 10.4018/979-8-3693-0782-3.ch015

ascertained to be linked with cognitive processes [6]. In terms of this, the application of computational thinking is viable through extensive digging into the cognitive systems which are responsible to help humans utilise natural problem-solving qualities. Evidence from earlier studies has strongly focused on the usage of the outermost circle which is the cognitive process which becomes the core fundamental for developing adequate computational thinking skills (Abdullah et al., 2019).

Following the above-stated notion, the term computational thinking typically defines the activity of witnessing a problem and approaching it through a systematic pattern. This systematic approach to solving issues is noted to be detrimental to inducing higher thinking skills among the masses, especially students (Alam, A., 2022). With respect to problem-solving qualities, the majority of corporate and business sector runs upon informed decision-making, which is a residue of analytical problem-solving, thereby, it is of utmost vitality to train students in recognising problems and issues through a specific manner and thus upscale their existing skill sets (Avila et al., 2021) and broaden their knowledge horizon.

CT is a problem-solving technique that depicts on thoughts and procedures from computer science to implements complex problems in an organised and effective manner. It encompasses collapsing down problems into subproblems, convenient components, recognizing patterns and perceptions, and originating algorithms to resolve them. Employing CT to learning can help to enhance students' ability to think significantly. Here are some ways to achieve that is Problem Decomposition, Pattern Recognition, Abstraction, Algorithm Design, Data Analysis, Logical Thinking, Problem Formulation, Iterative Approach, Debugging and Troubleshooting, Collaboration, Real-World Context, Multidisciplinary Connections, Ethical Considerations, Creativity and Reflection (Agbo et. Al., 2019). The brief description of each approach is given here.

Problem Decomposition: In this approach, students must teach about to break down large or complex problems into small sub-problems, which is more manageable and easier to handle. This benefits the students to recognise the problem's construction and recognize possible explanations. In a scholastic framework, problem decomposition helps students' approach to solve complex assignments, projects, or even exam questions more efficiently and effectively (Agbo et. Al., 2019, Abdullah et al., 2019). By demonstrating students to division of a problems, tutor can improve their fault-finding skills and thinking skills, as students learn to analyse, arrange, and disentangle each component one by one. This methodology also promotes coordinated thinking, organised planning, and better time administration. There are few advantages of problem decomposition like Clarity, Focus, Efficiency, Parallelism, Reusability and Debugging.

Steps involved in problem decomposition is given as:

- Understanding of the problem: Ahead of breaking down the problem into sub-problems, confirm a clear understanding of the overall problem's range and situation.
- Identification of sub-problems: Examine the problem and recognize different sections or portions that can be attended separately. Each sub-problem should be a self-contained, answerable problem, so that it would be helpful in solving the actual problem.
- Arrangement of sub-problems into a hierarchical order: Arrange the sub-problems in such a manner, where higher-level sub-problems may be determined by on the resolutions of lower-level sub-problems.
- Start solving subproblems: Start finding the solution of the sub-problems one by one and solve the entire sub-problems to get the solution of main problem. As students solved every sub-problem, so it contributed to solve the large problem.

- Integration of results: Once the sub-problems are solved, integrate their results to reach the solution of a large problem.

Pattern Recognition: The process of finding patterns in data or situations involves recognising patterns, regularities, and trends. Pattern identification is a key aspect of computational thinking. By identifying underlying patterns and exploiting them to gain insights and generate well-informed decisions, it enables people to make sense of complex material. Pattern recognition (Agbo et. Al., 2019) has extensive applications across various fields like Image and Speech Recognition, Financial Analysis, Natural Language Processing, Medical Diagnosis, Predictive Analytics, and security. A more detailed examination of pattern identification in the context of computational thinking is provided below:

Recognising Patterns: Pattern recognition involves identifying recurrent components, connections, or actions in data or issues. These patterns may be conceptual, visual, numerical, or sequential in character. Finding commonalities that offer useful information or insights is the aim.

Types of Patterns: There are several ways that patterns might appear, including:

- Visual Patterns: Recognition of forms, colours, textures, or designs in pictures or diagrams is known as visual patterning.
- Numerical Patterns: Finding sequences, trends, or correlations in numerical data is known as finding numerical patterns.
- Sequential Patterns: Observing patterns in sequential occurrences, behaviours, or process phases.
- Structural Patterns: Finding commonalities in how components are arranged or organised is known as structural pattern recognition.
- Categorical Patterns: Data is categorised using categorical patterns, or shared qualities.

Steps in Pattern Recognition: The steps involved in pattern recognition consists of five phases. The steps are given as:

- Data collection: It is the process of relevant data collection of information gathering that contains the pattern.
- Pre-processing: Data preprocessing is an important part of pattern recognition where data cleaning, and data normalization is performed. These processes are helpful in attaining high accuracy and consistency in results.
- Feature extraction: In this step, relevant and prominent features or attributes are extracted which is helpful in pattern recognition.
- Pattern detection: During this phase various machine learning and statistical algorithms are applied to detect patterns of the input data.
- Pattern interpretation: This step is used for the analysing the detected pattern to extract some meaningful information and insights.

Abstraction: In computational thinking (CT), abstraction is an important hypothesis that implies dividing a large problems or organisms into streamlined, manageable components or prototypes. It is an elementary aptitude used by computer experts, developers, and computer programmers, problem solvers to project a model and analyse results to various challenges (Agbo et. Al., 2019, Abdullah et al., 2019).

Abstraction permits individuals to concentrate on the principal aspects of a problem while pass over avoidable details, enabling them to create more effective and accessible solutions.

Abstraction in CT helps persons deliberate critically, model effective solutions, and manage complexity when handling large-scale systems and problems. It is an essential skill not only for hard core programmers or developer but also for anyone included in problem-solving system and decision-making system in the digital age. During the learning, students must concentrate on crucial details while disregarding irrelevant information. This will assist in making models that acquire the primary aspects of a problem and getting the optimized solutions.

Algorithm Design: In this, introduce students to designing step-by-step procedures (algorithms) to solve problems. This skill progresses systematic thinking and the capability to create efficient algorithms.

Data Analysis: During data analysis, teach students how to accumulate, combine, and analyse data to obtain insights. This skill benefits them supervise learned conclusions based on evidence and draw deductions from the data.

Logical Thinking: Computational thinking lay emphasis on logical reasoning, which is essential for assessing arguments, yielding deductions, and modelling coherent thoughts.

Problem Formulation: In CT, encourage learners or students to undoubtedly define problems before trying to answer the large or small piece of problems. This prevents prevent vagueness and guarantees that they are addressing the appropriate concerns.

Iterative Approach: In this approach, CT teaches learners that problem-solving is repeatedly an iterative approach. Learner should acquire to refine their solutions based on feedback and testing, improving their critical thinking and adaptability.

Debugging and Troubleshooting: Introduce students to the process of finding and fixing errors in their solutions. This fosters analytical skills and a willingness to learn from mistakes.

Collaboration: Promote collaborative learning and group problem-solving. Working with peers helps students gain different perspectives, learn from each other's methods, and upgrade their analytical thinking ability.

Real-World Context: These approaches focused on to relate computational thinking models towards real-world situations. This improves students monitor the practical presentations of their skills and assists them to think critically about how these perceptions can be used efficiently.

Multidisciplinary Connections: It shows how CT directs outside the computer science, such as in biology, economics, social sciences, biomedical engineering, astrology engineering, inter-domain engineering etc. This exhibits the adaptability of the approach and helps students think critically across several domains.

Ethical Considerations: It deals the ethical consequences of technology and computational decisions. This promotes students to think critically about the effects of their actions and the broader societal impacts.

Creativity: Emphasize that computational thinking is not just about following rigid steps but also about finding innovative solutions. Encourage students to think creatively while applying the principles of CT.

Reflection: Encourage students to reflect on their problem-solving processes and outcomes. This helps them refine their critical thinking skills over time.

By incorporating computational thinking into educational practices, you can foster students' analytical skills, logical reasoning, and ability to approach problems critically and systematically, equipping them with valuable skills for both academic and real-world challenges.

The observed issues with computational thinking strongly outline the disposition of a lack of adequate cognitive skills. In other words, the outermost portion of the human brain is responsible to help formulate decisions based on the data, however, most students differ in cognitive abilities (De Souza et al., 2019). This creates a significant gap in developing problem-solving skills among students which further affects their thinking capability. In addition, the inability to break down the problem is a highly observed issue (Fagerlund et al., 2021) which disrupts the process of computational problem-solving.

OBJECTIVES

The following objectives are formulated for this work:

- To recognize the grounds of the concept of computational thinking
- To explore the manners of using computational thinking
- To shed light on the correlation between CT and students' ability
- To discover the grounds of computational thinking
- To explore the manners of using computational thinking
- To shed light on the association between computational thinking and students' ability to think.
- To critically outline the influencing factors of computational thinking
- To highlight the benefits of computational thinking
- To measure the limitations of computational thinking

METHODOLOGY

In recent years, the approach to solving issues has become inherently linked with developing structured patterns followed by algorithms and other factors. Time and again it has been witnessed that students face severe mental blockage as a result of over-accumulation of information that has no path to follow or construct a pattern (Brating, K., & Kilhamn, C., 2021). In consideration of quantitative and data-related problems, through the implementation of computational thinking, issues solved in real-time that contain less amount of unrequited information (Cachero et al., 2020). In this section, it is seen that the primary notion of computational thinking is associated with identifying the key factors of a complex problem, evaluating the nature of the issue, and drafting certain probable solutions. Moreover, all this information related to the topic is studied by collecting online resources.

CONCEPT OF COMPUTATIONAL THINKING

The concept of CT is viewed in terms of an interconnected relationship between three major segments such as the field of computer science, computational thinking and branches of programming and coding as shown in Figure 1. Concerning these, the interrelated prospects of these mentioned factors are ascertained vital to help solve complicated problems and also advantageous in broadening knowledge horizons (Cutumisu, M., & Guo, Q., 2019). In accordance with pattern recognition, the aspect of computational thinking follows the path of abstraction wherein exclusive attention is laid on vital details and discarding irrelevant information (Diaz et al., 2020).

Figure 1. Patterns seen in the interrelationship between computational thinking, computer science

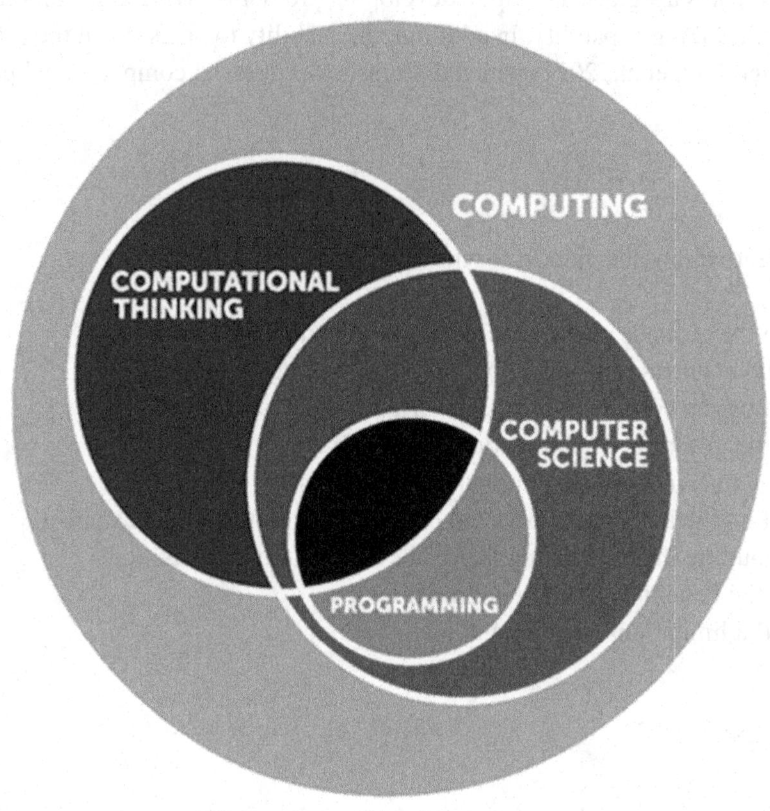

Similarly, one of the important aspects of empirical papers is related to pattern recognition. In addition, pattern recognition is an aspect of computational technology that helps to find patterns in different theoretical studies as well (Lin, P. H., & Chen, S. Y., 2020). The techniques of pattern recognition in theoretical studies are considered an improved thinking skill that helps to identify logical reasoning between different topics (Kallia et al., 2021).

Use of CT Model in Academics

Computational is a frequently growing aspect of academics and there are multiple uses depending on the subject. Mostly the use of computational thinking is seen in mathematical subjects (Fagerlund et al., 2021). Figure 2 describes the aspect which helps to increase the critical thinking ability of the students. Incorporating computational thinking in subjects that implies a practical implementation helps to get a theoretical framework for the subjects. For instance, aspects like automation, algorithmic thinking and visualisation have the most used scenarios for academic purposes (Fritz et al., 2022). In addition, computational thinking and mathematical expressions have a relationship in order to provide an improved concept of subjects. The aspect of abstraction, modelling and data analytics work as a link between computational thinking and mathematics. Additionally, such aspects of computational thinking help to understand the significance of computational thinking in academics (Kallia et al., 2021).

Figure 2. Use of different aspects of CT in empirical and theoretical studies

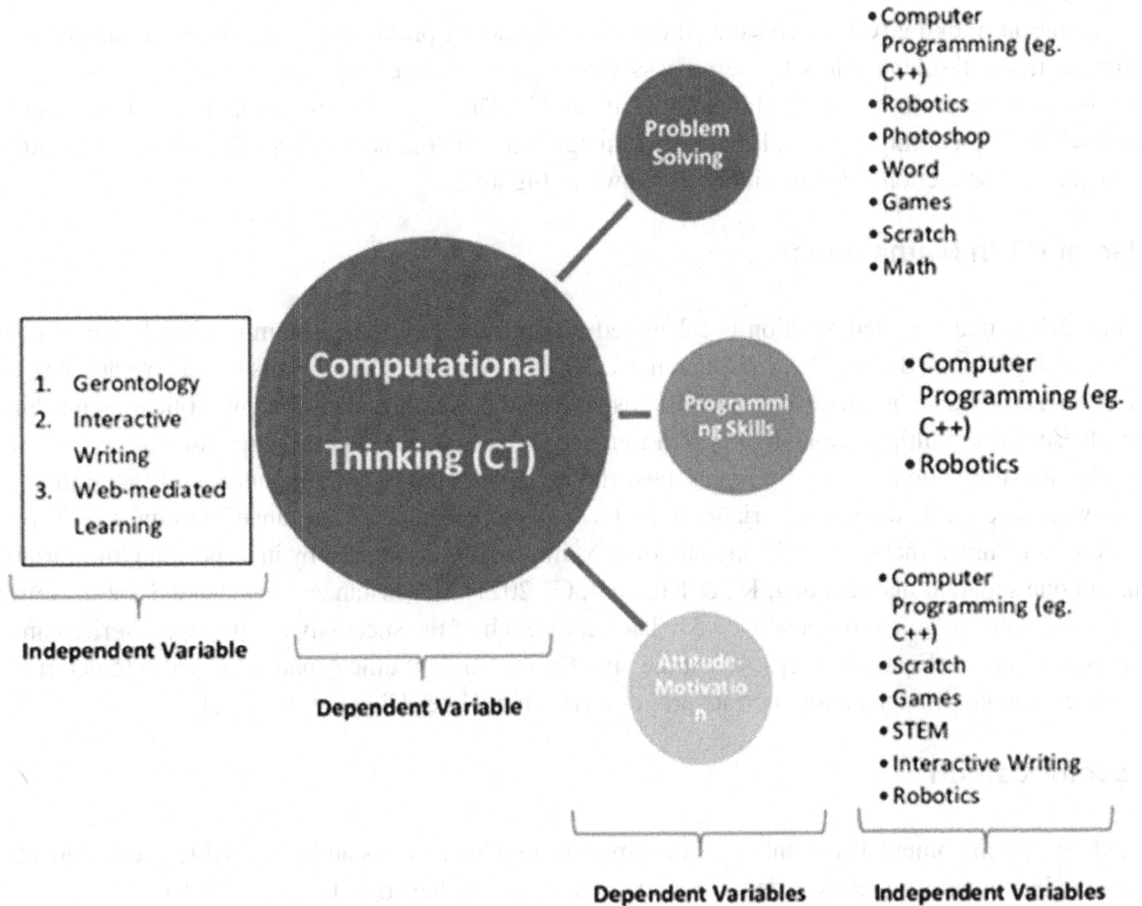

In academics, empirical papers are based on quantifiable verification and observation is related to the concept of using computational thinking. Computational thinking ability in order to describe subject increases the ability to critically thinking about a topic (Xu, F., & Zhang, S., 2021). On the other hand, subjects are related to theoretical expressions rarely computational thinking (Yang, D., & Feng, S., 2022). From an academic perspective visualisation and data analysis are part of scientific subjects. In addition, using computational thinking ability increases the capability of a student to critically analyse a topic and find patterns related to concepts. The incorporation of computational thinking ability further improves the capability of modelling such concepts in order to represent the relation of different concepts of the topic. Such computational thinking methods of empirical paper use vilifications tolls geometrical software in order to represent the relation of different aspects (Zapata-Cáceres et al., 2020). Moreover, in empirical papers, pattern recognition is based on visualisation tools that help in identifying different patterns in the output data. For instance, mathematical calculation and creating graphs using similar data help to visualise different patterns in empirical papers. In addition, visualization tools like excel and tableau are used in order to recognise patterns in empirical papers (Fagerlund et al., 2021).

WAYS OF USING COMPUTATIONAL THINKING

Computation thinking (CT) is considered the core element of problem-solving skills among students. Computational thinking refers to the process where problems are broken down into several parts to develop mitigation stapedius (El-Hamamsy et al., 2022). The ways of using computational thinking for improving the capability of critical analysis are integrated with four paths named decomposition, pattern recognition, abstraction, and algorithms as shown in Figure 3.

Use of CT in Mathematics

Considering, that repeated addition is calculated through the algebraic system of calculation of computational thinking (Brating, K., & Kilhamn, C., 2021). There are several watts to achieve at the result however a process is chosen tht corresponds closely with the overall process of a multiplication machine. For the scenario multiplication of 17.5 is chosen and the repeated addition is expressed as:

The above-mentioned scenario represented the variable count in correspondence of the number of lines which were calculated as a variable sum of a line is represented, for instance 34 in lines 2. Further, it needs to be noted that the sum in any other line is impossible to be used by incorporating the variable sum in one separate line (Brating, K., & Kilhamn, C., 2021). For instance, lines 2 and 3 are in contradiction as they are going to imply $34 = 51$. Therefore, each of the successive terms in the series can be represented as. <<Eqn015002.eps>>. At the end, the outcome of athe equation provides to determine a single variable through an appropriate program (Brating, K., & Kilhamn, C., 2021).

Decomposition

The first step in computation thinking is decomposition which assists an individual to break down the problem into smaller sections to understand the root cause (Echeverría, L., et al., 2019).

Abstraction

It is required to look for a generic solution to a problem without focusing on specific details. Abstraction in computational thinking incorporates the identification of generic issues and the reduction of extra details that complicated the problems (García, J. D. R., et al., 2019). Thus, the way is significant to focus on the basic part and root cause of the issues.

Algorithms

For solving the problem, a well-planned solution is needed and hence, the algorithms play a key role. Thus, algorithms focus on the strategies that are determined in a step-by-step process and help to solve the problem in a quicker manner.

Figure 3. Ways of using computational thinking

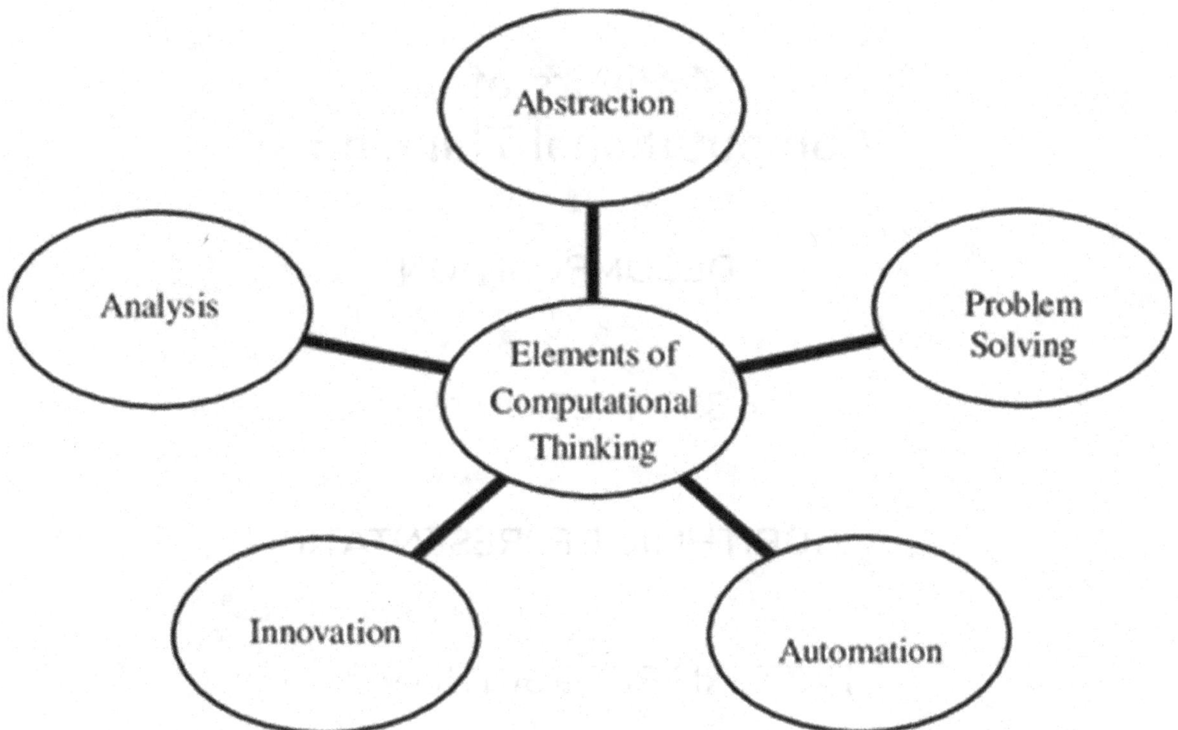

RELATION BETWEEN COMPUTATIONAL THINKING AND STUDENTS' ABILITY TO THINK

Computational thinking helps to derive solutions from a large size of problems and helps to derive students' capability of the thinking process. It has been observed computational thinking positively impacts the ability of the critical thinking process, systematic thinking process, holistic thinking, and creative thinking process (Gerosa, A., et al., 2021). With the help of CT, students can gain the ability to analyse terms by taking different viewpoints and resources into consideration which helps to develop critical thinking skills. Systematic thinking is developed with the help of CT which helps in executing the solution in a step-by-step manner (Fagerlund et al., 2021). The ability to understand the relationship between two issues along with the connection between challenges and solutions is required to develop holistic thinking processes as shown in Figure 4, that are integrated with computational thinking (Gardeli, A., & Vosinakis, S., 2019).

Creative thinking helps a student to develop innovative strategies for reducing the negative consequences of issues and improving their positive affect. Hence, it can be said that the ways of computation thinking help students to develop four dimensions of thinking capability. The decomposition helps to improve critical thinking ability and the systematic thinking process can be evolved by incorporating pattern recognition where similarities between issues are identified (Kazimoglu, C. 2020). The abstraction way helps to improve the holistic thinking process since it helps in deriving a generic solution. On the other hand, the creative thinking process is integrated with the algorithm design where innovations are made.

Figure 4. Holistic view of computational thinking

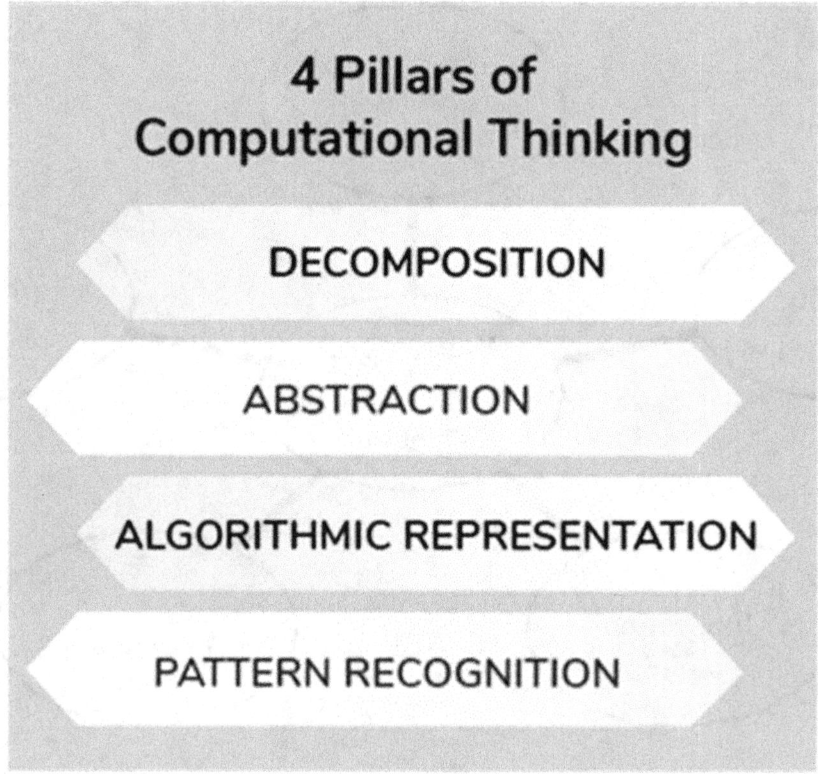

INFLUENCING FACTORS OF COMPUTATIONAL THINKING

Table 1 shows the list of factors of ICT.

It has been noticed that learning motivation, data collection, and data visualization are three core influencing factors that positively impact the computational thinking process. With the help of learning motivation, students can adopt different approaches in the context of computational thinking. On the

Table 1. Factors of CT

Factors	Description	Impact
Learning Motivation	Intrinsic motivation among student's triggers learning motivation.	Learning motivation encourages students to take other ways to enhance their computational thinking and derive the most effective solution to a problem.
Data collection	The factor assists a student to gather information from reliable resources.	Collected data impacts the extraction of solution since based on information the category of solution is derived.
Data visualization	Data visualization helps to overview the issues which are significant in breaking down the problems and deriving solutions.	The factor is significant in highlighting the roots of issues and improving the decision-making process.
(Source: Influenced by El-Hamamsy et al., 2022)		

Figure 5. Use of CT in academic

CT concept	Basic	Developing	Proficient
Abstraction and Problem decomposition	More than one script and more than one sprite	Make-a-blocks	Cloning
Parallelism	Two scripts start on "green flag"	Two scripts start on when key is pressed/when sprite is clicked on the same sprite	Two scripts start on "when I receive message", "create clone", "when %s is >%s" or "when backdrop change to" blocks
Logical thinking	"If" block	"If-else" block	Logic operations
Synchronisation	"Wait" block	"Broadcast", "when I receive message", "stop all", "stop program" or "stop programs sprite" blocks	"Wait until", "when backdrop change to" or "broadcast and wait" blocks
Flow control	Sequence of blocks	"Repeat" or "forever" blocks	"Repeat until" block
User interactivity	"Green flag" block	"Key pressed", "sprite clicked", "ask and wait" or mouse blocks	"When %s is >%s", video or audio blocks
Data representation	Modifiers of sprite properties	Operations on variables	Operations on lists

other hand, deriving effective solutions, collecting data and visualization play a major role. With the help of collected visualized data, a large-scale issue can be decomposed, and a solution can be derived.

BENEFITS OF COMPUTATIONAL THINKING

Enhancing Problem-Solving Skills

Problem-solving skills are one of the core elements of computation thinking and help in breaking down the problems into small portions which makes the extraction of the solution easier as compared to other processes (Kallia et al., 2021) shown in Figure 5.

Promotes and Boosts Efficiency

The involvement of the computational thinking process among students triggers creative thinking capability. It has been noticed that the integration of CT promotes working efficiency among students by involving innovation in solutions (El-Hamamsy et al., 2022).

Table 2. Limitation of computational thinking

Limitations	Description	Mitigation strategy
Lack of accurate prediction	The accuracy degree of the prediction derived with the help of CT is small as compared to another prediction process. The lower accuracy degree of the prediction in CT negatively impacts on the solution.	Gathered data about issues need to be sorted to increase accuracy degree.
Lack of confidence in using CT	Teachers play a major role in developing critical thinking skills with the help of CT among students. However, a lack of confidence in using CT skills among teachers negatively impacts on the development of students' thinking capabilities.	Teachers with confidence in using CT skills and experience in using CT skills are needed to be hired.

LIMITATION OF COMPUTATIONAL THINKING

The Table 2 highlights that the lack of prediction accuracy and the lack of confidence and skills in using computational thinking generates issues in developing the capability of thinking skills among students.

CONCLUSION

In conclusion, it can be stated that the initiation of computational thinking has a higher chance of constructing a developed skill set for students. The prospects of computational thinking are said to be detrimental since it aids students in enhancing their reasoning abilities as well as analytical qualities. Complex problems often require a deep-seated understanding of the issue and thereby create certain patterns and develop algorithms accordingly. The segment of CT is a kind of problem-solving skill that adopts the process of a computer's systematic manner. In response to this, computers are involved with deriving conclusions and solutions that are equipped with decomposing an issue, using analytical information and others. On the same hand, the application of computational thinking is deemed of high quality in inducing higher thinking capacities among students and strengthening their cognitive process which in turn brings forward advancing solutions.

REFERENCES

Abdullah, A. H., Othman, M. A., Ismail, N., Abd Rahman, S. N. S., Mokhtar, M., & Zaid, N. M. (2019, December). Development of mobile application for the concept of pattern recognition in computational thinking for mathematics subject. In *2019 IEEE International Conference on Engineering, Technology and Education (TALE)* (pp. 1-9). IEEE. 10.1109/TALE48000.2019.9225910

Agbo, F. J., Oyelere, S. S., Suhonen, J., & Adewumi, S. (2019, November). A systematic review of computational thinking approach for programming education in higher education institutions. In *Proceedings of the 19th Koli Calling International Conference on Computing Education Research* (pp. 1-10). IEEE. 10.1145/3364510.3364521

Alam, A. (2022, March). Educational robotics and computer programming in early childhood education: A conceptual framework for assessing elementary school students' computational thinking for designing powerful educational scenarios. In *2022 International Conference on Smart Technologies and Systems for Next Generation Computing (ICSTSN)* (pp. 1-7). IEEE. 10.1109/ICSTSN53084.2022.9761354

Avila, C. O., Foss, L., Bordini, A., Debacco, M. S., & da Costa Cavalheiro, S. A. (2019, July). Evaluation rubric for computational thinking concepts. In *2019 IEEE 19th International Conference on Advanced Learning Technologies (ICALT)* (*Vol. 2161*, pp. 279-281). IEEE.

Bråting, K., & Kilhamn, C. (2021). Exploring the intersection of algebraic and computational thinking. *Mathematical Thinking and Learning*, *23*(2), 170–185. doi:10.1080/10986065.2020.1779012

Caballero-González, Y. A., Muñoz, L., & Muñoz-Repiso, A. G. V. (2019, October). Pilot Experience: Play and Program with Bee-Bot to Foster Computational Thinking Learning in Young Children. In *2019 7th International Engineering, Sciences and Technology Conference (IESTEC)* (pp. 601-606). IEEE.

Cachero, C., Barra, P., Meliá, S., & López, O. (2020). Impact of programming exposure on the development of computational thinking capabilities: An empirical study. *IEEE Access : Practical Innovations, Open Solutions*, *8*, 72316–72325. https://ieeexplore.ieee.org/abstract/document/9063433/. doi:10.1109/ACCESS.2020.2987254

Cutumisu, M., & Guo, Q. (2019). Using topic modeling to extract pre-service teachers' understandings of computational thinking from their coding reflections. *IEEE Transactions on Education*, *62*(4), 325–332. doi:10.1109/TE.2019.2925253

De Souza, A. A., Barcelos, T. S., Munoz, R., Villarroel, R., & Silva, L. A. (2019). Data mining framework to analyze the evolution of computational thinking skills in game building Retrieved from: Workshops. *IEEE Access : Practical Innovations, Open Solutions*, *7*, 82848–82866. https://ieeexplore.ieee.org/abstract/document/8743397/. doi:10.1109/ACCESS.2019.2924343

Diaz, N. V. M., Meier, R., Trytten, D. A., & Yoon, S. Y. (2020, October). Computational thinking growth during a first-year engineering course. In *2020 IEEE Frontiers in Education Conference (FIE)* (pp. 1-7). IEEE. https://ieeexplore.ieee.org/abstract/document/9274250/

Echeverría, L., Cobos, R., & Morales, M. (2019). Improving the students computational thinking skills with collaborative learning techniques. *IEEE-RITA*, *14*(4), 196–206. doi:10.1109/RITA.2019.2952299

El-Hamamsy, L., Zapata-Cáceres, M., Marcelino, P., Zufferey, J. D., Bruno, B., Barroso, E. M., & Román-González, M. (2022). *Dataset for the comparison of two Computational Thinking (CT) test for upper primary school (grades 3-4): the Beginners' CT test (BCTt) and the competent CT test (cCTt)* No. DATASET.

Fagerlund, J., Häkkinen, P., Vesisenaho, M., & Viiri, J. (2021). Computational thinking in programming with Scratch in primary schools: A systematic review. *Computer Applications in Engineering Education*, *29*(1), 12–28. https://onlinelibrary.wiley.com/doi/pdfdirect/10.1002/cae.22255. doi:10.1002/cae.22255

Fritz, C., Bray, D., Lee, G., Julien, C., Burson, S., Castelli, D., Ramsey, C., & Payton, J. (2022). Project moveSMART: When Physical Education Meets Computational Thinking in Elementary Classrooms. *Computer, 55*(11), 29–39. doi:10.1109/MC.2022.3167600

García, J. D. R., León, J. M., González, M. R., & Robles, G. (2019, November). Developing computational thinking at school with machine learning: an exploration. In *2019 international symposium on computers in education (SIIE)* (pp. 1-6). IEEE.

Gardeli, A., & Vosinakis, S. (2019, September). ARQuest: A tangible augmented reality approach to developing computational thinking skills. In *2019 11th International Conference on Virtual Worlds and Games for Serious Applications (VS-Games)* (pp. 1-8). IEEE. 10.1109/VS-Games.2019.8864603

Gerosa, A., Koleszar, V., Tejera, G., Gómez-Sena, L., & Carboni, A. (2021). Cognitive abilities and computational thinking at age 5: Evidence for associations to sequencing and symbolic number comparison. *Computers and Education Open, 2,* 100043. https://www.sciencedirect.com/science/article/pii/S2666557321000148

Kallia, M., van Borkulo, S. P., Drijvers, P., Barendsen, E., & Tolboom, J. (2021). Characterising computational thinking in mathematics education: A literature-informed Delphi study. *Research in Mathematics Education, 23*(2), 159–187. doi:10.1080/14794802.2020.1852104

Kazimoglu, C. (2020). Enhancing confidence in using computational thinking skills via playing a serious game: A case study to increase motivation in learning computer programming. *IEEE Access : Practical Innovations, Open Solutions, 8,* 221831–221851. https://ieeexplore.ieee.org/abstract/document/9286386/. doi:10.1109/ACCESS.2020.3043278

Lin, P. H., & Chen, S. Y. (2020). Design and evaluation of a deep learning recommendation based augmented reality system for teaching programming and computational thinking. *IEEE Access : Practical Innovations, Open Solutions, 8,* 45689–45699. https://ieeexplore.ieee.org/abstract/document/9020189/. doi:10.1109/ACCESS.2020.2977679

Xu, F., & Zhang, S. (2021, March). Understanding the Source of Confusion with Computational Thinking: A Systematic Review of Definitions. In *2021 IEEE Integrated STEM Education Conference (ISEC)* (pp. 276-279). IEEE. Retrieved from: https://ieeexplore.ieee.org/abstract/document/9764144/

Yang, D., & Feng, S. (2022, March). A Collaborative Approach to Integrate Computational Thinking in an Integrated STEM Curriculum. In *2022 IEEE Integrated STEM Education Conference (ISEC)* (pp. 238-240). IEEE. 10.1109/ISEC54952.2022.10025038

Zapata-Cáceres, M., Martín-Barroso, E., & Román-González, M. (2020, April). Computational thinking test for beginners: Design and content validation. In *2020 IEEE global engineering education conference (EDUCON)* (pp. 1905-1914). IEEE.

Chapter 16
Wheat Crop Disease Detection and Classification Using Machine Learning

Nitin Dixit
https://orcid.org/0000-0002-2820-0536
ITM University, Gwalior, India

Rakhi Arora
ITM University, Gwalior, India

Deepak Gupta
ITM University, Gwalior, India

ABSTRACT

The fundamental significance of agriculture in preserving human life and providing necessary resources has led to frequent references to it as the "backbone of human civilization." In human cultures all over the world, the agricultural sector has played a significant role. In today's world, emerging technologies play a very important role in the agriculture sector. Machine learning is one of them; machine learning significantly contributes to the transformation of agriculture by facilitating data-driven decision-making, enhancing productivity, and optimizing numerous processes. There are so many areas where machine learning can be applied in agriculture processes like crop and soil monitoring, yield prediction, pest and disease management, etc. Pests and diseases are major obstacles to achieving higher productivity. The creation of effective techniques for the automatic detection and forecasting of pests and illnesses in crops is very much important. In this chapter, the authors use different machine-learning algorithms to detect early signs of pests, diseases, and weed infestations in crops.

DOI: 10.4018/979-8-3693-0782-3.ch016

1. INTRODUCTION

A new era of precision farming has begun with the use of machine learning in the identification and classification of wheat crop diseases. With the use of cutting-edge sensor technologies, including hyperspectral photography and multispectral drones, high-resolution data can now be collected, which, when combined with machine learning algorithms, enables the precise and prompt detection of wheat illnesses. Convolutional neural networks (CNNs), in particular, excel at identifying intricate patterns in images of leaves, stems, and grains, enabling the categorization and evaluation of the severity of diseases. With the help of real-time field monitoring devices, farmers can keep a close eye on their fields and get early alerts when disease outbreaks occur, allowing them to quickly take corrective action. Although the state of the art is quite encouraging, issues like data quality and user-friendly.

The main aim of machine learning techniques is to identify and detect early indicators of pests, illnesses, and weed infestations in crops so that timely treatments can be taken(Liakos et al., 2018). Early detection of these problems enables farmers to take swift action to lessen their effects, stop the spread of pests and diseases, and reduce crop losses. The primary goals of this research are as follows:

Early Detection: Machine learning algorithms are designed to find subtle and early symptoms of pests, illnesses, and weed infestations that might not be immediately obvious to the naked eye. Farmers can take action early and avoid substantial crop damage by spotting these problems before they spread(Sharma et al., 2021).

Accurate Identification: Machine learning models aim to identify precisely the particular pests, illnesses, or weed species harming the crops. For implementing focused therapies and choosing suitable control measures, this knowledge is essential. Farmers may prevent misdiagnosis and make sure the best remedies are used by having accurate identification(Pallathadka et al., 2022).

Rapid response: Machine learning facilitates the analysis of agricultural data in real-time or virtually real-time, allowing for quick responses and decisions. Farmers can take rapid action, such as applying targeted pesticides, modifying irrigation, or instituting cultural practices, to reduce damage and safeguard their crops as soon as pests, illnesses, or weeds are discovered.

Precision control: Machine learning approaches offer accurate and targeted control solutions by determining the precise location and extent of the infestation. This lessens the requirement for extensive resource and chemical application, lowering environmental impact and improving resource use.

Integrated pest management (IPM): IPM places a strong emphasis on the use of a variety of preventive actions, monitoring, and targeted interventions that are backed by machine learning. Machine learning makes it easier to apply IPM tactics that rely on biological control, cultural practices, and the prudent use of pesticides by precisely identifying and detecting pests and illnesses(Balducci et al., 2018).

Decision Support: Data-driven decision support tools are provided to farmers using machine learning. Machine learning models can offer suggestions on the best treatment alternatives, intervention timing, and resource allocation by examining historical data, weather patterns, and other pertinent variables. Farmers may now take informed judgments and active steps to safeguard their crops.

The goal of employing machine learning techniques in pest and disease detection is to encourage efficient and sustainable agricultural practices while also enhancing crop health and productivity(Katarya et al., 2020). Farmers can lessen the negative economic and environmental effects of pests and diseases on their crops by identifying and resolving problems early on(*Uow-Cs-Wp-1994-13*, n.d.).

The field of wheat crop disease detection and classification is being quickly transformed by the integration of wearable sensors and machine learning. Agriculture has strategically used wearable sensors, which were first developed for human applications, to collect a variety of information about crop health, behavior, and environmental factors. This data is processed by machine learning techniques like decision trees, random forests, and convolutional neural networks to find illness patterns, classify them, and even forecast the course of diseases. It should be noted that these sensors are not limited to static readings; they also allow for the monitoring of crop activities and behavioral changes, which can provide insight into the development of diseases. While issues like data management and user-friendly interfaces still need to be addressed, continuous research and development initiatives are prepared to do so.

Soybean, coffee, and wheat rust are the most harmful rust infections out of all of them. As a result, there are ongoing efforts being made to address this issue on a global scale. Along with rice and maize, wheat is a staple food crop. Around 220 million hectares are covered with wheat worldwide, and 772.64 million metric tonnes are produced there (2020–2021). Major fungal diseases, such as yellow rusts, stem rusts, and leaf rusts, impact the production of the wheat crop all over the world, but mainly in South Asian nations. The United Nations Food and Agricultural Organisation (FAO) predicts that due to high population increase, wheat supply may not be sufficient to meet demand in the near future(Sood et al., 2022).

2. LITERATURE SURVEY

Machine learning has been making significant contributions to the transformation of agriculture in recent years. Machine learning techniques have the potential to enhance several elements of agriculture, including crop management, pest control, yield prediction, and resource optimization(Benos et al., 2021). These improvements could be made possible by utilizing the power of data and sophisticated algorithms. To forecast future agricultural yields, machine learning algorithms can analyze previous data on weather patterns, soil quality, and crop yields. This aids farmers in selecting crops, allocating resources, and preparing for anticipated market changes in an educated manner(Anand et al., 2022). Machine learning algorithms can analyze images and sensor data from drones or IoT devices to detect early signs of pest infestations. This enables timely intervention and targeted use of pesticides, reducing environmental impact and minimizing crop losses (R. Khan et al., 2022). Machine learning techniques can analyze data from wearable sensors on livestock, providing insights into animal health, and behavior patterns, and predicting disease outbreaks. This helps farmers optimize feeding, breeding, and healthcare practices (Kamilaris et al., 2017). Machine learning and robotics can work together to create autonomous farming systems. These systems can handle tasks like planting, harvesting, and weed control, leading to increased efficiency and reduced labor costs(Tsouros et al., 2019). Traditional statistical regression approaches are used to build these models; more advanced machine-learning strategies have not been researched(Wang et al., 2020). The authors of this study proposed an efficient ML-based framework for different forms of wheat disease recognition and classification in order to automatically identify the brown- and yellow-rusted illnesses in wheat crops(H. Khan et al., 2022). To create highly predictive models, the author of this research said that the selection of the prediction algorithm is more crucial than the selection of the FS method. Additionally, the author came to the conclusion that gradient boosting techniques and random forests provide very accurate and reliable wheat grain yield GP models(Sirsat et al., 2022). Among the eight common machine learning models assessed in this study, Support Vector Machine, Gaussian Process Regression, and Random Forest are the top three most effective techniques for forecasting yields(Han et al., 2020). For generated datasets, the performance capability of machine learning models increased (Islam & Shehzad, 2022). In agriculture, machine learning improves yield prediction, crop quality, disease detection, species management, soil classification, and weed detection in addition to productivity gains(Sharma et al., 2021).

3. PESTS, ILLNESSES, AND WEED INFESTATIONS IN CROPS

A typical issue in agriculture that can have a big effect on crop health, productivity, and overall yield include pests, diseases, and weed infestations(Liakos et al., 2018). Let's examine each of these problems in turn:

1) Pests are living creatures that can damage crops by consuming plant tissues, spreading diseases, or even inflicting bodily harm. Insects like aphids, caterpillars beetles, mites, nematodes, rodents, and birds are among them. In addition to eating plant parts and acting as disease vectors for plants, pests can also decrease the growth and development of plants. Pest management is essential to keep crops healthy and stop financial losses.
2) A variety of pathogens such as fungi, bacteria, viruses, and other microorganisms, cause crop illnesses. These pathogens cause plants to get ill by interfering with normal cellular processes, which results in symptoms like wilting, yellowing, spots, lesions, and occasionally plant death. When weather circumstances are suitable, diseases can spread quickly in the field and significantly reduce agricultural output and quality. Crop rotation, the adoption of disease-resistant cultivars, and the administration of fungicides are among the methods that must be used to manage plant diseases efficiently.
3) Weed infestations: Infestations of undesired plants known as weeds coexist with farmed crops. They can negatively impact agricultural growth and productivity because they compete with crops for nutrients, water, and sunlight. Weeds are difficult to control because of their rapid growth and spread. Weed infestations can significantly reduce crop production and obstruct harvesting processes if they are not controlled. Herbicides, mechanized cultivation, mulching, and hand weeding are some of the methods used by farmers to control weeds(Dwivedi et al., 2021).

To ensure successful agricultural production, it is essential to control weed infestations, diseases, and pest infestations in crops(Pallathadka et al., 2023).

Like other crops, wheat is vulnerable to several diseases brought on by pathogens such as fungi, bacteria, and viruses. The following list of typical wheat illnesses includes:

1) infections known as rusts: Named for the distinctively rust-colored spores they generate, rusts are fungal infections that harm wheat. Three primary rust infections impact wheat:
   - (Puccinia graminis) Stem rot
   - (Puccinia triticina) Leaf rust
   - (Puccinia striiformis) Stripe rust
2) Fusarium head blight (FHB) or scab: The fungus Fusarium graminearum is the main culprit behind Fusarium head blight (FHB), often known as scab. It can dramatically lower grain quality and yield by infecting the wheat spikes, which results in shriveled, bleached, or pinkish grains.

3) Powdery mildew (Blumeria graminis f. sp. tritici): It attacks the plant's leaves, and stems, and occasionally spikes. It can decrease photosynthetic activity and yield and manifests as white, powdery patches on the surfaces of plants.
4) Septoria leaf blotch (Septoria tritici): In this Small, dark blotch with bright centers develops on wheat leaves. If not regulated, it may result in premature defoliation and yield loss.
5) Tan spot (Pyrenophora tritici-repentis): This fungus creates tan-colored lesions on wheat leaves, which decreases the plant's ability to photosynthesize and may reduce production.
6) Wheat Streak Mosaic Virus (WSMV): This virus can cause stunting, decreased production, and mosaic patterns on wheat leaves. It is spread by wheat curl mites.
7) Barley Yellow Dwarf Virus (BYDV): Wheat can be infected by the virus known as BYDV, although barley is its main target. It can result in stunted growth, discoloration, and lower yields and is spread by aphids.
8) Bacterial leaf streak (Xanthomonas translucent): This bacterial disease causes lengthy, yellow streaks on wheat plants' leaves with wet edges.
9) Gaeumannomyces graminis var. tritici: sometimes known as "take-all," is a soil-borne fungus disease that damages wheat roots. It causes poor root development and nutrient uptake, which stunts growth and reduces production.
10) Loose smut (Ustilago tritici): In this, the grain is replaced by dark, powdery masses of spores on the wheat inflorescence(H. Khan et al., 2022).

4. METHODOLOGY

By utilizing different methods and data sources, machine learning can be utilized to recognize and detect early indicators of pests, illnesses, and weed infestations in crops. The process of applying machine learning techniques follows the given steps:

4.1 Data collection: Useful information is gathered from a variety of sources, including field observations, sensors, drones, and satellite imaging. Images, spectral data, meteorological data, historical crop data, and records of pest occurrences can all be included in this data.

4.2 Preprocessing and Feature Extraction: To eliminate noise, account for environmental influences, and standardize the data format, the obtained data is first pre-processed. Feature extraction techniques are used to extract significant properties from the data, such as color, texture, shape, and spectral characteristics(Katarya et al., 2020).

4.3 Preparation of Training Data: After manually identifying and labeling instances of weeds, illnesses, and pests in the collected data by experts, a labeled dataset is created. The machine learning algorithm uses this labeled data as its training set.

4.4 Model Training: Machine learning algorithms are taught to recognize patterns and traits connected to pests, illnesses, and weeds using the labeled training dataset. Many algorithms can be employed, such as convolutional neural networks (CNN) for image processing, decision trees, and traditional machine learning approaches like support vector machines (SVM)(Jagtap et al., 2022).

4.5 Model Validation and Evaluation: To evaluate the trained model's performance, it is tested on a different validation dataset. To assess the precision and efficacy of the model in recognizing and detecting the target pests, illnesses, or weeds, evaluation metrics including precision, recall, and F1 score are computed(Sharma et al., 2021).

Figure 1. Methodology

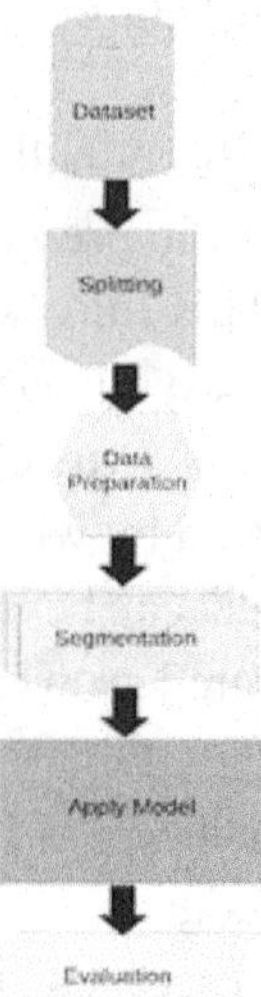

4.6 Deployment and monitoring: After the model has been verified, real-time applications can be used. For instance, farmers can use drones or mobile devices to take pictures of their crops, and a machine learning model can then evaluate those pictures to look for any indications of pest, disease, or weed infestation. Continuous surveillance enables early detection and rapid action.

4.7 Iterative Improvement: By gathering input and regularly updating the model, the machine learning model's performance can be progressively enhanced. The model can be retrained to improve accuracy and adaptability when new data become available and new patterns are found(Sahu et al., 2022).

Finally, we compare the results of different machine learning algorithms and find the best algorithm suitable for the early detection of pests and diseases in crops.

Methodology: The very first method used to identify any crop disease is depicted in Figure 1 We have provided step-by-step definitions of all the core methodology's necessary phases in this chapter.

A labeled dataset that is used to train and learn is necessary for every algorithm. The datasets are divided into two halves, one for model training and the other for model testing and performance evaluation(Vanitha et al., 2019).

Each experiment employed a dataset for model training and performance using accuracy or other performance criteria(Singh et al., 2021). Standard datasets include the CGIAR Dataset, the Large Wheat Disease Classification Dataset (LWDCD2020), the Wheat Fungal Diseases (WFD2020), and the Wheat Disease Dataset 2017.

5. ALGORITHM (COMPUTER VISION TECHNIQUES FOR IDENTIFYING PLANT DISEASES)

Machine Learning-based approaches:

The act of classifying involves adding labels to the dataset in order to divide it into several categories or groupings. These days, algorithm photos can be classified according to their category using machine

Figure 2. Description of machine learning used for plant disease detection

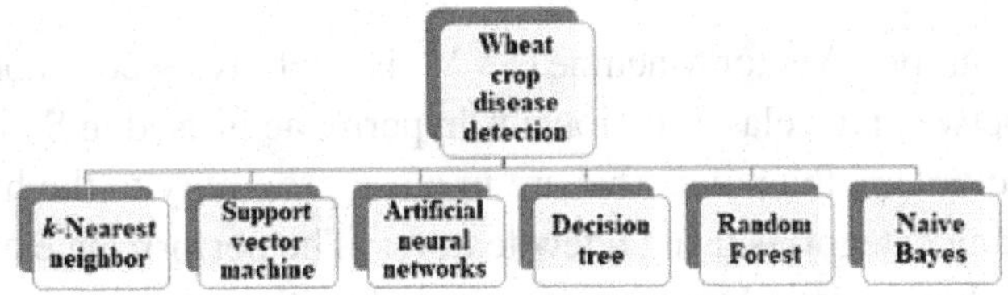

learning and deep learning methods. The supervised learning-based machine learning algorithms that are used to categorise plant diseases are listed below. In supervised learning, labels (categories of images) are provided together with the input images. Figure 2 depicts the machine-learning approaches for wheat crop disease detection.

In general, machine learning methodologies encompass a process of acquiring knowledge through the analysis of training data to carry out a specific task. In the field of machine learning, data is typically represented as a collection of examples. Typically, a singular instance is characterized by a collection of attributes, which are alternatively referred to as features or variables(Dwivedi et al., 2021).

Based on the signal from the learning system, machine learning tasks are separated into supervised and unsupervised learning. In the context of supervised learning, the process involves the presentation of data wherein example inputs are provided alongside their corresponding outputs(McQueen et al., 1995). The primary objective is to develop a comprehensive rule or model that effectively maps inputs to their respective outputs. In certain scenarios, inputs can be only partially accessible, with certain target outputs either missing or provided solely as feedback for actions within a dynamic environment, such as in the context of reinforcement learning. In the context of supervised learning, the acquired expertise, also known as the trained model, is employed to make predictions for the missing outputs, which are commonly referred to as labels, in the test data. In the context of unsupervised learning, it is important to note that the traditional distinction between training and test sets becomes irrelevant(H. Khan et al., 2022). This is primarily because the data used in unsupervised learning is typically unlabelled, meaning that there is no predetermined set of target values or labels associated with the data instances. Consequently, the absence of labels eliminates the need for a separate test set in the traditional sense, Since predictions or classifications based on labeled samples are not the goals of unsupervised learning, it instead seeks to identify innate patterns, structures, or relationships within the data itself. The process of learning involves the cognitive processing of input data to uncover latent patterns that may not be immediately apparent.

5.1 *k*-Nearest Neighbor

It is a machine learning algorithm that is calculated by k-neighbors and is used for classification. It is mostly used for statistical estimation, machine learning, and image processing. This technique operated on the basis of measuring the gap between various data points using the Manhattan and Euclidean distances. Getting data, defining k neighbours, calculating neighbour distance using Manhattan or Euclidean distance, and then assigning new instances to the majority of the neighbours are the stages that make it work.

5.2 Support Vector Machine

In statistical learning, the Support Vector Machine (SVM) is a relatively common classifier. The classifier seeks to distinguish between the classifications. A hyperplane is used in SVM to separate one class from another. Support vectors are the points that are in close proximity to the hyperplane. The SVM's job is to categorise the many categories using a few features. The performance of this approach is quite strong in extreme classes. Let's take a look at some characteristics of a specific plant, such as colour, texture, and shape. If we use two characteristics, like colour and texture, to distinguish between damaged and healthy leaves. The ideal choice boundary is needed to categorise them. For the new case, optimal decision bounds might lead to more misclassification.

5.3 Artificial Neural Networks

This particular machine learning algorithm is utilised for classification. Since the start of the 1980s, scientists have been working on artificial neural networks (ANNs). The construction of ANNs, a particular kind of classification system, was influenced by the human brain. Feature vector or pattern input from the outside world is used by ANNs. Each input value is multiplied by the appropriate weights, which add to the bias value. Additionally, the output is produced and the result is transferred to the activation function (binary, sigmoid).

5.4 Decision Tree

It is a machine learning algorithm used to address regression- and classification-based issues, and it falls under the category of supervised learning. The graphical depiction of pre-defined rules and the answer is called a decision tree. Decision nodes and leaf nodes are the two different sorts of nodes that make up the decision tree's graph. Additionally, leaf nodes contain the actual output while edges keep the information associated with the questions and their replies.

In addition to these methods, a number of other algorithms, including Random Forest and Naive Bayes, have demonstrated meaningful results in picture recognition.

6. EXPERIMENTAL RESULTS & DISCUSSION

This part covers the experimental setup, a dataset that was gathered, evaluation metrics, and performance assessment of our suggested framework. We also go into detail about how trained models perform in comparison.

6.1 Experimental Setup

The following were the experimental hardware platforms: A Ryzen 7 Quad-Core serves as the main processor. There was 32 GB of RAM. The GPU was a 16 GB NVIDIA GeForce GTX 1080. The following were the experimental software platforms: Windows 10(64-bit) was the operating system. The TensorFlow-GPU 1.14.0 backend was used with the Keras 2.2.4 frame for deep learning. Python 3.7 was used as the programming language. Spyder 3.3.3 and Anaconda 3 5.2.0 were used to construct the

Figure 3. Sample Images chosen from the selected data set

tool. Opencv was the third-party library, Cuda 9.2 was the development package, and CuDNN 7.0 was the deep-learning acceleration library.

6.2 Dataset

After gathering the data, the photos are pre-processed to reduce computation time using the inter-area interpolation technique. Additionally, there are three types of leaves in the distribution of the wheat leaf dataset (healthy, rusty, and yellow-rusted leaves)(Cravero et al., 2022).

6.2.1 Healthy Wheat Leaves: Healthy wheat leaves exhibit a vibrant green color with no visible signs of disease or damage. The leaf surface is uniform and smooth, with well-defined leaf edges. There are no discolorations, lesions, or abnormalities on the leaf surface. The leaf texture is consistent, with no rough patches or irregularities.

6.2.2 Rusted Wheat Leaves: Rusted wheat leaves show symptoms of rust disease, caused by fungal pathogens such as Puccinia spp.The leaf color may change to shades of yellow, orange, brown, or reddish-brown, depending on the severity of the infection. Small, powdery, rust-colored pustules (spore masses) appear on the leaf surface, primarily on the lower side.

6.2.3 Yellow-Rusted Wheat Leaves: Yellow-rusted wheat leaves represent a more advanced stage of rust disease infection. The leaf color turns predominantly yellow, with extensive areas of discoloration. The leaf may show signs of wilting or decreased turgidity due to the impact of the fungal infection on nutrient uptake and water transport. The pustules on the leaf surface may have ruptured, leading to the release of rust spores and further spreading of the disease.

Figure 3 also displays picture examples from several classes: the first row displays healthy samples, the second row displays rusted samples, and the final row depicts the yellow rust class of wheat disorders.

6.3 Evaluation Metrics

Based on the accuracy, precision, recall, and F1 score, the models are assessed. The confusion matrix is used to calculate each of these performance evaluation numbers(Shakoor et al., 2017). These equations contain various values that were derived from each model's confusion matrix. These magic numbers include True Positive (TP), which shows how often our model correctly identified the positive testing data sample as positive; True Negative (TN); False Positive (FP); and False Negative (FN); which shows how often our model identified different positive classes incorrectly. As a result, we discovered that DT had the highest accuracy, at 99.8% of our suggested framework. On the other side, LR's testing accuracy is found to be less accurate than all of the others which were 89.6%.

7. RESULTS

Various experiments on the training dataset are run to assess the performance. Each test image was chosen at random. The testing dataset had 225 photos for each group (healthy, rusted, and yellow-rusted), which made up 20% of our entire dataset.

Table 1 presents the results of our experiments on wheat disease detection using five different machine learning models: the proposed framework, DT LR, KNN, SVM, and NB. The models were evaluated on a testing dataset comprising 20% of the total dataset, with 225 images for each of the three categories (healthy, rusted, and yellow-rusted).

The accuracy of the suggested framework is 99.8%, which is higher than that of DT (99.2%) and KNN (99.0%). In contrast, the LR's accuracy is 89.6%, which is lower than that of all the models discussed. SVM and NB, on the other hand, were accurate to up to 94.0% and 97.7%, respectively.

Figures 4 and 5 show a Graphical representation of the outcome

The proposed framework demonstrated superior performance in accurately detecting wheat diseases, achieving the highest accuracy, precision, recall, and F1 score values among the three models. The Decision Tree (DT) model also performed well, but its performance was slightly lower than the proposed framework. However, the Logistic Regression (LR) model exhibited lower accuracy and F1 score values, suggesting that it may not be as effective as the other models in distinguishing between healthy and diseased wheat leaves.

Overall, the results indicate the effectiveness of the proposed framework for wheat disease detection, showcasing its potential for practical applications in agriculture and crop management.

Additionally, it is noted from the testing set that other evaluation graphs, such as confusion matrix, overall accuracy, and performance evaluation graph, exist. These graphs show the F1-score, recall, and precision. In these tests, the combined extracted features of the colors, shapes, and textures (Local Binary Patterns) of wheat leaves are analysed in comparison to the performance of ML models.

We split the whole sample size from the dataset chosen for testing by the total accuracy of predictions provided by trained models.

Table 1. Results

S.No	Algorithm	Accuracy
1	Proposed Framework	99.8
2	Decision Tree	99.2
3	Logistic regression	89.6
4	KNN	99.0
5	SVM	94.0
6	NB	97.7

Figure 4. Accuracy graph

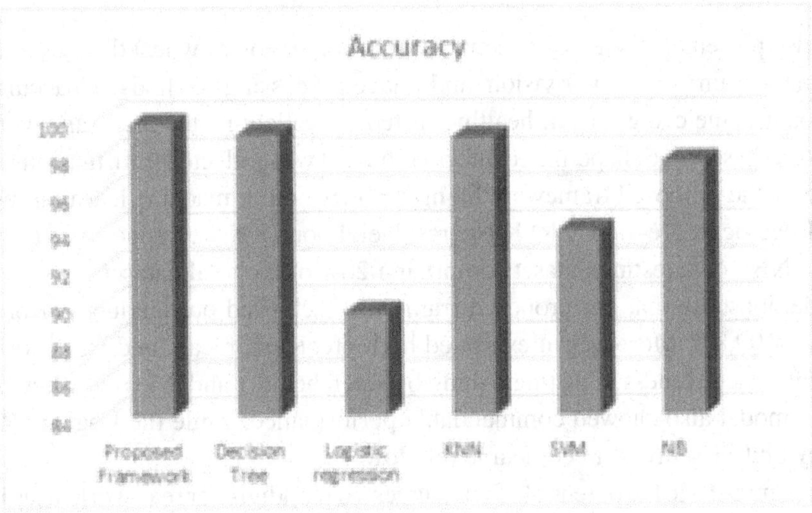

Figure 5. Performance metrics

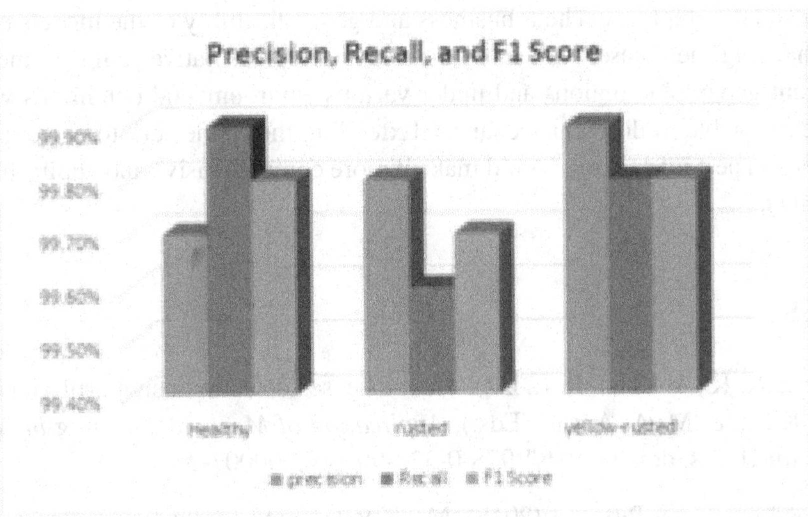

The proposed framework's accuracy is 99.8%, which is greater than the accuracy of DT (99.2%) and KNN (99.0%). The LR's accuracy, on the other hand, is the lowest of all the models discussed at 89.6%. SVM and NB, however, have accuracy levels as high as 94.0% and 97.7%, respectively.

CONCLUSION AND FUTURE WORK

In this research, we presented a unique framework for the detection of wheat diseases utilizing a combination of machine learning, computer vision, and image processing methods. To accurately categorize wheat leaf photos into the categories of healthy, rusted, or yellow-rusted, the framework makes use of color-based, texture-based, and shape-based information that was collected from the images. We evaluated the performance of the proposed framework along with five other machine learning models, Decision Tree (DT), and Logistic Regression (LR) K-Nearest Neighbour(KNN), Support vector machine(SVM), and Naïve Bayes(NB), on a testing dataset comprising 20% of the total dataset.

The results demonstrate that our proposed framework achieved outstanding performance, with an overall accuracy of 99.8%. Moreover, it exhibited high precision, recall, and F1 score values for each class, indicating its effectiveness in distinguishing between healthy and diseased wheat leaves. The Decision Tree (DT) model also showed commendable performance, while the Logistic Regression (LR) model's accuracy and F1 score were comparatively lower.

These results demonstrate the potential of our suggested paradigm for real-world uses in crop management and agriculture. Farmers can avoid future spread and reduce crop losses by quickly and precisely identifying wheat illnesses at an early stage.

Although the results of our suggested framework are encouraging, there are still many opportunities for further study and development. The robustness and generalizability of the models could be further improved by enhancing the dataset with more diverse and representative samples. Incorporating images from different geographic regions and under various environmental conditions would make the framework more adaptable to different scenarios. Extending the framework to detect additional types of wheat diseases and pest infestations would make it more comprehensive and applicable in real-world agricultural settings.

REFERENCES

Anand, R., Mishra, R. K., & Khan, R. (2022). Plant diseases detection using artificial intelligence. In M. A. Khan, R. Khan, & M. A. Ansari (Eds.), *Application of Machine Learning in Agriculture* (pp. 173–190). Academic Press. doi:10.1016/B978-0-323-90550-3.00007-2

Balducci, F., Impedovo, D., & Pirlo, G. (2018). Machine learning applications on agricultural datasets for smart farm enhancement. *Machines*, *6*(3), 38. doi:10.3390/machines6030038

Benos, L., Tagarakis, A. C., Dolias, G., Berruto, R., Kateris, D., & Bochtis, D. (2021). Machine learning in agriculture: A comprehensive updated review. In Sensors (Vol. 21, Issue 11). MDPI AG. doi:10.339021113758

Cravero, A., Pardo, S., Sepúlveda, S., & Muñoz, L. (2022). Challenges to Use Machine Learning in Agricultural Big Data: A Systematic Literature Review. In Agronomy (Vol. 12, Issue 3). MDPI. doi:10.3390/agronomy12030748

Dwivedi, P., Kumar, S., Vijh, S., & Chaturvedi, Y. (2021). Study of Machine Learning Techniques for Plant Disease Recognition in Agriculture. *2021 11th International Conference on Cloud Computing, Data Science & Engineering (Confluence)*, (pp. 752–756). IEEE. 10.1109/Confluence51648.2021.9377186

Han, J., Zhang, Z., Cao, J., Luo, Y., Zhang, L., Li, Z., & Zhang, J. (2020). Prediction of winter wheat yield based on multi-source data and machine learning in China. *Remote Sensing (Basel), 12*(2), 236. doi:10.3390/rs12020236

Islam, M., & Shehzad, F. (2022). A Prediction Model Optimization Critiques through Centroid Clustering by Reducing the Sample Size, Integrating Statistical and Machine Learning Techniques for Wheat Productivity. *Scientifica, 7271293*, 1–11. doi:10.1155/2022/7271293 PMID:35310811

Jagtap, S. T., Phasinam, K., Kassanuk, T., Jha, S. S., Ghosh, T., & Thakar, C. M. (2022). Towards application of various machine learning techniques in agriculture. *Materials Today: Proceedings, 51*, 793–797. doi:10.1016/j.matpr.2021.06.236

Kamilaris, A., Kartakoullis, A., & Prenafeta-Boldú, F. X. (2017). A review on the practice of big data analysis in agriculture. *Computers and Electronics in Agriculture, 143*, 23–37. doi:10.1016/j.compag.2017.09.037

Katarya, R., Raturi, A., Mehndiratta, A., & Thapper, A. (2020). Impact of Machine Learning Techniques in Precision Agriculture. *2020 3rd International Conference on Emerging Technologies in Computer Engineering: Machine Learning and Internet of Things (ICETCE)*, (pp. 1–6). IEEE. 10.1109/ICETCE48199.2020.9091741

Khan, H., Haq, I. U., Munsif, M., Mustaqeem, Khan, S. U., & Lee, M. Y. (2022). Automated Wheat Diseases Classification Framework Using Advanced Machine Learning Technique. *Agriculture, 12*(8), 1226. doi:10.3390/agriculture12081226

Khan, R., Khan, M. A., Ansari, M. A., Dhingra, N., & Bhati, N. (2022). Machine learning-based agriculture. In M. A. Khan, R. Khan, & M. A. Ansari (Eds.), *Application of Machine Learning in Agriculture* (pp. 3–27). Academic Press. doi:10.1016/B978-0-323-90550-3.00003-5

Liakos, K. G., Busato, P., Moshou, D., Pearson, S., & Bochtis, D. (2018). Machine learning in agriculture: A review. In Sensors (Switzerland) (Vol. 18, Issue 8). MDPI AG. doi:10.339018082674

McQueen, R. J., Garner, S. R., Nevill-Manning, C. G., & Witten, I. H. (1995). Applying machine learning to agricultural data. *Computers and Electronics in Agriculture, 12*(4), 275–293. doi:10.1016/0168-1699(95)98601-9

Pallathadka, H., Jawarneh, M., Sammy, F., Garchar, V., Sanchez, D. T., & Naved, M. (2022). A Review of Using Artificial Intelligence and Machine Learning in Food and Agriculture Industry. *2022 2nd International Conference on Advance Computing and Innovative Technologies in Engineering (ICACITE)*, (pp. 2215–2218). IEEE. 10.1109/ICACITE53722.2022.9823427

Pallathadka, H., Mustafa, M., Sanchez, D. T., Sekhar Sajja, G., Gour, S., & Naved, M. (2023). IMPACT OF MACHINE learning ON Management, healthcare AND AGRICULTURE. *Materials Today: Proceedings, 80*, 2803–2806. doi:10.1016/j.matpr.2021.07.042

Sahu, P., Singh, A. P., Chug, A., & Singh, D. (2022). A Systematic Literature Review of Machine Learning Techniques Deployed in Agriculture: A Case Study of Banana Crop. *IEEE Access : Practical Innovations, Open Solutions, 10*, 87333–87360. doi:10.1109/ACCESS.2022.3199926

Shakoor, M. T., Rahman, K., Rayta, S. N., & Chakrabarty, A. (2017). Agricultural production output prediction using Supervised Machine Learning techniques. *2017 1st International Conference on Next Generation Computing Applications (NextComp)*, (pp. 182–187). IEEE. 10.1109/NEXTCOMP.2017.8016196

Sharma, A., Jain, A., Gupta, P., & Chowdary, V. (2021). Machine Learning Applications for Precision Agriculture: A Comprehensive Review. In IEEE Access (Vol. 9, pp. 4843–4873). Institute of Electrical and Electronics Engineers Inc. doi:10.1109/ACCESS.2020.3048415

Singh, G., Sethi, G. K., & Singh, S. (2021). Survey on Machine Learning and Deep Learning Techniques for Agriculture Land. *SN Computer Science*, *2*(6), 487. doi:10.100742979-021-00929-6

Sirsat, M. S., Oblessuc, P. R., & Ramiro, R. S. (2022). Genomic Prediction of Wheat Grain Yield Using Machine Learning. *Agriculture*, *12*(9), 1406. doi:10.3390/agriculture12091406

Sood, S., Singh, H., & Jindal, S. (2022). *Rust Disease Classification Using Deep Learning Based Algorithm: The Case of Wheat*. doi:10.5772/intechopen.104426

Tsouros, D. C., Bibi, S., & Sarigiannidis, P. G. (2019). A review on UAV-based applications for precision agriculture. In *Information (Switzerland)* (*Vol. 10*, Issue 11). MDPI AG.

Vanitha, C. N., Archana, N., & Sowmiya, R. (2019). Agriculture Analysis Using Data Mining And Machine Learning Techniques. *2019 5th International Conference on Advanced Computing & Communication Systems (ICACCS)*, (pp. 984–990). IEEE. 10.1109/ICACCS.2019.8728382

Wang, Y., Zhang, Z., Feng, L., Du, Q., & Runge, T. (2020). Combining multi-source data and machine learning approaches to predict winter wheat yield in the conterminous United States. *Remote Sensing (Basel)*, *12*(8), 1232. doi:10.3390/rs12081232

Compilation of References

A. P. G. V., Sree, Meera, & Kalpana. (2020). Automated Irrigation System and Detection of Nutrient Content in the Soil. *International Conference on Power, Energy, Control and Transmission Systems (ICPECTS)*, 1-3. 10.1109/ICPECTS49113.2020.9336990

Abdelaziz, A., Elhoseny, M., Salama, A. S., & Riad, A. M. (2018). A machine learning model for improving healthcare services on cloud computing environment. *Measurement, 119*, 117–128. doi:10.1016/j.measurement.2018.01.022

Abdullah, A. H., Othman, M. A., Ismail, N., Abd Rahman, S. N. S., Mokhtar, M., & Zaid, N. M. (2019, December). Development of mobile application for the concept of pattern recognition in computational thinking for mathematics subject. In *2019 IEEE International Conference on Engineering, Technology and Education (TALE)* (pp. 1-9). IEEE. 10.1109/TALE48000.2019.9225910

Abirami, R., & Padmakumar, M. (2022). Pandemic War Natural Calamities and Sustainability: Industry 4.0 Technologies to Overcome Traditional and Contemporary Supply Chain Challenges. *Logistics 6*(4) 81-10.3390/logistics6040081

Acemoglu, D. (2002). Technical change, inequality, and the labor market. *Journal of Economic Literature, 40*(1), 7–72. doi:10.1257/jel.40.1.7

Adel, A. (2020). Future of industry 5.0 in society: Human-centric solutions, challenges and prospective research areas. *Journal of Cloud Computing (Heidelberg, Germany), 11*(1), 40. doi:10.1186/s13677-022-00314-5 PMID:36101900

Adhikari, R., Li, C., Kalbaugh, K., & Nemali, K. (2020). A low-cost smartphone controlled sensor based on image analysis for estimating whole-plant tissue nitrogen (N) content in floriculture crops. *Computers and Electronics in Agriculture, 169*, 105173. doi:10.1016/j.compag.2019.105173

Agarwal, V., & Shendure, J. (2020). Predicting mRNA abundance directly from genomic sequence using deep convolutional neural networks. *Cell Reports, 31*(7), 107663. doi:10.1016/j.celrep.2020.107663 PMID:32433972

Agbo, F. J., Oyelere, S. S., Suhonen, J., & Adewumi, S. (2019, November). A systematic review of computational thinking approach for programming education in higher education institutions. In *Proceedings of the 19th Koli Calling International Conference on Computing Education Research* (pp. 1-10). IEEE. 10.1145/3364510.3364521

Ahirwar, S., Swarnker, R., Bhukya, S., & Namwade, G. (2019). Application of Drone in Agriculture. *International Journal of Current Microbiology and Applied Sciences, 8*(1), 2500–2505. doi:10.20546/ijcmas.2019.801.264

Ahmad, M. A., Eckert, C., & Teredesai, A. (2018). Interpretable machine learning in healthcare. *Proceedings of the 2018 ACM International Conference on Bioinformatics, Computational Biology, and Health Informatics,* 559-560.

Ahuja, A. S. (2020). Artificial Intelligence in Industry 5.0: The Next Revolution in Productivity and Working Conditions. *International Journal of Research and Analytical Reviews, 7*(3), 331–335.

AI in Education Market Size, Share, Trends and Industry Analysis 2018 - 2030. (n.d.). Markets and Markets. https://www.marketsandmarkets.com/Market-Reports/ai-in-education-market-200371366.html

Akbari, M., Foroudi, P., Shahmoradi, M., Padash, H., Parizi, Z. S., Khosravani, A., Ataei, P., & Cuomo, M. T. (2022). The Evolution of Food Security: Where Are We Now, Where Should We Go Next? *Sustainability (Basel), 14*(6), 3634. doi:10.3390/su14063634

Akundi, A., Euresti, D., Luna, S., Ankobiah, W., Lopes, A., & Edinbarough, I. (2022). State of Industry 5.0 – Analysis and Identification of Current Research Trends. *Applied System Innovation, 5*(1), 1–14. doi:10.3390/asi5010027

Alam, A. (2022, March). Educational robotics and computer programming in early childhood education: A conceptual framework for assessing elementary school students' computational thinking for designing powerful educational scenarios. In *2022 International Conference on Smart Technologies and Systems for Next Generation Computing (ICSTSN)* (pp. 1-7). IEEE. 10.1109/ICSTSN53084.2022.9761354

Al-Emran, M., & Al-Sharafi, M. A. (2022). Revolutionizing Education with Industry 5.0: Challenges and Future Research Agendas. *International Journal of Information Technology and Language Studies, 6*(3), 1–5.

Alharbi, W. S., & Rashid, M. (2022). A review of deep learning applications in human genomics using next-generation sequencing data. *Human Genomics, 16*(1), 26. doi:10.1186/s40246-022-00396-x PMID:35879805

Al-Naji, A., Fakhri, A. B., Gharghan, S. K., & Chahl, J. (2021). Soil color analysis based on a RGB camera and an artificial neural network towards smart irrigation: A pilot study. *Heliyon, 7*(1), e06078. Advance online publication. doi:10.1016/j.heliyon.2021.e06078 PMID:33537493

Alojaiman, B. (2023). Technological Modernizations in the Industry 5.0 Era: A Descriptive Analysis and Future Research Directions. *Processes (Basel, Switzerland), 11*(5), 1318. doi:10.3390/pr11051318

Alvarez-Aros, E. L., & Bernal-Torres, C. A. (2021). Technological competitiveness and emerging technologies in Industry 4.0 and Industry 5.0. *Anais da Academia Brasileira de Ciências, 93*(1), 1–20. doi:10.1590/0001-3765202120191290 PMID:33886700

Alves, J., Lima, T. M., & Gaspar, P. D. (2023). Is Industry 5.0 a Human-Centred Approach? A Systematic Review. *Processes (Basel, Switzerland), 11*(1), 193. doi:10.3390/pr11010193

Amara, J., Bouaziz, B., & Algergawy, A. (2017). A deep learning-based approach for banana leaf diseases classification. *Datenbanksysteme für Business, Technologie und Web (BTW 2017)-Workshopband.*

Anand, R., Mishra, R. K., & Khan, R. (2022). Plant diseases detection using artificial intelligence. In M. A. Khan, R. Khan, & M. A. Ansari (Eds.), *Application of Machine Learning in Agriculture* (pp. 173–190). Academic Press. doi:10.1016/B978-0-323-90550-3.00007-2

Andres, B., Sempere-Ripoll, F., Esteso, A., & Alemany, M. (2022). Mapping between Industry 5.0 and Education 5.0. *EDULEARN22 Proceedings*, 2921–2926. 10.21125/edulearn.2022.0739

Angermueller, C., Lee, H. J., Reik, W., & Stegle, O. (2017). DeepCpG: Accurate prediction of single-cell DNA methylation states using deep learning. *Genome Biology, 18*(1), 67–67. doi:10.1186/s13059-017-1189-z PMID:28395661

Apan, A., Held, A., Phinn, S., & Markley, J. (2004). Detecting sugarcane 'orange rust'disease using EO-1 Hyperion hyperspectral imagery. *International Journal of Remote Sensing, 25*(2), 489–498. doi:10.1080/01431160310001618031

Arno, J., Casasnovas, M. J., Ribes-Dasi, M., Rosell-Polo, J. R., (2009). Review. Precision Viticulture. Research topics, challenges and opportunitiesin site-specific vineyard management. *Spanish Journal ofAgricultural Research, 7*, 779-790.

Aslam, F., Aimin, W., Li, M., & Ur Rehman, K. (2020). Innovation in the era of IoT and industry 5.0: Absolute innovation management (AIM) framework. *Information (Basel)*, *11*(2), 124. doi:10.3390/info11020124

Avila, C. O., Foss, L., Bordini, A., Debacco, M. S., & da Costa Cavalheiro, S. A. (2019, July). Evaluation rubric for computational thinking concepts. In *2019 IEEE 19th International Conference on Advanced Learning Technologies (ICALT)* (*Vol. 2161*, pp. 279-281). IEEE.

Ávila-Gutiérrez, M. J., Aguayo-González, F., & Lama-Ruiz, J. R. (2021). Framework for the development of affective and smart manufacturing systems using sensorised surrogate models. *Sensors (Basel)*, *21*(7), 2274. Advance online publication. doi:10.3390/s21072274 PMID:33805015

Ayaz, M., Ammad-Uddin, M., Sharif, Z., Mansour, A., & Aggoune, E. M. (2019). Internet-of-Things (IoT)-based smart agriculture: Toward making the fields talk. *IEEE Access : Practical Innovations, Open Solutions*, *7*, 129551–129583. doi:10.1109/ACCESS.2019.2932609

Badsha, M. B., Li, R., Liu, B., Li, Y. I., Xian, M., Banovich, N. E., & Fu, A. Q. (2020). Imputation of single-cell gene expression with an autoencoder neural network. *Quantitative Biology*, *8*(1), 78–94. doi:10.1007/s40484-019-0192-7 PMID:32274259

Baig, M. I., & Yadegaridehkordi, E. (2023). Industry 5.0 applications for sustainability: A systematic review and future research directions. *Sustainable Development (Bradford)*, sd.2699. Advance online publication. doi:10.1002/sd.2699

Bajwa, S. G., Rupe, J. C., & Mason, J. (2017). Soybean disease monitoring with leaf reflectance. *Remote Sensing (Basel)*, *9*(2), 127. doi:10.3390/rs9020127

Baker, T., Smith, L., & Anissa, N. (2019). *Educ-AI-tion rebooted? Exploring the future of artificial intelligence in schools and colleges*. Academic Press.

Baker, R. S., & Inventado, P. S. (2014). Educational data mining and learning analytics. In J. Larusson & B. White (Eds.), *Learning Analytics: From Research to Practice* (pp. 61–75). Springer. doi:10.1007/978-1-4614-3305-7_4

Baker, R. S., & Yacef, K. (2009). The State of Educational Data Mining in 2009: A Review and Future Visions. *Journal of Educational Data Mining*, *1*(1), 3–17.

Bakir, A., & Dahlan, M. (2023). Higher education leadership and curricular design in industry 5.0 environment: A cursory glance. *Development and Learning in Organizations*, *37*(3), 15–17. doi:10.1108/DLO-08-2022-0166

Balducci, F., Impedovo, D., & Pirlo, G. (2018). Machine learning applications on agricultural datasets for smart farm enhancement. *Machines*, *6*(3), 38. doi:10.3390/machines6030038

Balleda, K., Satyanvesh, D., Sampath, N. V. S. S. P., Varma, K. T. N., & Baruah, P. K. (2014, January). Agpest: An efficient rule-based expert system to prevent pest diseases of rice & wheat crops. In *2014 IEEE 8th International Conference on Intelligent Systems and Control (ISCO)* (pp. 262-268). IEEE.

Barata, J., & Kayser, I. (2023). Industry 5.0 – Past, Present, and Near Future. *Procedia Computer Science*, *219*, 778–788. doi:10.1016/j.procs.2023.01.351

Barbosa, A., Trevisan, R., Hovakimyan, N., & Martin, N. F. (2020). Modeling yield response to crop management using convolutional neural networks. *Computers and Electronics in Agriculture*, *170*, 105197. doi:10.1016/j.compag.2019.105197

Barker, T. (2010). An automated feedback system based on adaptive testing: Extending the model. *International Journal of Emerging Technologies in Learning*, *5*(2), 11–14. doi:10.3991/ijet.v5i2.1235

Barker, T. (2011). An Automated Individual Feedback and Marking System: An Empirical Study. *Electronic Journal of e-Learning*, *9*(1), 1–14.

Barnes, E. M., Clarke, T. R., Richards, S. E., Colaizzi, P. D., Haberland, J., Kostrzewski, M., . . . Moran, M. S. (2000, July). Coincident detection of crop water stress, nitrogen status and canopy density using ground based multispectral data. In *Proceedings of the Fifth International Conference on Precision Agriculture, Bloomington, MN, USA* (*Vol. 1619*, p. 6). Academic Press.

Baruah, I., & Baruah, G. (2020). Improvement of Seed Quality: A Biotechnological Approach. In A. K. Tiwari (Ed.), *Advances in Seed Production and Management*. Springer. doi:10.1007/978-981-15-4198-8_26

Bazoobandi, A., Emamgholizadeh, S., & Ghorbani, H. (2022). Estimating the amount of cadmium and lead in the polluted soil using artificial intelligence models. *European Journal of Environmental and Civil Engineering*, *26*(3), 933–951. doi:10.1080/19648189.2019.1686429

Begum, S. (2021). A Study on growth in Technology and Innovation across the globe in the Field of Education and Business. *International Research Journal on Advanced Science Hub, 3*(6S), 148-156.

Behrens, T., Förster, H., Scholten, T., Steinrücken, U., Spies, E. D., & Goldschmitt, M. (2005). Digital soil mapping using artificial neural networks. *Journal of Plant Nutrition and Soil Science*, *168*(1), 21–33. doi:10.1002/jpln.200421414

Bengio, Y. (2009). Learning Deep Architectures for AI. *Foundations and Trends in Machine Learning*, *2*(1), 1–127. doi:10.1561/2200000006

Benincasa, P., Antognelli, S., Brunetti, L., Fabbri, C. A., Natale, A., Sartoretti, V., Modeo, G., Guiducci, M., Tei, F., & Vizzari, M. (2018). Reliability of NDVI derived by high resolution satelliteand UAV compared to in-field methods for the evaluation of early crop N status and grain yield in wheat. *Experimental Agriculture*, *54*(4), 604–622. doi:10.1017/S0014479717000278

Benos, L., Tagarakis, A. C., Dolias, G., Berruto, R., Kateris, D., & Bochtis, D. (2021). Machine learning in agriculture: A comprehensive updated review. In Sensors (Vol. 21, Issue 11). MDPI AG. doi:10.3390/s21113758

Benyezza, H., Bouhedda, M., & Rebouh, S. (2021). Zoning irrigation smart system based on fuzzy control technology and IoT for water and energy saving. *Journal of Cleaner Production*, *302*, 127001. Advance online publication. doi:10.1016/j.jclepro.2021.127001

Bhattacharya, S., Maddikunta, R., Somayaji, S. R., Lakshmanna, K., Kaluri, R., Hussein, A., & Gadekallu, T. R. (2020). Load balancing of energy cloud using wing driven and firefly algorithms in internet of everything. *Journal of Parallel and Distributed Computing*, 16–26.

Bhojani, S. H., & Bhatt, N. (2020). Wheat crop yield prediction using new activation functions in neural network. *Neural Computing & Applications*, *32*(17), 13941–13951. doi:10.1007/s00521-020-04797-8

Bibri, S. E., & Krogstie, J. (2017). Smart sustainable cities of the future: An extensive interdisciplinary literature review. *Sustainable Cities and Society*, *31*, 183–212. doi:10.1016/j.scs.2017.02.016

Bida, M. N., Mosito, S. M., Miya, T. V., Demetriou, D., Blenman, K. R., & Dlamini, Z. (2023). Transformation of the Healthcare Ecosystem in the Era of Society 5.0. In *Society 5.0 and Next Generation Healthcare: Patient-Focused and Technology-Assisted Precision Therapies* (pp. 223–248). Springer Nature Switzerland. doi:10.1007/978-3-031-36461-7_10

Bigan, C. (2022). *Trends in Teaching Artificial Intelligence for Industry 5.0.* Advances in Sustainability Science and Technology. doi:10.1007/978-981-16-7365-8_10

Birtchnell, T., & Urry, J. (2016). *A New Industrial Future? 3D Printing and the Reconfiguring of Production, Distribution, and Consumption.* Routledge. doi:10.4324/9781315776798

Blikstein, P. (2013). Multimodal Learning Analytics and Education Data Mining: Using Computational Technologies to Measure Complex Learning Tasks. *Journal of Learning Analytics, 1*(2), 185–189.

Bochtis, D. D., Sorensen, C. G., & Busato, P. (2014). Advances in agricultural machinery management: A review. *Biosystems Engineering, 126*, 69–81. doi:10.1016/j.biosystemseng.2014.07.012

Boursianis, A. D., Papadopoulou, M. S., Diamantoulakis, P., Liopa-Tsakalidi, A., Barouchas, P., Salahas, G., Karagiannidis, G., Wan, S., & Goudos, S. K. (2020). Internet of Things (IoT) and agricultural unmanned aerial vehicles (UAVs) in smart farming: A comprehensive review. *Internet of Things : Engineering Cyber Physical Human Systems*, (100187). Advance online publication. doi:10.1016/j.iot.2020.100187

Box, G. E., Jenkins, G. M., & Reinsel, G. C. (1994). *Time Series Analysis: Forecasting and Control.* John Wiley & Sons.

Bråting, K., & Kilhamn, C. (2021). Exploring the intersection of algebraic and computational thinking. *Mathematical Thinking and Learning, 23*(2), 170–185. doi:10.1080/10986065.2020.1779012

Brazeau, M. (2018). *Fighting weeds: Can we reduce, or even eliminate, herbicides by utilizing robotics and AI.* Genetic Literacy Project, North Wales.

Breque, M., De Nul, L., & Petridis, A. (2021). *Industry 5.0: towards a sustainable, human-centric and resilient European industry.* European Commission, Directorate-General for Research and Innovation.

Brinker, T. J., Hekler, A., Enk, A. H., Klode, J., Hauschild, A., Berking, C., Schilling, B., Haferkamp, S., Schadendorf, D., Fröhling, S., Utikal, J. S., von Kalle, C., Ludwig-Peitsch, W., Sirokay, J., Heinzerling, L., Albrecht, M., Baratella, K., Bischof, L., Chorti, E., ... Schrüfer, P. (2019). A convolutional neural network trained with dermoscopic images performed on par with 145 dermatologists in a clinical melanoma image classification task. *European Journal of Cancer (Oxford, England), 111*, 148–154. doi:10.1016/j.ejca.2019.02.005 PMID:30852421

Broo, D. G., Kaynak, O., & Sait, S. M. (2022). Rethinking engineering education at the age of industry 5.0. *Journal of Industrial Information Integration, 25*, 100311. doi:10.1016/j.jii.2021.100311

Broussard, D. M., Rahman, Y., Kulshreshth, A. K., & Borst, C. W. (2021). An Interface for Enhanced Teacher Awareness of Student Actions and Attention in a VR Classroom. *2021 IEEE Conference on Virtual Reality and 3D User Interfaces Abstracts and Workshops (VRW)*, 284-290. 10.1109/VRW52623.2021.00058

Brown, L., & Miller, C. (2020). *Ethical and Legal Considerations of AI in Healthcare.* HealthTech Magazine.

Brunelli, D., Albanese, A., Acunto, D., & Nardello, M. (2019). Energy neutral machine learning based IoT device for pest detection in precision agriculture. *IEEE Internet Things Mag., 2*(4), 10–13. doi:10.1109/IOTM.0001.1900037

Bryant, J., Heitz, C., Sanghvi, S., & Wagle, D. (2020). How artificial intelligence will impact K-12 teachers. *Retrieved*, (May), 12.

Bullock, D. G., Bullock, D. S., Nafziger, E. D., Doerge, T. A., Paszkiewicz, S. R., Carter, P. R., & Peterson, T. A. (1998). Does Variable Rate Seeding of Corn Pay? *Agronomy Journal, 90*(6), 830-836. doi:10.2134/agronj1998.00021962009000060019x

Bungărdean, R. M., Şerbănescu, M. S., Streba, C. T., & Crişan, M. (2021). Deep learning with transfer learning in pathology. Case study: Classification of basal cell carcinoma. *Romanian Journal of Morphology and Embryology, 62*(4), 1017–1028. doi:10.47162/RJME.62.4.14 PMID:35673821

Caballero-González, Y. A., Muñoz, L., & Muñoz-Repiso, A. G. V. (2019, October). Pilot Experience: Play and Program with Bee-Bot to Foster Computational Thinking Learning in Young Children. In *2019 7th International Engineering, Sciences and Technology Conference (IESTEC)* (pp. 601-606). IEEE.

Cachero, C., Barra, P., Meliá, S., & López, O. (2020). Impact of programming exposure on the development of computational thinking capabilities: An empirical study. *IEEE Access : Practical Innovations, Open Solutions*, *8*, 72316–72325. https://ieeexplore.ieee.org/abstract/document/9063433/. doi:10.1109/ACCESS.2020.2987254

Callaway, D. (2021). *The Impact of Personalized Learning on Achievement in an Elementary School Mathematics Classroom.* Wilkes University.

Cao, Y., Liu, C., Liu, B., Brunette, M. J., Zhang, N., & Sun, T. (2016). Improving tuberculosis diagnostics using deep learning and mobile health technologies among resource-poor and marginalized communities. In *2016 IEEE First International Conference on Connected Health: Applications, Systems and Engineering Technologies (CHASE).* New York, NY: IEEE. 10.1109/CHASE.2016.18

Carayannis, E. G. (2021a). THE NEW EARTH INITIATIVE (NEI) VIA INDUSTRY 5.0 PLUS. *Reducing the average earth temperature not just the increase rate of it.* Springer. https://www.springer. com/journal/13132/updates/19860990.

Carayannis, E. G. (2021b). *Sustainable Development Goal 8 - an Interview with Elias Carayannis.* Springer. https://www.springer.com/journal/13132/updates/19771722.

Carayannis, E. G. (2021c). Introducing the notions of Industry 5.0 & Society 5.0. *Discover the way a Greek Scientifc Paradigm may develop Innovation Ecosystem 5.0.* The Greek Scientists Society. https://www.youtube.com/watch?v=nul_yXWuWqI.

Caturegli, L., Corniglia, M., Gaetani, M., Grossi, N., Magni, S., Migliazzi, M., ... Raffaelli, M. (2016). Unmanned aerial vehicle to estimate nitrogen status ofturfgrasses. *PLoS One*, *11*(6), e0158268. doi:10.1371/journal.pone.0158268 PMID:27341674

Chander, B., Pal, S., De, D., & Buyya, R. (2022). Artificial intelligence-based internet of things for industry 5.0. *Artificial intelligence-based internet of things systems*, 3-45.

Char, D. S., Abràmoff, M. D., & Feudtner, C. (2020). Identifying ethical considerations for machine learning healthcare applications. *The American Journal of Bioethics*, *20*(11), 7–17. doi:10.1080/15265161.2020.1819469 PMID:33103967

Cheng, J. Z., Ni, D., Chou, Y. H., Qin, J., Tiu, C. M., Chang, Y. C., Huang, C.-S., Shen, D., & Chen, C.-M. (2016). Computer-Aided diagnosis with deep learning architecture: Applications to breast lesions in US images and pulmonary nodules in CT scans. *Scientific Reports*, *6*(1), 24454–24454. doi:10.1038/srep24454 PMID:27079888

Cheng, Y., Wang, F., Zhang, P., & Hu, J. (2016). Risk prediction with electronic health records: a deep learning approach. In *Proceedings of the 2016 SIAM International Conference on Data Mining.* SIAM. 10.1137/1.9781611974348.49

Chen, I. Y., Joshi, S., Ghassemi, M., & Ranganath, R. (2020). Probabilistic machine learning for healthcare. *Annual Review of Biomedical Data Science*, 4. PMID:34465179

Chen, L., Chen, P., & Lin, Z. (2020). Artificial intelligence in education: A review. *IEEE Access : Practical Innovations, Open Solutions*, *8*, 75264–75278. doi:10.1109/ACCESS.2020.2988510

Chen, S., Wang, L., & Zhang, H. (2023). Time Series Analysis for Remote Patient Monitoring using Wearable Devices. *Journal of Medical Sensors*, *10*(3), 156–168.

Chin, S. T. S. (2021). Influence of emotional intelligence on the workforce for industry 5.0. IBIMA Business Review. *Article*, *882278*, 1–7. doi:10.5171/2021.882278

Choi, Y., Chiu, C. Y.-I., & Sontag, D. A. (2016). Learning low-dimensional representations of medical concepts. *AMIA Joint Summits on Translational Science Proceedings AMIA Summit on Translational Science*, 41–50. PMID:27570647

Choi, Y., Ms, C. Y.-I. C., & Sontag, D. (2016). Learning low-dimensional representations of medical concepts. *Proceedings of the AMLA Summit on Clinical Research Informatics.*

Choudhary, S., Gaurav, V., Singh, A., & Agarwal, S. (2019). Autonomous crop irrigation system using artificial intelligence. *International Journal of Engineering and Advanced Technology*, *8*(5), 46–51. doi:10.35940/ijeat.E1010.0585S19

Ciasullo, M. V., Orciuoli, F., Douglas, A., & Palumbo, R. (2022). Putting Health 4.0 at the service of Society 5.0: Exploratory insights from a pilot study. *Socio-Economic Planning Sciences, 80*. doi:10.1016/j.seps.2021.101163

Cicero, M., Bilbily, A., Colak, E., Dowdell, T., Gray, B., Perampaladas, K., & Barfett, J. (2017). Training and validating a deep convolutional neural network for computer-aided detection and classification of abnormalities on frontal chest radiographs. *Investigative Radiology*, *52*(5), 281–287. doi:10.1097/RLI.0000000000000341 PMID:27922974

Cillis, D., Pezzuolo, A., Marinello, F., & Sartori, L. (2018). Field-scale electrical resistivity profiling mapping for delineating soil condition in a nitrate vulnerable zone. *Applied Soil Ecology*, *123*, 780–786. doi:10.1016/j.apsoil.2017.06.025

Cioffi, R., Travaglioni, M., Piscitelli, G., Petrillo, A., & Parmentola, A. (2020). Smart manufacturing systems and applied industrial technologies for a sustainable industry: A systematic literature revie. *Applied Sciences (Basel, Switzerland)*, *10*(8), 2897. doi:10.3390/app10082897

Coelho, P., Bessa, C., Landeck, J., & Silva, C. (2023). Industry 5.0: The Arising of a Concept. *Procedia Computer Science*, *217*, 1137–1144. doi:10.1016/j.procs.2022.12.312

Coronado, E., Kiyokawa, T., Ricardez, G. A. G., Ramirez-Alpizar, I. G., Venture, G., & Yamanobe, N. (2022). Evaluating quality in human-robot interaction: A systematic search and classification of performance and human-centered factors, measures and metrics towards an industry 5.0. *Journal of Manufacturing Systems*, *63*(9), 392–410. doi:10.1016/j.jmsy.2022.04.007

Correndo, A., McArtor, B., Prestholt, A., Hernandez, C., Kyveryga, P. M., & Ciampitti, I. A. (2022). Interactive soybean variable-rate seedingsimulator for farmers. *Agronomy Journal*, *114*(6), 3554–3565. doi:10.1002/agj2.21181

Cravero, A., Pardo, S., Sepúlveda, S., & Muñoz, L. (2022). Challenges to Use Machine Learning in Agricultural Big Data: A Systematic Literature Review. In Agronomy (Vol. 12, Issue 3). MDPI. doi:10.3390/agronomy12030748

Cruz, A., Ampatzidis, Y., Pierro, R., Materazzi, A., Panattoni, A., De Bellis, L., & Luvisi, A. (2019). Detection of grapevine yellows symptoms in Vitis vinifera L. with artificial intelligence. *Computers and Electronics in Agriculture*, *157*, 63–76. doi:10.1016/j.compag.2018.12.028

Cutumisu, M., & Guo, Q. (2019). Using topic modeling to extract pre-service teachers' understandings of computational thinking from their coding reflections. *IEEE Transactions on Education*, *62*(4), 325–332. doi:10.1109/TE.2019.2925253

Czyba, R., Lemanowicz, M., Gorol, Z., & Kudala, T. (2018). Construction Prototyping, Flight Dynamics Modeling, and Aerodynamic Analysis of Hybrid VTOLUnmanned Aircraft. *Journal of Advanced Transportation, 2018*, 1–15. doi:10.1155/2018/7040531

Dai, L., Fang, R., Li, H., Hou, X., Sheng, B., Wu, Q., & Jia, W. (2018). Clinical report guided retinal microaneurysm detection with multi-sieving deep learning. *IEEE Transactions on Medical Imaging*, *37*(5), 1149–1161. doi:10.1109/TMI.2018.2794988 PMID:29727278

Dalamagkidis, K., Valavanis, K. P., & Piegl, L. A. (2012). Aviation history and unmanned flight. In *On integrating unmanned aircraft systems into the nationalairspace system* (pp. 11–42). Springer. doi:10.1007/978-94-007-2479-2_2

Daponte, P., Vito, L. D., Glielmo, L., Iannelli, L., Liuzza, D., Picariello, F., & Silano, G. (2019). A review on the use of drones for precisionagriculture. *IOP Conference Series. Earth and Environmental Science*, *275*(1), 1–10. doi:10.1088/1755-1315/275/1/012022

Davenport, T. H., & Kalakota, R. (2019). The Potential for Artificial Intelligence in Healthcare. *Future Healthcare Journal*, *6*(2), 94–98. doi:10.7861/futurehosp.6-2-94 PMID:31363513

De Souza, A. A., Barcelos, T. S., Munoz, R., Villarroel, R., & Silva, L. A. (2019). Data mining framework to analyze the evolution of computational thinking skills in game building Retrieved from: Workshops. *IEEE Access : Practical Innovations, Open Solutions*, *7*, 82848–82866. https://ieeexplore.ieee.org/abstract/document/8743397/. doi:10.1109/ACCESS.2019.2924343

Deivanayagampillai, N., Jacob, K., Manohar, G. V., & Broumi, S. (2023). Investigation of industry 5.0 hurdles and their mitigation tactics in emerging economies by TODIM arithmetic and geometric aggregation operators in single value neutrosophic environment. Facta Universitatis, Series. *Mechanical Engineering (New York, N.Y.)*.

Dela Cruz, J. R., Baldovino, R. G., Bandala, A. A., & Dadios, E. P. (2017, May). Water usage optimization of Smart Farm Automated Irrigation System using artificial neural network. In *2017 5th International Conference on Information and Communication Technology (ICoIC7)* (pp. 1-5). IEEE. 10.1109/ICoICT.2017.8074668

Demir, K. A., Doven, G., & Sezen, B. (2019). Industry 5.0 and human-robot coworking. *Procedia Computer Science*, *158*, 688–695. doi:10.1016/j.procs.2019.09.104

Demoncourt, F., Lee, J. Y., & Uzuner, O. (2016). De-identification of patient notes with recurrent neural networks. *Journal of the American Medical Informatics Association : JAMIA*, *24*(3), 596–606. doi:10.1093/jamia/ocw156 PMID:26644398

Dhar, K. K., Bhattacharya, P., Kumar, N., & Mitra, A. (2023). Abnormality Detection in Chest Diseases Using a Convolutional Neural Network, Journal of Information and Optimization Sciences. *Taru Publication*, *44*(1), 97–111. doi:10.47974/JIOS-1298

Dhirani, L. L., Mukhtiar, N., Chowdhry, B. S., & Newe, T. (2023). Ethical dilemmas and privacy issues in emerging technologies: A review. *Sensors (Basel)*, *23*(3), 1151. doi:10.3390/s23031151 PMID:36772190

Diaz, N. V. M., Meier, R., Trytten, D. A., & Yoon, S. Y. (2020, October). Computational thinking growth during a first-year engineering course. In *2020 IEEE Frontiers in Education Conference (FIE)* (pp. 1-7). IEEE. https://ieeexplore.ieee.org/abstract/document/9274250/

Du, B., Chai, Y., Huangfu, W., Zhou, R., & Ning, H. (2021). Undergraduate University Education in Internet of Things Engineering in China: A Survey. *Education Sciences*, *11*(5), 202. doi:10.3390/educsci11050202

Duggal, A. S., Malik, P. K., Gehlot, A., Singh, R., Gaba, G. S., Masud, M., & Al-Amri, J. F. (2021). A sequential roadmap to Industry 6.0: Exploring future manufacturing trends. *IET Communications*, *16*(5), 521–531. doi:10.1049/cmu2.12284

Dwivedi, P., Kumar, S., Vijh, S., & Chaturvedi, Y. (2021). Study of Machine Learning Techniques for Plant Disease Recognition in Agriculture. *2021 11th International Conference on Cloud Computing, Data Science & Engineering (Confluence)*, (pp. 752–756). IEEE. 10.1109/Confluence51648.2021.9377186

Echeverría, L., Cobos, R., & Morales, M. (2019). Improving the students computational thinking skills with collaborative learning techniques. *IEEE-RITA*, *14*(4), 196–206. doi:10.1109/RITA.2019.2952299

ElFar, O. A., Chang, C. K., Leong, H. Y., Peter, A. P., Chew, K. W., & Show, P. L. (2021). Prospects of Industry 5.0 in algae: Customization of production and new advance technology for clean bioenergy generation. *Energy Conversion and Management: X, 10*, 100048.

El-Hamamsy, L., Zapata-Cáceres, M., Marcelino, P., Zufferey, J. D., Bruno, B., Barroso, E. M., & Román-González, M. (2022). *Dataset for the comparison of two Computational Thinking (CT) test for upper primary school (grades 3-4): the Beginners' CT test (BCTt) and the competent CT test (cCTt)* No. DATASET.

Eli-Chukwu, N. C. (2019). Applications of artificial intelligence in agriculture: A review. *Engineering, Technology &. Applied Scientific Research, 9*(4), 4377–4383.

ElShawi, R., Sherif, Y., Al-Mallah, M., & Sakr, S. (2021). Interpretability in healthcare: A comparative study of local machine learning interpretability techniques. *Computational Intelligence, 37*(4), 1633–1650. doi:10.1111/coin.12410

Embarak, O. (2019). *Demolish falsy ratings in recommendation systems. In 2019 Sixth HCT Information Technology Trends*. ITT. doi:10.1109/ITT48889.2019.9075130

Embarak, O. H. (2018). *Three Layered Factors Model for Mining Students Academic Performance. In 2018 Fifth HCT Information Technology Trends*. ITT. doi:10.1109/CTIT.2018.8649491

Esmaeilzadeh, P., Mirzaei, T., & Dharanikota, S. (2021). Patients' perceptions toward human–artificial intelligence interaction in health care: Experimental study. *Journal of Medical Internet Research, 23*(11), e25856. doi:10.2196/25856 PMID:34842535

Fagerlund, J., Häkkinen, P., Vesisenaho, M., & Viiri, J. (2021). Computational thinking in programming with Scratch in primary schools: A systematic review. *Computer Applications in Engineering Education, 29*(1), 12–28. https://onlinelibrary.wiley.com/doi/pdfdirect/10.1002/cae.22255. doi:10.1002/cae.22255

Fan, M., Shen, J., Yuan, L., Jiang, R., Chen, X., Davies, W. J., & Zhang, F. (2012). Improving crop productivity and resource use efficiency to ensure food security and environmental quality in China. *Journal of Experimental Botany, 63*(1), 13–24. doi:10.1093/jxb/err248 PMID:21963614

Faraj, S., Pachidi, S., & Sayegh, K. (2018). Working and organizing in the age of the learning algorithm. *Information and Organization, 28*(1), 62–70. doi:10.1016/j.infoandorg.2018.02.005

Farid, H. U., Mustafa, B., Khan, Z. M., Anjum, N. M., Ahmad, I., Mubeen, M., & Shahzad, H. (2023). An Overview of Precision Agricultural Technologies for Crop Yield Enhancement and Environmental Sustainability. In *Climate Change Impacts on Agriculture* (pp. 239–257). Springer. doi:10.1007/978-3-031-26692-8_14

Farooq, M. S., Riaz, S., Abid, A., Abid, K., & Naeem, M. A. (2019). A survey on the role of IoT in agriculture for the implementation of smart farming. *IEEE Access : Practical Innovations, Open Solutions, 7*, 156237–156271. doi:10.1109/ACCESS.2019.2949703

Frankel, F.R. (1971). *India's Green Revolution. Economic Gains and Politir.* I Costs Bombay.

Fritz, C., Bray, D., Lee, G., Julien, C., Burson, S., Castelli, D., Ramsey, C., & Payton, J. (2022). Project moveSMART: When Physical Education Meets Computational Thinking in Elementary Classrooms. *Computer, 55*(11), 29–39. doi:10.1109/MC.2022.3167600

Furstenau, L. B., Sott, M. K., Kipper, L. M., Machado, E. L., Lopez-Robles, J. R., Dohan, M. S., Cobo, M. J., Zahid, A., Abbasi, Q. H., & Imran, M. A. (2020). Link between Sustainability and Industry 4.0: Trends, Challenges and New Perspectives. *IEEE Access : Practical Innovations, Open Solutions, 8*, 140079–140096. doi:10.1109/ACCESS.2020.3012812

Gago, J., Douthe, C., Coopman, R., Gallego, P., Ribas-Carbo, M., Flexas, J., & Medrando, H. (2015). UAVs challenge to assess water stress forsustainable agriculture. *Agricultural Water Management, 153*, 9–19. doi:10.1016/j.agwat.2015.01.020

Gai, J., Tang, L., & Steward, B. L. (2020). Automated crop plant detection based on the fusion of color and depth images for robotic weed control. *Journal of Field Robotics, 37*(1), 35–52. doi:10.1002/rob.21897

Gaitán-Cremaschi, D., Klerkx, L., Duncan, J., Trienekens, J. H., Huenchuleo, C., Dogliotti, S., Contesse, M. E., & Rossing, W. A. H. (2018). Characterizing diversity of food systems in view of sustainability transitions. A review. *Agronomy for Sustainable Development*, *39*(1), 1–22. doi:10.1007/s13593-018-0550-2 PMID:30881486

Gandhi, M., Kamdar, J., & Shah, M. (2020). Preprocessing of non-symmetrical images for edge detection. *Augmented Human Research*, *5*(1), 1–10. doi:10.1007/s41133-019-0030-5

Gandomi, A., & Haider, M. (2015). Beyond the hype: Big data concepts, methods, and analytics. *International Journal of Information Management*, *35*(2), 137–144. doi:10.1016/j.ijinfomgt.2014.10.007

Gao, S., Rehman, J., & Dai, Y. (2022). Assessing comparative importance of DNA sequence and epigenetic modifications on gene expression using a deep convolutional neural network. *Computational and Structural Biotechnology Journal*, *20*, 3814–3823. doi:10.1016/j.csbj.2022.07.014 PMID:35891778

García, J. D. R., León, J. M., González, M. R., & Robles, G. (2019, November). Developing computational thinking at school with machine learning: an exploration. In *2019 international symposium on computers in education (SIIE)* (pp. 1-6). IEEE.

García, S., Luengo, J., & Herrera, F. (2015). *Data Preprocessing in Data Mining*. Springer. doi:10.1007/978-3-319-10247-4

Gardeli, A., & Vosinakis, S. (2019, September). ARQuest: A tangible augmented reality approach to developing computational thinking skills. In *2019 11th International Conference on Virtual Worlds and Games for Serious Applications (VS-Games)* (pp. 1-8). IEEE. 10.1109/VS-Games.2019.8864603

Garg, P. K. (2022). The future healthcare technologies: a roadmap to society 5.0. In *Geospatial Data Science in Healthcare for Society 5.0* (pp. 305–318). Springer Singapore. doi:10.1007/978-981-16-9476-9_14

Garre, P., & Harish, A. (2018, December). Autonomous agricultural pesticide spraying uav. *IOP Conference Series. Materials Science and Engineering*, *455*, 012030. doi:10.1088/1757-899X/455/1/012030

Gerhards, R., & Christensen, S. (2003). Real-time weed detection, decision making and patch spraying in maize, sugarbeet, winter wheat and winter barley. *Weed Research*, *43*(6), 385–392. doi:10.1046/j.1365-3180.2003.00349.x

Gerosa, A., Koleszar, V., Tejera, G., Gómez-Sena, L., & Carboni, A. (2021). Cognitive abilities and computational thinking at age 5: Evidence for associations to sequencing and symbolic number comparison. *Computers and Education Open, 2,* 100043. https://www.sciencedirect.com/science/article/pii/S2666557321000148

Ghobakhloo. (2023). *Behind the definition of Industry 5.0: a systematic review of technologies, principles, components, and values*. doi:10.1080/21681015.2023.2216701

Ghobakhloo, M., Fathi, M., Iranmanesh, M., Maroufkhani, P., & Morales, M. E. (2021). Industry 4.0 ten years on: A bibliometric and systematic review of concepts, sustainability value drivers, and success determinants. *Journal of Cleaner Production*, *302*, 127052. doi:10.1016/j.jclepro.2021.127052

Ghobakhloo, M., Iranmanesh, M., Foroughi, B., Tirkolaee, F. B., Asadi, S., & Amran, A. (2023, September). Industry 5.0 implications for inclusive sustainable manufacturing: An evidence-knowledge-based strategic roadmap. *Journal of Cleaner Production*, *417*, 138023. Advance online publication. doi:10.1016/j.jclepro.2023.138023

Ghobakhloo, M., Iranmanesh, M., Mubarak, M., Mubarik, M., Rejeb, A., & Nilashi, M. (2022). Identifying industry 5.0 contributions to sustainable development: A strategy roadmap for delivering sustainability values. *Sustainable Production and Consumption*, *33*, 716–737. doi:10.1016/j.spc.2022.08.003

Ghosh, P., & Kumpatla, S. P. (2022). GIS Applications in Agriculture. In Geographis Information Systems and Applications in Coastal Studies. Intechopen. doi:10.5772/intechopen.104786

Ghosh, D., Anand, A., Gautam, S. S., & Vidyarthi, A. (2022). Soil Fertility Monitoring with Internet of Underground Things: A Survey. *IEEE Micro*, *42*(1), 8–16. doi:10.1109/MM.2021.3121496

Ghosh, S., Majumder, S., & Peng, S. L. (2023). An Empirical Study on Adoption of Artificial Intelligence in Human Resource Management. In *Artificial Intelligence Techniques in Human Resource Management* (pp. 29–85). Apple Academic Press. doi:10.1201/9781003328346-3

Goel, D., Gulati, P., Jha, S. K., Kumar, N., Khan, A., & Kumari, P. (2022). *Evaluation of Machine Learning Techniques for Crop Yield Prediction, Artificial Intelligence Applications in Agriculture and Food Quality Improvement*. IGI Global. doi:10.4018/978-1-6684-5141-0.ch007

Gomathi, L., Mishra, A. K., & Tyagi, A. K. (2023). Industry 5.0 for Healthcare 5.0: Opportunities, Challenges and Future Research Possibilities. *2023 7th International Conference on Trends in Electronics and Informatics (ICOEI)*, 204-213. 10.1109/ICOEI56765.2023.10125660

Gonot-Schoupinsky, F. N., Garip, G., & Sheffield, D. (2022). Facilitating the planning and evaluation of narrative intervention reviews: Systematic Transparency Assessment in Intervention Reviews (STAIR). *Evaluation and Program Planning*, *91*, 102043. doi:10.1016/j.evalprogplan.2021.102043 PMID:34839113

Gonzalez, G., & Williams, M. (2021). IoT-Driven Smart Healthcare: A Comprehensive Survey. *IEEE Access : Practical Innovations, Open Solutions*, *9*, 78902–78925.

Government of India. (2011). *State of the Economy and Prospects Economic Survey*. http:// exim.indiamart.com/economic-survey10-11/pdfs/ echap-01.pdf

Goyal. (2016). Indian Agriculture and Farmers-Problems and Reforms. Academic Press.

Goyal, D. T. (2019). A Look at Top 35 Problems in the Computer Science Field for the Next Decade. *Proceedings of 4th international conference on information and communication technology for competitive strategies(ICTCS)*, 379 – 396.

Grabowska, S., Saniuk, S., & Gajdzik, B. (2022). Industry 5.0: Improving humanization and sustainability of Industry 4.0. *Scientometrics*, *127*(6), 3117–3144. doi:10.1007/s11192-022-04370-1 PMID:35502439

Grant, M. J., & Booth, A. (2009). A typology of reviews: An analysis of 14 review types and associated methodologies. *Health Information and Libraries Journal*, *26*(2), 91–108. doi:10.1111/j.1471-1842.2009.00848.x PMID:19490148

Green, B. N., Johnson, C. D., & Adams, A. (2006). Writing narrative literature reviews for peer-reviewed journals: Secrets of the trade. *Journal of Chiropractic Medicine*, *5*(3), 101–117. doi:10.1016/S0899-3467(07)60142-6 PMID:19674681

Gulshan, V., Peng, L., Coram, M., Stumpe, M. C., Wu, D., Narayanaswamy, A., Venugopalan, S., Widner, K., Madams, T., Cuadros, J., Kim, R., Raman, R., Nelson, P. C., Mega, J. L., & Webster, D. R. (2016). Development and validation of a deep learning algorithm for detection of diabetic retinopathy in retinal fundus photographs. *Journal of the American Medical Association*, *316*(22), 2402–2410. doi:10.1001/jama.2016.17216 PMID:27898976

Gupta, A., & Katarya, R. (2020). Social media based surveillance systems for healthcare using machine learning: a systematic review. *J. Biomed. Inf.*, 103500.

Gupta, S., Tyagi, S., & Kishor, K. (2022). Study and Development of Self Sanitizing Smart Elevator. In D. Gupta, Z. Polkowski, A. Khanna, S. Bhattacharyya, & O. Castillo (Eds.), *Proceedings of Data Analytics and Management. Lecture Notes on Data Engineering and Communications Technologies* (Vol. 90). Springer. doi:10.1007/978-981-16-6289-8_15

Gürdür Broo, D., Kaynak, O., & Sait, S. M. (2021). Rethinking engineering education at the age of industry 5.0. *Journal of Industrial Information Integration*, *25*(8).

Han, J., Zhang, Z., Cao, J., Luo, Y., Zhang, L., Li, Z., & Zhang, J. (2020). Prediction of winter wheat yield based on multi-source data and machine learning in China. *Remote Sensing (Basel), 12*(2), 236. doi:10.3390/rs12020236

Hannan, E., & Liu, S. (2023). AI: New source of competitiveness in higher education. *Competitiveness Review, 33*(2), 265–279. doi:10.1108/CR-03-2021-0045

Hanson, B., Orloff, S., & Sanden, B. (2007). *Monitoring soil moisture for irrigation water management* (Vol. 21635). University of California, Agriculture and Natural Resources.

Hanson, B., Orloff, S., & Peters, D. (2000). Monitoring soil moisture helps refine irrigation management. *California Agriculture, 54*(3), 38–42. doi:10.3733/ca.v054n03p38

Heinrichs, W. (2005). Do it anywhere. Electronics Systems and Software, 3(4). doi:10.1049/ess:20050405

Higginbotham, S. (2020). What 5G hype gets wrong - [Internet of everything]. *IEEE Spectrum, 57*, 22–24.

Hinton, G. E., Osindero, S., & Teh, Y.-W. (2006). A fast learning algorithm for deep belief nets. *Neural Computation, 18*(7), 1527–1554. doi:10.1162/neco.2006.18.7.1527 PMID:16764513

Hu, C., Ju, R., Shen, Y., Zhou, P., & Li, Q. (2016). Clinical decision support for Alzheimer's disease based on deep learning and brain network. In *2016 IEEE International Conference on Communications (ICC)*. New York, NY: IEEE. 10.1109/ICC.2016.7510831

Hu, G., Wu, H., Zhang, Y., & Wan, M. (2019). A low shot learning method for tea leaf's disease identification. *Computers and Electronics in Agriculture, 163*, 104852. doi:10.1016/j.compag.2019.104852

Hu, P., Huang, Y., You, Z., Li, S., Chan, K. C. C., & Leung, H. (2019). Learning from deep representations of multiple networks for predicting drug-target interactions. In Lecture Notes in Computer Science: Vol. 11644. *Intelligent Computing Theories, ICIC 2019*. Springer. doi:10.1007/978-3-030-26969-2_14

Islam, M., & Shehzad, F. (2022). A Prediction Model Optimization Critiques through Centroid Clustering by Reducing the Sample Size, Integrating Statistical and Machine Learning Techniques for Wheat Productivity. *Scientifica, 7271293*, 1–11. doi:10.1155/2022/7271293 PMID:35310811

Ivanov, D. (2023). The Industry 5.0 framework: Viability-based integration of the resilience, sustainability, and human-centricity perspectives. *International Journal of Production Research, 61*(5), 1683–1695. doi:10.1080/00207543.2022.2118892

Jagtap, S. T., Phasinam, K., Kassanuk, T., Jha, S. S., Ghosh, T., & Thakar, C. M. (2022). Towards application of various machine learning techniques in agriculture. *Materials Today: Proceedings, 51*, 793–797. doi:10.1016/j.matpr.2021.06.236

Jain, A., Saify, A., & Kate, V. (2020). *Prediction of Nutrients (N, P, K) in soil using Colour Sensor (TCS3200). International Journal of Innovative Technology and Exploring Engineering.*

Javaid, M., & Haleem, A. (2020). Critical components of industry 5.0 towards a successful adoption in the field of manufacturing. *Journal of Industrial Integration and Management, 5*(03), 327–348. doi:10.1142/S2424862220500141

Javaid, M., Haleem, A., Singh, R. P., Haq, M. I., Raina, A., & Suman, R. (2020). "Industry 5.0: Potential applications in COVID-19", (2020). *Journal of Industrial Integration and Management, 05*(04), 507–530. doi:10.1142/S2424862220500220

Javed, A. R., Shahzad, F., & Rehman, S. (2022). *Future smart cities requirements, emerging technologies, applications, challenges, and future aspects* (Vol. 129). Cities.

Jha, D., & Kumar, S. (2006). *Research resource allocation in Indian agriculture, Policy Paper 23*. National Centre for Agricultural Economics and Policy Research.

Jha, K., Doshi, A., Patel, P., & Shah, M. (2019). A comprehensive review on automation in agriculture using artificial intelligence. *Artificial Intelligence in Agriculture, 2*, 1–12. doi:10.1016/j.aiia.2019.05.004

Jha, P., Kumar, V., Godara, R. K., & Chauhan, B. S. (2017). Weed management using crop competition in the United States: A review. *Crop Protection (Guildford, Surrey), 95*, 31–37. doi:10.1016/j.cropro.2016.06.021

Jiayu, Z., Shiwei, X., Zhemin, L., Wei, C., & Dongjie, W. (2015, May). Application of intelligence information fusion technology in agriculture monitoring and early-warning research. In *2015 International Conference on Control, Automation and Robotics* (pp. 114-117). IEEE. 10.1109/ICCAR.2015.7166013

Jibran, S., & Mufti, A. (2019). Issues and challenges in Indian agriculture. *International Journal of Commerce and Business Management, 12*(2), 85–88. doi:10.15740/HAS/IJCBM/12.2/85-88

Johnson, A. B., Kramer, A. A., & Clifford, G. D. (2023). Predictive Analytics in the Intensive Care Unit: A Survey. *IEEE Transactions on Biomedical Engineering, 70*, 63–76.

Jones, D., & Barnes, E. (2000). Fuzzy composite programming to combineremote sensing and crop models for decision support in precisioncrop management. *Agricultural Systems, 65*(3), 137–158. doi:10.1016/S0308-521X(00)00026-3

Jože. (2023). Human-centric artificial intelligence architecture for industry 5.0 applications. International Journal of Production Research, 61(20), 6847–6872. doi:10.1080/00207543.2022.2138611

Jr, P. P., Hatfield, J. L., Schepers, J. S., Barnes, E. M., Moran, S. M., Daughtry, C. S., & Upchurch, D. R. (2003). Remote Sensing for Crop Management. *Photogrammetric Engineering and Remote Sensing, 18*(6), 647–664. doi:10.14358/PERS.69.6.647

Kallia, M., van Borkulo, S. P., Drijvers, P., Barendsen, E., & Tolboom, J. (2021). Characterising computational thinking in mathematics education: A literature-informed Delphi study. *Research in Mathematics Education, 23*(2), 159–187. doi:10.1080/14794802.2020.1852104

Kamalov, F., Santandreu Calonge, D., & Gurrib, I. (2023). New Era of Artificial Intelligence in Education: Towards a Sustainable Multifaceted Revolution. *Sustainability (Basel), 15*(16), 12451. doi:10.3390/su151612451

Kamble, P. (2018). *Sustainability of Indian Agriculture of Indian Agriculture: Challenges and Opportunities*. Academic Press.

Kamilaris, A., Kartakoullis, A., & Prenafeta-Boldú, F. X. (2017). A review on the practice of big data analysis in agriculture. *Computers and Electronics in Agriculture, 143*, 23–37. doi:10.1016/j.compag.2017.09.037

Kamilaris, A., & Prenafeta-Boldú, F. X. (2018). Deep learning in agriculture: A survey. *Computers and Electronics in Agriculture, 147*, 70–90. doi:10.1016/j.compag.2018.02.016

Kanaka, K. K., Sukhija, N., Sivalingam, J., Goli, R. C., Rathi, P., Jaglan, K., & Raj, C. (2023). Deep Learning in Neural Networks and their Application in Genomics. *Acta Scientific Veterinary Sciences, 5*(7), 21–26. doi:10.31080/ASVS.2023.05.0683

Kang, H., & Chen, C. (2020). Fast implementation of real-time fruit detection in apple orchards using deep learning. *Computers and Electronics in Agriculture, 168*, 105108. doi:10.1016/j.compag.2019.105108

Karaca-Atik, A., Meeuwisse, M., Gorgievski, M., & Smeets, G. (2023). Uncovering important 21st-Century skills for sustainable career development of social sciences graduates: A systematic review. *Educational Research Review, 39*, 100528. doi:10.1016/j.edurev.2023.100528

Karimi, Y., Prasher, S. O., Patel, R. M., & Kim, S. H. (2006). Application of support vector machine technology for weed and nitrogen stress detection in corn. *Computers and Electronics in Agriculture*, *51*(1-2), 99–109. doi:10.1016/j.compag.2005.12.001

Kashyap, B., & Kumar, R. (2021). Sensing Methodologies in Agriculture for Soil Moisture and Nutrient Monitoring. *IEEE Access : Practical Innovations, Open Solutions*, *9*, 14095–14121. doi:10.1109/ACCESS.2021.3052478

Katariya, S. S., Gundal, S. S., Kanawade, M. T., & Mazhar, K. (2015). Automation in agriculture. *International Journal of Recent Scientific Research*, *6*(6), 4453–4456.

Katarya, R., Raturi, A., Mehndiratta, A., & Thapper, A. (2020). Impact of Machine Learning Techniques in Precision Agriculture. *2020 3rd International Conference on Emerging Technologies in Computer Engineering: Machine Learning and Internet of Things (ICETCE)*, (pp. 1–6). IEEE. 10.1109/ICETCE48199.2020.9091741

Kazimoglu, C. (2020). Enhancing confidence in using computational thinking skills via playing a serious game: A case study to increase motivation in learning computer programming. *IEEE Access : Practical Innovations, Open Solutions*, *8*, 221831–221851. https://ieeexplore.ieee.org/abstract/document/9286386/. doi:10.1109/ACCESS.2020.3043278

Kermany, D. S., Goldbaum, M., Cai, W., Valentim, C. C. S., Liang, H., Baxter, S. L., McKeown, A., Yang, G., Wu, X., Yan, F., Dong, J., Prasadha, M. K., Pei, J., Ting, M. Y. L., Zhu, J., Li, C., Hewett, S., Dong, J., Ziyar, I., ... Zhang, K. (2018). Identifying medical diagnoses and treatable diseases by image-based deep learning. *Cell*, *172*(5), 1122–1131. doi:10.1016/j.cell.2018.02.010 PMID:29474911

Khaki, S., Wang, L., & Archontoulis, S. V. (2020). A cnn-rnn framework for crop yield prediction. *Frontiers in Plant Science*, *10*, 1750. doi:10.3389/fpls.2019.01750 PMID:32038699

Khan & Amjad. (2019). Mutation Based Genetic Algorithm for Efficiency Optimization of Unit Testing. *International Journal of Advanced Intelligence Paradigms, 12*(3-4), 254-265. . doi:10.1504/IJAIP.2019.098563

Khan, Kumar, Dhingra, & Bhati. (2021). The Use of Different Image Recognition Techniques in Food Safety: A Study. *Journal of Food Quality*. doi:10.1155/2021/7223164

Khan, Kumar, Srivastava, Dhingra, Gupta, Bhati, & Kumari. (2021). Machine Learning and IoT Based Waste Management Model. *Computational Intelligence and Neuroscience*. doi:10.1155/2021/5942574

Khan, H., Haq, I. U., Munsif, M., Mustaqeem, Khan, S. U., & Lee, M. Y. (2022). Automated Wheat Diseases Classification Framework Using Advanced Machine Learning Technique. *Agriculture*, *12*(8), 1226. doi:10.3390/agriculture12081226

Khan, M., Gupta, B., Verma, A., Praveen, P., & Peoples, C. J. (Eds.). (2023). *Smart Village Infrastructure and Sustainable Rural Communities*. IGI Global. doi:10.4018/978-1-6684-6418-2

Khan, R., Khan, M. A., Ansari, M. A., Dhingra, N., & Bhati, N. (2022). Machine learning-based agriculture. In M. A. Khan, R. Khan, & M. A. Ansari (Eds.), *Application of Machine Learning in Agriculture* (pp. 3–27). Academic Press. doi:10.1016/B978-0-323-90550-3.00003-5

Khan, R., Shabaz, M., Hussain, S., Ahmed, F., & Mishra, P. (2021). *Early Flood Detection and Rescue Using Bioinformatic devices, Internet of Things (IoT) and Android App. Word Journal of Engineering Journal of Emerald Publishing*. doi:10.1108/WJE-05-2021-0269

Khoa, T. A., Man, M. M., Nguyen, T. Y., Nguyen, V., & Nam, N. H. (2019). Smart agriculture using IoT multi-sensors: A novel watering management system. *Journal of Sensor and Actuator Networks*, *8*(3), 45. doi:10.3390/jsan8030045

Kim, J., Kim, S., Ju, C., & Son, H. I. (2019). Unmanned aerial vehicles in agriculture: A review of perspective of platform, control, and applications. *IEEE Access: Practical Innovations, Open Solutions*, *7*, 105100–105115. doi:10.1109/ACCESS.2019.2932119

Kim, N., Ha, K. J., Park, N. W., Cho, J., Hong, S., & Lee, Y. W. (2019). A comparison between major artificial intelligence models for crop yield prediction: Case study of the midwestern United States, 2006–2015. *ISPRS International Journal of Geo-Information*, *8*(5), 240. doi:10.3390/ijgi8050240

Kim, S.-H., Lee, C., & Youn, C.-H. (2020). An Accelerated Edge Cloud System for Energy Data Stream Processing Based on Adaptive Incremental Deep Learning Scheme. *IEEE Access : Practical Innovations, Open Solutions*, *8*, 195341–195358. doi:10.1109/ACCESS.2020.3033771

Kim, Y., Evans, R. G., & Iversen, W. M. (2008). Remote sensing and control of an irrigation system using a distributed wireless sensor network. *IEEE Transactions on Instrumentation and Measurement*, *57*(7), 1379–1387. doi:10.1109/TIM.2008.917198

Kishor, K. (2023). Cloud Computing in Blockchain. In Cloud-based Intelligent Informative Engineering for Society 5.0 (pp. 79-105). Chapman and Hall/CRC. doi:10.1201/9781003213895-5

Kishor, K. (2023). Impact of Cloud Computing on Entrepreneurship, Cost, and Security. In Cloud-based Intelligent Informative Engineering for Society 5.0 (pp. 171-191). CRC Press. doi:10.1201/9781003213895-10

Kishor, K., & Nand, P. (2023). Wireless Networks Based in the Cloud That Support 5G. In Cloud-based Intelligent Informative Engineering for Society 5.0 (pp. 23-40). Chapman and Hall/CRC. doi:10.1201/9781003213895-2

Kishor, K., & Pandey, D. (2022). Study and Development of Efficient Air Quality Prediction System Embedded with Machine Learning and IoT. In *Proceeding International Conference on Innovative Computing and Communications. Lect. Notes in Networks, Syst.* (Vol. 471). Springer. 10.1007/978-981-19-2535-1_24

Kishor, K., Saxena, N., & Pandey, D. (Eds.). (2023). Cloud-based Intelligent Informative Engineering for Society 5.0. Chapman and Hall/CRC. doi:10.1201/9781003213895

Kishor, K., Tyagi, R., Bhati, R., & Rai, B. K. (2023). Develop Model for Recognition of Handwritten Equation Using Machine Learning. In *Proceedings of International Conference on Recent Trends in Computing. Lecture Notes in Networks and Systems* (vol. 600). Springer. 10.1007/978-981-19-8825-7_23

Kishor, K. (2022). Communication-Efficient Federated Learning. In S. P. Yadav, B. S. Bhati, D. P. Mahato, & S. Kumar (Eds.), *Federated Learning for IoT Applications. EAI/Springer Innovations in Communication and Computing*. Springer. doi:10.1007/978-3-030-85559-8_9

Kishor, K. (2022). Personalized Federated Learning. In S. P. Yadav, B. S. Bhati, D. P. Mahato, & S. Kumar (Eds.), *Federated Learning for IoT Applications. EAI/Springer Innovations in Communication and Computing*. Springer. doi:10.1007/978-3-030-85559-8_3

Kishor, K. (2023). *10 Study of quantum computing for data analytics of predictive and prescriptive analytics models*. In S. P. Yadav, R. Singh, V. Yadav, F. Al-Turjman, & S. A. Kumar (Eds.), *Quantum-Safe Cryptography Algorithms and Approaches: Impacts of Quantum Computing on Cybersecurity* (pp. 121–146). De Gruyter. doi:10.1515/9783110798159-010

Kishor, K. (2023). *12 Review and significance of cryptography and machine learning in quantum computing*. In S. P. Yadav, R. Singh, V. Yadav, F. Al-Turjman, & S. A. Kumar (Eds.), *Quantum-Safe Cryptography Algorithms and Approaches: Impacts of Quantum Computing on Cybersecurity* (pp. 159–176). De Gruyter. doi:10.1515/9783110798159-012

Kishor, K. (2023). *17 Application of quantum computing for digital forensic investigation.* In S. P. Yadav, R. Singh, V. Yadav, F. Al-Turjman, & S. A. Kumar (Eds.), *Quantum-Safe Cryptography Algorithms and Approaches: Impacts of Quantum Computing on Cybersecurity* (pp. 231–248). De Gruyter., doi:10.1515/9783110798159-017

Kishor, K., Nand, P., & Agarwal, P. (2018). Notice of Retraction Design adaptive Subnetting Hybrid Gateway MANET Protocol on the basis of Dynamic TTL value adjustment. *Aptikom Journal on Computer Science and Information Technologies, 3*(2), 59–65. doi:10.11591/APTIKOM.J.CSIT.115

Kishor, K., Nand, P., & Agarwal, P. (2018). Secure and Efficient Subnet Routing Protocol for MANET. *Indian Journal of Public Health Research & Development, 9*(12), 200. doi:10.5958/0976-5506.2018.01830.2

Kishor, K., Sharma, R., & Chhabra, M. (2022). Student Performance Prediction Using Technology of Machine Learning. In D. K. Sharma, S. L. Peng, R. Sharma, & D. A. Zaitsev (Eds.), *Micro-Electronics and Telecommunication Engineering. Lecture Notes in Networks and Systems* (Vol. 373). Springer. doi:10.1007/978-981-16-8721-1_53

Kishor, K., Singh, P., & Vashishta, R. (2023). Develop Model for Malicious Traffic Detection Using Deep Learning. In D. K. Sharma, S. L. Peng, R. Sharma, & G. Jeon (Eds.), *Micro-Electronics and Telecommunication Engineering. Lecture Notes in Networks and Systems* (Vol. 617). Springer. doi:10.1007/978-981-19-9512-5_8

Kitchen, D. B., Decornez, H., Furr, J. R., & Bajorath, J. (2004). Docking and Scoring in Virtual Screening for Drug Discovery: Methods and Applications. *Nature Reviews. Drug Discovery, 3*(11), 935–949. doi:10.1038/nrd1549 PMID:15520816

Knight, S., Wise, A. F., & Chen, B. (2017). Seeing through the glass: Using automated feedback to uncover learners' feedback literacy. *Computers in Human Behavior, 71*, 275–285.

Koedinger, K. R., & Corbett, A. T. (2006). Cognitive tutors: Technology bringing learning science to the classroom. In R. K. Sawyer (Ed.), *The Cambridge Handbook of the Learning Sciences* (pp. 61–78). Cambridge University Press.

Kohavi, R., & Quinlan, J. R. (2002). Data Mining Tasks and Methods: Classification: Decision Trees. Encyclopedia of Machine Learning, 12(1), 63-72.

Kolasa, K. (2023). *The Digital Transformation of the Healthcare System: Healthcare 5.0.* Taylor & Francis. doi:10.4324/b23291

Koller, D., & Ng, A. (2016). The Online Revolution: Education for Everyone. *Daedalus, 145*(1), 62–69.

Kothari, J. D. (2018). Plant Disease Identification using Artificial Intelligence: Machine Learning Approach. *International Journal of Innovative Research in Computer and Communication Engineering, 7*(11), 11082–11085.

Krishna Kumar, K., Rupa Kumar, K., Ashrit, R. G., Deshpande, N. R., & Hansen, J. W. (2004). Climate impacts on Indian agriculture. *International Journal of Climatology, 24*(11), 1375–1393. doi:10.1002/joc.1081

Kumar, N., Siddiqi, A. H., & Alam, K. (2017). Wavelet Based EEG Signal Classification. *Biomedical and Pharmacology Journal, 10*(4), 2061-2069.

Kumar, N., Tripathi, P., & Alam, K. (2016). Non-Negative Factorization Based EEG Signal Classification. *Indian Journal of Industrial and Applied Mathematics, 7*(2), 2012-219.

Kumar, I., Mishra, Z., Rajput, A. S., & Parmar, O. (2022). IoT based Motor Pump Control System. *2022 IEEE International Conference on Current Development in Engineering and Technology (CCET),* 1-5. 10.1109/CCET56606.2022.10080813

Kuyper, M. C., & Balendonck, J. (1997, August). Application of dielectric soil moisture sensors for real-time automated irrigation control. In *III International Symposium on Sensors in Horticulture 562* (pp. 71-79). Academic Press.

Lal, H., Jones, J. W., Peart, R. M., & Shoup, W. D. (1992). FARMSYS—A whole-farm machinery management decision support system. *Agricultural Systems*, *38*(3), 257–273. doi:10.1016/0308-521X(92)90069-Z

Lamm, R. D., Slaughter, D. C., & Giles, D. K. (2002). Precision weed control system for cotton. *Transactions of the ASAE. American Society of Agricultural Engineers*, *45*(1), 231.

Lee, Lee, & Chung. (2001). Face recognition using Fisherface algorithm and elastic graph matching. *Proceedings 2001 International Conference on Image Processing*, 998-1001.

Lee, I., & Lee, K. (2015). The Internet of Things (IoT): Applications, investments, and challenges for enterprises. *Business Horizons*, *58*(4), 431–440. doi:10.1016/j.bushor.2015.03.008

Lee, J., Wang, J., Crandall, D., Šabanović, S., & Fox, G. (2017, April). Real-time, cloud-based object detection for unmanned aerial vehicles. In *2017 First IEEE International Conference on Robotic Computing (IRC)* (pp. 36-43). IEEE. 10.1109/IRC.2017.77

Leena, U., Premasudha, B., & Basavaraja, P. (2016). Sensible approach for soil fertility management using GIS cloud. *International Conference on Advances in Computing, Communications and Informatics (ICACCI)*, 2776-2781. 10.1109/ICACCI.2016.7732483

Lee, W., Alchanatis, V., Yang, C., Hirafuji, M., Moshou, D., & Li, C. (2010). Sensing technologies for precision specialty crop production. *Computers and Electronics in Agriculture*, *74*(1), 2–33. doi:10.1016/j.compag.2010.08.005

Li, Q., Li, R., Ji, K., & Dai, W. (2015). Kalman Filter and Its Application. *2015 8th International Conference on Intelligent Networks and Intelligent Systems (ICINIS)*, 74-77. 10.1109/ICINIS.2015.35

Liakos, K. G., Busato, P., Moshou, D., Pearson, S., & Bochits, D. (2018). Machine Learning in Agriculture: A Review. *Sensors*.

Liakos, K. G., Busato, P., Moshou, D., Pearson, S., & Bochtis, D. (2018). Machine learning in agriculture: A review. *Sensors (Basel)*, *18*(8), 2674. doi:10.3390/s18082674 PMID:30110960

Liao, Z., Yin, Q., Huang, Y., & Sheng, L. (2014). Management and application of mobile big data International. *J. Embedded Syst.*, *7*(1), 63–70. doi:10.1504/IJES.2015.066143

Li, H., Liu, D. Y., & Wu, X. (2020). Artificial intelligence in education: A review. *Journal of Computer Assisted Learning*, *36*(1), 6–27.

Li, J., Zhang, L., & Liu, S. (2022). Deep Learning for Diabetic Retinopathy Detection Using Retinal Fundus Images. *IEEE Transactions on Medical Imaging*, *41*(2), 323–331.

Lin, P. H., & Chen, S. Y. (2020). Design and evaluation of a deep learning recommendation based augmented reality system for teaching programming and computational thinking. *IEEE Access : Practical Innovations, Open Solutions*, *8*, 45689–45699. https://ieeexplore.ieee.org/abstract/document/9020189/. doi:10.1109/ACCESS.2020.2977679

Liong-Rung, L., Hung-Wen, C., Ming-Yuan, H., Shu-Tien, H., Ming-Feng, T., Chia-Yu, C., & Kuo-Song, C. (2022). Using Artificial Intelligence to Establish Chest X-Ray Image Recognition Model to Assist Crucial Diagnosis in Elder Patients With Dyspnea. *Frontiers in Medicine*, *9*, 893208. doi:10.3389/fmed.2022.893208 PMID:35721050

Liu, R. (2018). B., Yang, E., Zio, X., Chen: Artificial intelligence forfault diagnosis of rotating machinery: A review. *Mechanical Systems and Signal Processing*, *108*, 33–47. doi:10.1016/j.ymssp.2018.02.016

Liu, Y., Zhang, Y., & Hu, Z. (2021). Personalized Chemotherapy Selection for Advanced Gastric Cancer Patients using Machine Learning. *Clinical Cancer Research*, *27*(4), 1037–1045. PMID:33272982

Longo, F., Padovano, A., & Umbrella, S. (2020). Value-oriented and ethical technology engineering in industry 5.0: A human-centric perspective for the design of the factory of the future. *Applied Sciences (Basel, Switzerland)*, *10*(12), 4182. doi:10.3390/app10124182

Lotfy Abdrabou, E. A. M., & Salem, A.-B. M. (2010). A breast cancer classifier based on a combination of case-based reasoning and ontology approach. *Proceedings of the International Multiconference on Computer Science and Information Technology*, 3-10. 10.1109/IMCSIT.2010.5680045

Luciani, R., Laneve, G., & JahJah, M. (2019). Agricultural monitoring, an automatic procedure for crop mapping and yield estimation: The great rift valley of Kenya case. *IEEE Journal of Selected Topics in Applied Earth Observations and Remote Sensing*, *12*(7), 2196–2208. doi:10.1109/JSTARS.2019.2921437

Luckin, R., & Holmes, W. (2016). *Intelligence unleashed: An argument for AI in education.* Academic Press.

Luckin, R. (2018). Machine Learning and Human Intelligence: The Future of Education for the 21st Century. *Zeitschrift für Psychologie mit Zeitschrift für Angewandte Psychologie*, *226*(2), 82–93.

Lu, Y. (2017)."Industry 4.0: A survey on technologies, applications and open research issues. *Journal of Industrial Information Integration*, *6*, 110. doi:10.1016/j.jii.2017.04.005

Lu, Y., Adrados, J. S., Chand, S. S., & Wang, L. (2021). Humans are not machines—Anthropocentric human–machine symbiosis for ultra-flexible smart manufacturing. *Engineering (Beijing)*, *7*(6), 734–737. doi:10.1016/j.eng.2020.09.018

Lu, Y., Liu, C., Kevin, L., Wang, K., Huang, H., & Xu, X. (2020). Digital twin - driven smart manufacturing: Connotations, reference model, applications, and research issues. *Robotics and Computer-integrated Manufacturing*, *61*, 101837. doi:10.1016/j.rcim.2019.101837

Lu, Y., Yi, S., Zeng, N., Liu, Y., & Zhang, Y. (2017). Identification of rice diseases using deep convolutional neural networks. *Neurocomputing*, *267*, 378–384. doi:10.1016/j.neucom.2017.06.023

Lykov & Razumowsky. (2023). Industry 5.0 and human capital. *E3S Web of Conferences, 376.* doi:10.1051/e3sconf/202337605053

Maddikunta, P. K., Pham, Q. V., Prabadevi, B., Deepa, N., Dev, K., Gadekallu, T. R., Ruby, R., & Liyanage, M. (2021). Industry 5.0: A survey on enabling technologies and potential applications. *Journal of Industrial Information Integration*, *26*, 100257. doi:10.1016/j.jii.2021.100257

Madsen, D. O., & Berg, T. (2021). An exploratory bibliometric analysis of the birth and emergence of industry 5.0. *Applied System Innovation*, *4*(4), 87. doi:10.3390/asi4040087

Manek, A. H., & Singh, P. K. (2016, July). Comparative study of neural network architectures for rainfall prediction. In 2016 IEEE Technological Innovations in ICT for Agriculture and Rural Development (TIAR) (pp. 171-174). IEEE doi:10.1109/TIAR.2016.7801233

Mann, S., Pathak, N., Sharma, N., Kumar, R., Porwal, R., Sharma, S. K., & Aung, S. M. Y. (2022). Study of Energy-Efficient Optimization Techniques for High-Level Homogeneous Resource Management. *Wireless Communications and Mobile Computing*. doi:10.1155/2022/1953510

Martinez, E., Johnson, A., & Garcia, M. (2023). Personalized Treatment Plans using Machine Learning Algorithms. *Journal of Personalized Medicine*, *13*(4), 257–270.

Martinez-Murcia, F. J., Ortiz, A., Gorriz, J. M., Ramirez, J., & Castillo-Barnes, D. (2020). Studying the Manifold Structure of Alzheimer's Disease: A Deep Learning Approach Using Convolutional Autoencoders. *IEEE Journal of Biomedical and Health Informatics*, *24*(1), 17–26. doi:10.1109/JBHI.2019.2914970 PMID:31217131

Martos, V., Ahmad, A., Cartujo, P., & Ordonez, J. (2021). Ensuring agricultural sustainability through remote sensing in the era of agriculture 5.0. *Applied Sciences (Basel, Switzerland), 11*(13), 5911. doi:10.3390/app11135911

Masood, T., & Sonntag, P. (2020). Industry 4.0: Adoption challenges and benefits for SMEs. *Computers in Industry, 121*, 103261. doi:10.1016/j.compind.2020.103261

Ma, T., Xiao, C., & Wang, F. (2018). Health-ATM: A deep architecture for multifaceted patient health record representation and risk prediction. In *Proceedings of the 2018 SIAM International Conference on Data Mining (SDM).* SIAM. 10.1137/1.9781611975321.30

Mbunge, E., Muchemwa, B., & Batani, J. (2021). Sensors and healthcare 5.0: Transformative shift in virtual care through emerging digital health technologies. *Global Health Journal (Amsterdam, Netherlands), 5*(4), 169–177. doi:10.1016/j.glohj.2021.11.008

McCarthy, J., Minsky, M., Rochester, N., & Shannon, C. E. (2006). A Proposal for the Dartmouth Summer Research Project on Artificial Intelligence, August 31, 1955. *AI Magazine, 27*, 12–14.

McGann, M. (2011). FRED Pose Prediction and Virtual Screening Accuracy. *Journal of Chemical Information and Modeling, 51*(3), 578–596. doi:10.1021/ci100436p PMID:21323318

McQueen, R. J., Garner, S. R., Nevill-Manning, C. G., & Witten, I. H. (1995). Applying machine learning to agricultural data. *Computers and Electronics in Agriculture, 12*(4), 275–293. doi:10.1016/0168-1699(95)98601-9

Mehdiabadi, A., Shahabi, V., Shamsinejad, S., Amiri, M., Spulbar, C., & Birau, R. (2022). Investigating Industry 5.0 and Its Impact on the Banking Industry: Requirements, Approaches and Communications. *Applied Sciences (Basel, Switzerland), 12*(10), 5126. doi:10.3390/app12105126

Merzon, E. E., & Ibatullin, R. R. (2016). Architecture of smart learning courses in higher education. *2016 IEEE 10th International Conference on Application of Information and Communication Technologies (AICT),* 1-5. 10.1109/ICAICT.2016.7991809

Miao, F., Holmes, W., Huang, R., & Zhang, H. (2021). *AI and education: A guidance for policymakers.* UNESCO Publishing.

Miotto, R., Li, L., Kidd, B. A., & Dudley, J. T. (2016). Deep patient: An unsupervised representation to predict the future of patients from the electronic health records. *Scientific Reports, 6*(1), 26094–26094. doi:10.1038/srep26094 PMID:27185194

Mitchell, J., & Guile, D. (2022). Fusion skills and industry 5.0: conceptions and challenges. *Insights Into Global Engineering Education After the Birth of Industry 5.0*, 53.

Moghaddam, M. T., Muccini, H., Dugdale, J., & Kjægaard, M. B. (2022). Designing Internet of Behaviors Systems. *2022 IEEE 19th International Conference on Software Architecture (ICSA),* 124-134. 10.1109/ICSA53651.2022.00020

Mohanta, Das, & Patnaik. (2019). Healthcare 5.0: A Paradigm Shift in Digital Healthcare System Using Artificial Intelligence, IOT and 5G Communication. *2019 International Conference on Applied Machine Learning (ICAML),* 191-196. 10.1109/ICAML48257.2019.00044

Mohanty, S. P., Hughes, D. P., & Salathé, M. (2016). Using deep learning for image-based plant disease detection. *Frontiers in Plant Science, 7*, 1419. doi:10.3389/fpls.2016.01419 PMID:27713752

Monteiro, A. L., de Freitas Souza, M., Lins, H. A., da Silva Teófilo, T. M., Júnior, A. P. B., Silva, D. V., & Mendonça, V. (2021). A new alternative to determine weed control in agricultural systems based on artificial neural networks (ANNs). *Field Crops Research, 263*, 108075. doi:10.1016/j.fcr.2021.108075

Mooney, D. F., Roberts, R. K., English, B. C., Lambert, D. M., Larson, J. A., Velandia, M., & Reeves, J. M. (2009). Precision Farming by Cotton Producers in Twelve Southern States: Results from the 2009 Southern Cotton Precision Farming Survey. *Research in Agriculture and Applied Economics.*

Moore, I. D., Gessler, P. E., Nielsen, G. A., & Peterson, G. A. (1993, May). Terrain analysis for soil specific crop management. In *Proceedings of Soil Specific Crop Management: A Workshop on Research and Development Issues* (pp. 27-55). Madison, WI, USA: American Society of Agronomy, Crop Science Society of America, Soil Science Society of America. 10.2134/1993.soilspecificcrop.c3

Moran, M. S., Inoue, Y., & Barnes, E. M. (1997). Opportunities and limitations for image-based remote sensing in precision crop management. *Remote Sensing of Environment, 61*(3), 319–346. doi:10.1016/S0034-4257(97)00045-X

Moshou, D., Bravo, C., West, J., Wahlen, S., McCartney, A., & Ramon, H. (2004). Automatic detection of 'yellow rust' in wheat using reflectance measurements and neural networks. *Computers and Electronics in Agriculture, 44*(3), 173–188. doi:10.1016/j.compag.2004.04.003

Mostavi, M., Chiu, Y.-C., Huang, Y., & Chen, Y. (2020). Convolutional neural network models for cancer type prediction based on gene expression. *BMC Medical Genomics, 13*(S5), 44. doi:10.1186/s12920-020-0677-2 PMID:32241303

Muthuswamy, P. (2022). Artificial intelligence based tool condition monitoring for digital twins and industry 4.0 applications. *International Journal on Interactive Design and Manufacturing (IJIDeM),* 1-21.

Nahavandi, S. (2019). Industry 5.0—A Human-Centric Solution. *Sustainability (Basel), 11*(16), 4371. doi:10.3390/su11164371

Nair, M. M., Tyagi, A. K., & Sreenath, N. (2021). The Future with Industry 4.0 at the Core of Society 5.0: Open Issues, Future Opportunities and Challenges. *International Conference on Computer Communication and Informatics (ICCCI),* 1–7. 10.1109/ICCCI50826.2021.9402498

Nguyen, P., Tran, T., Wickramasinghe, N., & Venkatesh, S. (2017). Deepr: A Convolutional Net for Medical Records. *IEEE Journal of Biomedical and Health Informatics, 21*(1), 22–30. doi:10.1109/JBHI.2016.2633963 PMID:27913366

Nickerson, P., Tighe, P., Shickel, B., & Rashidi, P. (2016). Deep neural network architectures for forecasting analgesic response. In *2016 38th Annual International Conference of the IEEE Engineering in Medicine and Biology Society (EMBC).* New York, NY: IEEE. 10.1109/EMBC.2016.7591352

Nithin Reddy, Danish, Babu, & Koperundevi. (2018). Automatic Irrigation and Soil Quality Testing. *International Conference on Recent Innovations in Electrical, Electronics & Communication Engineering.*

Nourmohammadi, A., Fathi, M., & Ng, A. H. C. (2022). Balancing and scheduling assembly lines with human-robot collaboration tasks. *Computers & Operations Research, 140,* 105674. doi:10.1016/j.cor.2021.105674

Nyéki, A., Kerepesi, C., Daróczy, B., Benczúr, A., Milics, G., Nagy, J., Harsányi, E., Kovács, A. J., & Neményi, M. (2021). Application of spatio-temporal data in site-specific maize yield prediction with machine learning methods. *Precision Agriculture, 22*(5), 1397–1415. doi:10.1007/s11119-021-09833-8

Olszewski, M. (2020). Modern industrial robotics. *Pomiary Automatyka Robotyka, 24*(1), 5–20. doi:10.14313/PAR_235/5

Ortiz, A., Munilla, J., Górriz, J. M., & Ramírez, J. (2016). Ensembles of deep learning architectures for the early diagnosis of the Alzheimer's disease. *International Journal of Neural Systems, 26*(7), 1650025. doi:10.1142/S0129065716500258 PMID:27478060

Ortuani, B., Facchi, A., Mayer, A., Bianchi, D., Bianchi, A., & Brancadoro, L. (2019). Assessing the Effectiveness of Variable-Rate Drip Irrigation on Water Use Efficiency in a Vineyard in Northern Italy. *Water (Basel), 11*(10), 1964. Advance online publication. doi:10.3390/w11101964

Oshaughnessy, S., Evett, S. R., Colaizzi, P. D., Andrade, M. A., & Marek, T. H. (2019). *Identifying Advantages and Disadvantages of Variable Rate Irrigation-An Updated Review*. Biological Systems Engineering.

Østergaard, E. H. (2018). Welcome to industry 5.0. *Retrieved Febr, 5*, 2020.

Özdemir, V., & Hekim, N. (2018). Birth of Industry 5.0: Making Sense of Big Data with Artificial Intelligence. *OMICS: A Journal of Integrative Biology, 22*(Jan), 65–76. doi:10.1089/omi.2017.0194 PMID:29293405

Pallathadka, H., Jawarneh, M., Sammy, F., Garchar, V., Sanchez, D. T., & Naved, M. (2022). A Review of Using Artificial Intelligence and Machine Learning in Food and Agriculture Industry. *2022 2nd International Conference on Advance Computing and Innovative Technologies in Engineering (ICACITE)*, (pp. 2215–2218). IEEE. 10.1109/ICACITE53722.2022.9823427

Pallathadka, H., Mustafa, M., Sanchez, D. T., Sekhar Sajja, G., Gour, S., & Naved, M. (2023). IMPACT OF MACHINE learning ON Management, healthcare AND AGRICULTURE. *Materials Today: Proceedings, 80*, 2803–2806. doi:10.1016/j.matpr.2021.07.042

Pallevada, Potu, Munnangi, Rayapudi, Gadde, & Chinta. (2021). Real-time Soil Nutrient detection and Analysis. *International Conference on Advance Computing and Innovative Technologies in Engineering (ICACITE)*.

Pal, S., & Jha, D. (2007). Public-private partnerships in Agricultural R&D: Challenges and Prospects. In V. Ballabh (Ed.), *Institutional Alternatives and Governance of Agriculture*. Academic Foundation.

Pan, M. L. (2016). *Preparing literature reviews: Qualitative and quantitative approaches*. Taylor & Francis. doi:10.4324/9781315265872

Pan, S. J., & Yang, Q. (2010). A survey on transfer learning. *IEEE Transactions on Knowledge and Data Engineering, 22*(10), 1345–1359. doi:10.1109/TKDE.2009.191

Pappas, I. O., Mikalef, P., Giannakos, M. N., Krogstie, J., & Lekakos, G. (2018). Big data and business analytics ecosystems: Paving the way towards digital transformation and sustainable societies. *Information Systems and e-Business Management, 16*(3), 479–491. doi:10.1007/s10257-018-0377-z

Parmar, J., Palav, P., Nagda, T., & Lopes, H. (2018). IOT Based Weather Intelligence. *International Conference on Smart City and Emerging Technology (ICSCET)*.

Partel, V., Kakarla, S. C., & Ampatzidis, Y. (2019). Development and evaluation of a low-cost and smart technology for precision weed management utilizing artificial intelligence. *Computers and Electronics in Agriculture, 157*, 339–350. doi:10.1016/j.compag.2018.12.048

Patel, R., Sharma, A., & Williams, J. (2023). Machine Learning Models for Drug Discovery and Development. *Journal of Pharmaceutical Sciences, 22*(1), 45–57.

Pathak, Pal, & Mohapatra. (2020). Mahatma Gandhi's Vision of Agriculture: Achievements of ICAR. Indian Council of Agricultural Research.

Paymode, A. S., Shinde, J. N. M. U. B., & Malode, V. B. (2021). Artificial intelligence for agriculture: A technique of vegetables crop onion sorting and grading using deep learning. *International Journal, 6*(4).

Pereira, A. G., Lima, T. M., & Santos, F. C. (2020). Industry 4.0 and society 5.0: Opportunities and threats. *International Journal of Recent Technology and Engineering*, *8*(5), 3305–3308. doi:10.35940/ijrte.D8764.018520

Perez, S., Massey-Allard, J., Butler, D., Ives, J., Bonn, D., Yee, N., & Roll, I. (2017). Identifying productive inquiry in virtual labs using sequence mining. *Artificial Intelligence in Education: 18th International Conference, AIED 2017, Wuhan, China, June 28–July 1, 2017 Proceedings*, *18*, 287–298.

Perin, D., & Lauterbach, M. (2018). Assessing text-based writing of low-skilled college students. *International Journal of Artificial Intelligence in Education*, *28*(1), 56–78. doi:10.1007/s40593-016-0122-z

Pernapati, K. (2018). IoT based low cost smart irrigation system. *Proc. 2nd Int. Conf. Inventive Commun. Comput. Technol. (ICICCT)*, 1312–1315.

Pianykh, O. S., Guitron, S., Parke, D., Zhang, C., Pandharipande, P., Brink, J., & Rosenthal, D. (2020). Improving healthcare operations management with machine learning. *Nature Machine Intelligence*, *2*(5), 266–273. doi:10.1038/s42256-020-0176-3

Pilarski, T., Happold, M., Pangels, H., Ollis, M., Fitzpatrick, K., & Stentz, A. (2002). The demeter system for automated harvesting. *Autonomous Robots*, *13*(1), 9–20. doi:10.1023/A:1015622020131

Planning Commission. (2011). *Draft on Faster, Sustainable and More Inclusive Growth – An approach to Twelfth Five Year Plan*. Author.

Poblete, T., Ortega-Farías, S., Moreno, M. A., & Bardeen, M. (2017). Artificial neural network to predict vine water status spatial variability using multispectralinformation obtained from an unmanned aerial vehicle (UAV). *Sensors (Basel)*, *17*(11), 2488. doi:10.3390/s17112488 PMID:29084169

Pokhrel, B. K., Paudel, K. P., & Segarra, E. (2018). Factors Affecting the Choice, Intensity, and Allocation of Irrigation Technologies by U.S. Cotton Farmers. *Water (Basel)*, *10*(6), 706. Advance online publication. doi:10.3390/w10060706

Popenici, S. A., & Kerr, S. (2017). Exploring the impact of artificial intelligence on teaching and learning in higher education. *Research and Practice in Technology Enhanced Learning*, *12*(1), 1–13. doi:10.1186/s41039-017-0062-8 PMID:30595727

Poplin, R., Varadarajan, A. V., Blumer, K., Liu, Y., McConnell, M. V., Corrado, G. S., Peng, L., & Webster, D. R. (2018). Prediction of cardiovascular risk factors from retinal fundus photographs via deep learning. *Nature Biomedical Engineering*, *2*(3), 158–164. doi:10.1038/s41551-018-0195-0 PMID:31015713

Popoola, S. I., Adebisi, B., Hammoudeh, M., Gui, G., & Gacanin, H. (2020). Hybrid deep learning for botnet attack detection in the internet-of-things networks. *IEEE Internet of Things Journal*, *8*(6), 4944–4956. doi:10.1109/JIOT.2020.3034156

Pramod, Naicker, & Tyagi. (2022). Emerging Innovations shortly Using Deep Learning Techniques. *Advanced Analytics and Deep Learning Models*.

Praveen, K. R. M., Pham, Q.-V., & Prabadevi, B. (2022). Industry 5.0: A survey on enabling technologies and potential applications. *Journal of Industrial Information*, *26*, 1–8.

Praveen, P., Khan, A., Verma, A. R., Kumar, M., & Peoples, C. (2023). *Smart Village Infrastructure and Rural Communities*. IGI Global. doi:10.4018/978-1-6684-6418-2.ch001

Pujahari, Yadav, & Khan. (2022). Intelligence Farming System through Weather Forecast Support and Crop Production. In *Application of Machine Learning in Smart Agriculture*. Elsevier. . doi:10.1016/B978-0-323-90550-3.00009-6

Pyingkodi, M., Thenmozhi, K., Karthikeyan, M., Kalpana, T., & Suresh Palarimath, G. (2022). IoT based Soil Nutrients Analysis and Monitoring System for Smart Agriculture. *Proceedings of the Third International Conference on Electronics and Sustainable Communication Systems.*

Quinlan, J. R. (1986). Induction of Decision Trees. *Machine Learning, 1*(1), 81–106. doi:10.1007/BF00116251

Rachmawati, I., Multisari, W., Triyono, T., Simon, I. M., & da Costa, A. (2021). Prevalence of Academic Resilience of Social Science Students in Facing the Industry 5.0 Era. *International Journal of Evaluation and Research in Education, 10*(2), 676–683. doi:10.11591/ijere.v10i2.21175

Radoglou-Grammatikis, P., Sarigiannidis, P., Lagkas, T., & Moscholios, I. (2020). A compilation of UAVapplications for precision agriculture. Computer Networks, 172, 107148.

Rahutomo, R., Perbangsa, A. S., Lie, Y., Cenggoro, T. W., & Pardamean, B. (2019, August). Artificial intelligence model implementation in web-based application for pineapple object counting. In *2019 International Conference on Information Management and Technology (ICIMTech)* (Vol. 1, pp. 525-530). IEEE. 10.1109/ICIMTech.2019.8843741

Rai, B. K., Sharma, S., Kumar, G., & Kishor, K. (2022). Recognition of Different Bird Category Using Image Processing. *International Journal of Online and Biomedical Engineering, 18*(07), 101–114. doi:10.3991/ijoe.v18i07.29639

Rajkomar, A., Oren, E., Chen, K., Dai, A. M., Hajaj, N., & Hardt, M. (2018). Scalable and accurate deep learning with electronic health records. *NPJ Digital Medicine, 1*, 18. PMID:31304302

Rajotte, E. G., Bowser, T., Travis, J. W., Crassweller, R. M., Musser, W., Laughland, D., & Sachs, C. (1992, February). Implementation and adoption of an agricultural expert system: The Penn State Apple Orchard Consultant. In *III International Symposium on Computer Modelling in Fruit Research and Orchard Management 313* (pp. 227-232). 10.17660/ActaHortic.1992.313.28

Rajput, S., Khanna, L., & Kumari, P. (2023). *Artificial Intelligence and Machine Learning-Based Agriculture.* doi:10.4018/978-1-6684-6418-2.ch002

Raju, D., & Schumacker, R. (2015). Exploring student characteristics of retention that lead to graduation in higher education using data mining models. *Journal of College Student Retention, 16*(4), 563–591. doi:10.2190/CS.16.4.e

Rajumesh. (2023). Promoting sustainable and human-centric industry 5.0: a thematic analysis of emerging research topics and opportunities. *Journal of Business and Socio-Economic Development.* doi:10.1108/JBSED-10-2022-0116

Ramos, P. J., Prieto, F. A., Montoya, E. C., & Oliveros, C. E. (2017). Automatic fruit count on coffee branches using computer vision. *Computers and Electronics in Agriculture, 137*, 9–22. doi:10.1016/j.compag.2017.03.010

Rane, N. (2023). ChatGPT and Similar Generative Artificial Intelligence (AI) for Smart Industry: Role, Challenges and Opportunities for Industry 4.0, Industry 5.0 and Society 5.0. *Challenges and Opportunities for Industry, 4.*

Ray, P. P. (2017). Internet of Things for smart agriculture: Technologies, practices and future direction. *Journal of Ambient Intelligence and Smart Environments, 9*(4), 395–420. doi:10.3233/AIS-170440

Redmon, J., Divvala, S., Girshick, R., & Farhadi, A. (2016). You Only Look Once: Unified, Real-Time Object Detection. *2016 IEEE Conference on Computer Vision and Pattern Recognition (CVPR),* 779-788. 10.1109/CVPR.2016.91

Ruppert, Darányi, Medvegy, & Csereklei. (2022). Demonstration Laboratory of Industry 4.0 Retrofitting and Operator 4.0 Solutions: Education towards Industry 5.0. *Sensors, 23*(1), 283.

Sahani, Srivastava, & Khan. (2021). Modelling Techniques to Improve the Quality of Food using Artificial Intelligence. *Artificial Intelligence in Food Quality Improvement.* doi:10.1155/2021/2140010

Sahu, P., Singh, A. P., Chug, A., & Singh, D. (2022). A Systematic Literature Review of Machine Learning Techniques Deployed in Agriculture: A Case Study of Banana Crop. *IEEE Access : Practical Innovations, Open Solutions, 10*, 87333–87360. doi:10.1109/ACCESS.2022.3199926

Samavedham, L., & Ragupathi, K. (2012). Facilitating 21st century skills in engineering students. *Journal of Engineering Education, 26*(1), 38–49.

Saniuk, S., & Grabowska, S. (2022). *Development of knowledge and skills of engineers and managers in the era of Industry 5.0 in the light of expert research.* Zeszyty Naukowe. Organizacja i Zarządzanie/Politechnika Śląska. doi:10.29119/1641-3466.2022.158.35

Sannakki, S. S., Rajpurohit, V. S., Nargund, V. B., Kumar, A., & Yallur, P. S. (2011). Leaf disease grading by machine vision and fuzzy logic. *International Journal (Toronto, Ont.), 2*(5), 1709–1716.

Santosh, K. C., & Gaur, L. (2022). *Artificial intelligence and machine learning in public healthcare: Opportunities and societal impact.* Springer Nature.

Sarmadian, F., & Keshavarzi, A. (2010). Developing pedotransfer functions for estimating some soil properties using artificial neural network and multivariate regression approaches. *Int. J. Environ. Earth Sci, 1*, 31–37.

Saxena, A., Pant, D., Saxena, A., & Patel, C. (2020). Emergence of Educators for Industry 5.0 - An Indological Perspective. *International Journal of Innovative Technology and Exploring Engineering, 9*(12), 359–363. doi:10.35940/ijitee.L7883.1091220

Schmidhuber, J. (2015). Deep Learning in Neural Networks: An Overview. *Neural Networks, 61*, 85–117. doi:10.1016/j.neunet.2014.09.003 PMID:25462637

Schmid, T., Rodríguez-Rastrero, M., Escribano, P., Palacios-Orueta, A., Ben-Dor, E., Plaza, A., Milewski, R., Huesca, M., Bracken, A., Cicuendez, V., Pelayo, M., & Chabrillat, S. (2015). Characterization of soil erosion indicators using hyperspectral data from a Mediterranean rainfed cultivated region. *IEEE Journal of Selected Topics in Applied Earth Observations and Remote Sensing, 9*(2), 845–860. doi:10.1109/JSTARS.2015.2462125

Schwab, K. (2022). *The Fourth Industrial Revolution: what it means, how to respond.* World Economic Forum.

Sekhon, A., Singh, R., & Qi, Y. (2018). DeepDiff: DEEP-learning for predicting DIFFerential gene expression from histone modifications. *Bioinformatics (Oxford, England), 34*(17), i891–i900. doi:10.1093/bioinformatics/bty612 PMID:30423076

Sendak, M. P., D'Arcy, J., Kashyap, S., Gao, M., Nichols, M., Corey, K., & Balu, S. (2020). A path for translation of machine learning products into healthcare delivery. *EMJ Innov, 10*, 19–172.

Shakoor, M. T., Rahman, K., Rayta, S. N., & Chakrabarty, A. (2017). Agricultural production output prediction using Supervised Machine Learning techniques. *2017 1st International Conference on Next Generation Computing Applications (NextComp)*, (pp. 182–187). IEEE. 10.1109/NEXTCOMP.2017.8016196

Sharma, A., Jain, A., Gupta, P., & Chowdary, V. (2021). Machine Learning Applications for Precision Agriculture: A Comprehensive Review. In IEEE Access (Vol. 9, pp. 4843–4873). Institute of Electrical and Electronics Engineers Inc. doi:10.1109/ACCESS.2020.3048415

Sharma, S. K., Sharma, N. K., & Potter, P. P. (2020, December). Fusion approach for document classification using random forest and svm. In *2020 9th International Conference System Modeling and Advancement in Research Trends (SMART)* (pp. 231-234). IEEE. 10.1109/SMART50582.2020.9337131

Sharma, S. K., & Sharma, N. K. (2019a). Text Document Categorization using Modified K-Means Clustering Algorithm. *International Journal of Recent Technology and Engineering, 8*(2), 508–511.

Sharma, S. K., & Sharma, N. K. (2019b). Text Classification using Ensemble of Non-Linear Support Vector Machines. *International Journal of Innovative Technology and Exploring Engineering*, *8*(10), 3170–3174. doi:10.35940/ijitee.J9520.0881019

Sharma, S. K., & Sharma, N. K. (2019c). Text Classification using LSTM based Deep Neural Network Architecture. *International Journal on Emerging Technologies.*, *10*(4), 38–42.

Sharma, S. K., Sharma, N. K., & Singh, G. (2019). Unified framework for deep learning based text classification. *International Journal of Scientific Technology Research*, *8*(10), 1479–1483.

SharmaV. (2020). Major Agricultural Problems of India and Various Government Initiatives. doi:10.13140/RG.2.2.33455.76968

Shengde, C., Lan, Y., Jiyu, L., Zhiyan, Z., Aimin, L., & Yuedong, M. (2017). Effect of wind field below unmanned helicopter on droplet deposition distribution of aerial spraying. *International Journal of Agricultural and Biological Engineering*, *10*(3), 67–77.

Shin, H.-C., Roth, H. R., Gao, M., Lu, L., Xu, Z., Nogues, I., Yao, J., Mollura, D., & Summers, R. M. (2016). Deep convolutional neural networks for computer-aided detection: Cnn architectures, dataset characteristics and transfer learning. *IEEE Transactions on Medical Imaging*, *35*(5), 1285–1298. doi:10.1109/TMI.2016.2528162 PMID:26886976

Siddique, S., & Chow, J. C. (2021). Machine learning in healthcare communication. *Encyclopedia*, *1*(1), 220–239. doi:10.3390/encyclopedia1010021

Siemens, G. (2013). Learning Analytics: The Emergence of a Discipline. *The American Behavioral Scientist*, *57*(10), 1380–1400. doi:10.1177/0002764213498851

Siemens, G., & Baker, R. (Eds.). (2019). *Educational Data Mining: Applications and Trends*. Springer.

Silver, D., & Huang, A. (2021). AI in Drug Discovery and Development: Challenges and Opportunities. *Frontiers in Artificial Intelligence*, *4*, 52.

Singh, A. K., Praveen, P., Tripathi, D., Pandey, V. P., & Verma, P. (2023). *Revolutionizing Agriculture with Cloud-Connected Irrigation Technology. In Smart Village Infrastructure and Sustainable Rural Communities*. IGI Global. doi:10.4018/978-1-6684-6418-2.ch010

Singh, G., Sethi, G. K., & Singh, S. (2021). Survey on Machine Learning and Deep Learning Techniques for Agriculture Land. *SN Computer Science*, *2*(6), 487. doi:10.1007/s42979-021-00929-6

Singh, R., Lanchantin, J., Robins, G., & Qi, Y. (2016). Deep chrome: Deep-learning for predicting gene expression from histone modifications. *Bioinformatics (Oxford, England)*, *32*(17), 639–648. doi:10.1093/bioinformatics/btw427 PMID:27587684

Singh, S., Yang, Y., Póczos, B., & Ma, J. (2019). Predicting enhancer-promoter interaction from genomic sequence with deep neural networks. *Quantitative Biology*, *7*(2), 122–137. doi:10.1007/s40484-019-0154-0 PMID:34113473

Sirsat, M. S., Oblessuc, P. R., & Ramiro, R. S. (2022). Genomic Prediction of Wheat Grain Yield Using Machine Learning. *Agriculture*, *12*(9), 1406. doi:10.3390/agriculture12091406

Sishodia, R. P., Ray, R. L., & Singh, S. K. (2020). Applications of Remote Sensing in Precision Agriculture: A Review. *Remote Sensing (Basel)*, *12*(19), 3136. Advance online publication. doi:10.3390/rs12193136

Smith, A., & Johnson, B. (2022). Machine Learning in Healthcare: Current Applications and Future Trends. *Journal of Medical Technology*, *45*(3), 132–148.

Smith, R., Brown, A., & Johnson, M. (2022). Deep Learning for Disease Diagnosis Using Convolutional Neural Networks. *Journal of Medical Imaging (Bellingham, Wash.)*, *48*(5), 205–214.

Smith, R., Brown, A., & Johnson, M. (2023). Predicting Drug-Protein Interactions using Machine Learning and Molecular Descriptors. *Journal of Chemical Information and Modeling*, *63*(4), 1456–1467.

Sofia, M., Fraboni, F., De Angelis, M., Puzzo, G., Giusino, D., & Pietrantoni, L. (2023). The impact of artificial intelligence on workers' skills: Upskilling and reskilling in organisations. *Informing Science*, *26*, 39–68. doi:10.28945/5078

Sood, S., Singh, H., & Jindal, S. (2022). *Rust Disease Classification Using Deep Learning Based Algorithm: The Case of Wheat*. doi:10.5772/intechopen.104426

Sousa, M. J., & Wilks, D. (2018). Sustainable skills for the world of work in the digital age. *Systems Research and Behavioral Science*, *35*(4), 399–405. doi:10.1002/sres.2540

Stiglic, G., Kocbek, P., Fijacko, N., Zitnik, M., Verbert, K., & Cilar, L. (2020). Interpretability of machine learning-based prediction models in healthcare. *Wiley Interdisciplinary Reviews. Data Mining and Knowledge Discovery*, *10*(5), 1379. doi:10.1002/widm.1379

Stine, M. A., & Weil, R. R. (2002). The relationship between soil quality and crop productivity across three tillage systems in south central Honduras. *American Journal of Alternative Agriculture*, *17*(1), 2–8.

Suganyadevi, S., & Seethalakshmi, V. (2022). CVD-HNet: Classifying Pneumonia and COVID-19 in Chest X-ray Images Using Deep Network. *Wireless Personal Communications*, *19*(4), 1–25. doi:10.1007/s11277-022-09864-y PMID:35756172

Syers, J. K. (1997). Managing soils for long-term productivity. *Philosophical Transactions of the Royal Society of London. Series B, Biological Sciences*, *352*(1356), 1011–1021. doi:10.1098/rstb.1997.0079

Tagung, T., Singh, S. K., Singh, P., Kashiwar, S. R., & Singh, S. K. (2022). GPS and GIS based Soil Fertility Assessment and Mapping in Blocks of Muzaffarpur District of Bihar. *Biological Forum : An International Journal*, *14*(3), 1663–1671.

Taj, I., & Zaman, N. (2022). Towards industrial revolution 5.0 and explainable artificial intelligence: Challenges and opportunities. *International Journal of Computing and Digital Systems*, *12*(1), 295–320. doi:10.12785/ijcds/120128

Tang, Y. (2023). Deep learning in drug discovery: Applications and limitations. *Frontiers in Computing and Intelligent Systems*, *3*(2), 118–123. doi:10.54097/fcis.v3i2.7575

Tapalova, O., & Zhiyenbayeva, N. (2022). Artificial Intelligence in Education: AIEd for Personalised Learning Pathways. *Electronic Journal of e-Learning*, *20*(5), 639–653. doi:10.34190/ejel.20.5.2597

Teimouri, N., Dyrmann, M., & Jørgensen, R. N. (2019). A novel spatio-temporal FCN-LSTM network for recognizing various crop types using multi-temporal radar images. *Remote Sensing (Basel)*, *11*(8), 990. doi:10.3390/rs11080990

Teoh, C., Mohamad, B., Radzi, F. F., Najib, M. M., Zamzuri, C. F., Hassan, D., & Nordin, M. n. (2016). Variable rate application of fertilizer in riceprecision farming. *International Conference on Agricultural andFood Engineering (Cafei2016)*, 23-25.

Tian, T., Min, M. R., & Wei, Z. (2021). Model-based autoencoders for imputing discrete single-cell RNA-seq data. *Methods (San Diego, Calif.)*, *192*, 112–119. doi:10.1016/j.ymeth.2020.09.010 PMID:32971193

Tiulpin, A., & Saarakkala, S. (2020). Automatic Grading of Individual Knee Osteoarthritis Features in Plain Radiographs Using Deep Convolutional Neural Networks. *Diagnostics (Basel)*, *10*(11), 932. doi:10.3390/diagnostics10110932 PMID:33182830

Tiulpin, A., Thevenot, J., Rahtu, E., Lehenkari, P., & Saarakkala, S. (2018). Automatic knee osteoarthritis diagnosis from plain radiographs: A deep learning-based approach. *Scientific Reports, 8*(1), 1727. doi:10.1038/s41598-018-20132-7 PMID:29379060

Tiwari, S., Bahuguna, P. C., & Walker, J. (2022). Industry 5.0: A macroperspective approach. In Handbook of Research on Innovative Management Using AI in Industry 5.0 (pp. 59-73). IGI Global.

Tobal, A., & Mokhtar, S. A. (2014). Weeds identification using Evolutionary Artificial Intelligence Algorithm. *Journal of Computational Science, 10*(8), 1355–1361. doi:10.3844/jcssp.2014.1355.1361

Torres-Sánchez, J., Peña, J. M., de Castro, A. I., & López-Granados, F. (2014). Multi-temporal mapping of the vegetation fraction in early-season wheat fields using images from UAV. *Computers and Electronics in Agriculture, 103*, 104–113. doi:10.1016/j.compag.2014.02.009

Tóth, A., Nagy, L., Kennedy, R., Bohuš, B., Abonyi, J., & Ruppert, T. (2023). The human-centric Industry 5.0 collaboration architecture. *MethodsX, 11*(December), 102260. doi:10.1016/j.mex.2023.102260 PMID:37388166

Tran, T., Nguen, T. D., & Phung, D. (2015). Learning vector representation of medical objects via EMR-driven non-negative restricted Boltzmann machines (eNRBM). *Journal of Biomedical Informatics, 54*, 96–105. doi:10.1016/j.jbi.2015.01.012 PMID:25661261

Tripathi, P., Kumar, N., Rai, M., Shukla, P. K., & Verma, K. N. (2023). *Applications of Machine Learning in Agriculture.* doi:10.4018/978-1-6684-6418-2.ch006

Tripathi, P. (2020). Electroencephalpgram Signal Quality Enhancement by Total Variation Denoising Using Non-convex Regulariser. *International Journal of Biomedical Engineering and Technology, 33*(2), 134–145. doi:10.1504/IJBET.2020.107709

Tripathi, P., Kumar, N., Rai, M., & Khan, A. (2022). Applications of Deep Learning in Agriculture. In *Artificial Intelligence Applications in Agriculture and Food Quality Improvement* (pp. 17–28). IGI Global. doi:10.4018/978-1-6684-5141-0.ch002

Tripathi, P., Kumar, N., Rai, M., & Khan, A. (2022). *Applications of Deep Learning in Agriculture. In Artificial Intelligence Applications in Agriculture and Food Quality Improvement.* IGI Global.

Tripathi, P., Kumar, N., & Siddiqi, A. H. (2020). De-noising Raman spectra using total variation de-noising with iterative clipping algorithm. In *Computational Science and its Applications, Taylor and Francis Group* (pp. 225–231). CRC Press. doi:10.1201/9780429288739-14

Tripathi, P., & Siddiqi, A. H. (2016). Solution of Inverse Problem for de-noising Raman Spectral Data with Total variation using Majorization-Minimization Algorithm. *Int. J. Computing Science and Mathematics, 7*(3), 274–282. doi:10.1504/IJCSM.2016.077855

Tripathi, P., & Siddiqi, A. H. (2017). De-noising EEG signal using iterative clipping algorithm. *Biosciences Biotechnology Research Asia, 14*(1), 497–502. doi:10.13005/bbra/2470

Tripathy, H. P., & Pattanaik, P. (2020). Birth of industry 5.0: "the internet of things" and next-generation technology policy. *International Journal of Advanced Research in Engineering and Technology, 11*(11), 1904–1910.

Tsai, C. C., Cheng, W. C., Taur, J. S., & Tao, C. W. (2006). Face Detection Using Eigenface and Neural Network. *2006 IEEE International Conference on Systems, Man and Cybernetics,* 4343-4347. 10.1109/ICSMC.2006.384817

Tsouros, D. C., Bibi, S., & Sarigiannidis, P. G. (2019). A review on UAV-based applications for precision agriculture. In *Information (Switzerland)* (*Vol. 10*, Issue 11). MDPI AG.

Tsouros, D. C., Bibi, S., & Sarigiannidis, P. G. (2019). A review on UAV-based applications for precision agriculture. *Information (Basel)*, *10*(11), 349. doi:10.3390/info10110349

Tucker, A., Wang, Z., Rotalinti, Y., & Myles, P. (2020). Generating high-fidelity synthetic patient data for assessing machine learning healthcare software. *NPJ Digital Medicine*, *3*(1), 1–13. doi:10.1038/s41746-020-00353-9 PMID:33299100

Tyagi, A. K., Dananjayan, S., Agarwal, D., & Thariq Ahmed, H. F. (2023). Blockchain—Internet of Things Applications: Opportunities and Challenges for Industry 4.0 and Society 5.0. *Sensors (Basel)*, *23*(2), 947. doi:10.3390/s23020947 PMID:36679743

Tyagi, N., Rai, M., Sahw, P., Tripathi, P., & Kumar, N. (2022). *Methods for the Recognition of Human Emotions Based on Physiological Response: Facial Expressions in Smart Healthcare for Sustainable Urban Development*. IGI Global. doi:10.4018/978-1-6684-2508-4.ch013

Ullah, F., Salam, A., Abrar, M., Ahmad, M., Ullah, F., Khan, A., Alharbi, A., & Alosaimi, W. (2022). Machine health surveillance system by using deep learning sparse autoencoder. *Soft Computing*, *26*(16), 7737–7750. doi:10.1007/s00500-022-06755-z

Usmaedi, U. (2021). Education curriculum for society 5.0 in the next decade. *Jurnal Pendidikan Dasar Setiabudhi*, *4*(2), 63–79.

Uzal, L. C., Grinblat, G. L., Namías, R., Larese, M. G., Bianchi, J. S., Morandi, E. N., & Granitto, P. M. (2018). Seed-per-pod estimation for plant breeding using deep learning. *Computers and Electronics in Agriculture*, *150*, 196–204. doi:10.1016/j.compag.2018.04.024

Valous, N., & Sun, D.-W. (2012). Image processing techniques for computer vision in the food and beverage industries. In D.-W. Sun (Ed.), *Computer Vision Technology in the Food and Beverage Industries* (pp. 97–129). Woodhad Publishing. doi:10.1533/9780857095770.1.97

Vanitha, C. N., Archana, N., & Sowmiya, R. (2019). Agriculture Analysis Using Data Mining And Machine Learning Techniques. *2019 5th International Conference on Advanced Computing & Communication Systems (ICACCS)*, (pp. 984–990). IEEE. 10.1109/ICACCS.2019.8728382

Vanmathi, C., Mangayarkarasi, R., & Jaya Subalakshmi, R. (2020). Real Time Weather Monitoring using Internet of Things. *2020 International Conference on Emerging Trends in Information Technology and Engineering*.

Varsha, R., Nair, S. M., Tyagi, A. K., Aswathy, S. U., & RadhaKrishnan, R. (2020). The Future with Advanced Analytics: A Sequential Analysis of the Disruptive Technology's Scope. *Hybrid Intelligent Systems*, *1375*, 565–579. doi:10.1007/978-3-030-73050-5_56

Vassilakopoulou, P., Haug, A., Salvesen, L. M., & Pappas, I. O. (2023). Developing Human/AI interactions for chat-based-customer-services: Lessons learned from the norwegian government. *European Journal of Information Systems*, *32*(1), 10–22. doi:10.1080/0960085X.2022.2096490

Vaswani, A., Shazeer, N., Parmar, N., Uszkoreit, J., Jones, L., Gomez, A. N., Kaiser, Ł., & Polosukhin, I. (2017). *Attention is all you need.* In *Proceedings of the 31st Conference on Neural Information Processing Systems (NIPS 2017)*, Long Beach, CA, USA.

Villa-Henriksen, A., Edwards, G. T. C., Pesonen, L. A., Green, O., & Sørensen, C. A. G. (2020). Internet of Things in arable farming: Implementation, applications, challenges and potential. *Biosystems Engineering*, *191*, 60–84. doi:10.1016/j.biosystemseng.2019.12.013

Waleed, M., Um, T. W., Kamal, T., Khan, A., & Iqbal, A. (2020). Determining the precise work area of agriculture machinery using internet of things and artificial intelligence. *Applied Sciences (Basel, Switzerland)*, *10*(10), 3365. doi:10.3390/app10103365

Wang, H., Chen, X., & Zhang, Z. (2023). Continuous Heart Rate Variability Monitoring Using Wearable Sensors and Machine Learning. *Journal of Biomedical Informatics*, *115*, 103738.

Wang, Y., Zhang, Z., Feng, L., Du, Q., & Runge, T. (2020). Combining multi-source data and machine learning approaches to predict winter wheat yield in the conterminous United States. *Remote Sensing (Basel)*, *12*(8), 1232. doi:10.3390/rs12081232

Waring, J., Lindvall, C., & Umeton, R. (2020). Automated machine learning: Review of the state-of-the-art and opportunities for healthcare. *Artificial Intelligence in Medicine*, *104*, 101822. doi:10.1016/j.artmed.2020.101822 PMID:32499001

Wazid, M., Das, A. K., Mohd, N., & Park, Y. (2022). Healthcare 5.0 Security Framework: Applications, Issues and Future Research Directions. *IEEE Access : Practical Innovations, Open Solutions*, *10*, 129429–129442. doi:10.1109/ACCESS.2022.3228505

Wojke, N., Bewley, A., & Paulus, D. (2017). Simple online and realtime tracking with a deep association metric. *2017 IEEE International Conference on Image Processing (ICIP)*, 3645-3649. 10.1109/ICIP.2017.8296962

Xu, F., & Zhang, S. (2021, March). Understanding the Source of Confusion with Computational Thinking: A Systematic Review of Definitions. In *2021 IEEE Integrated STEM Education Conference (ISEC)* (pp. 276-279). IEEE. Retrieved from: https://ieeexplore.ieee.org/abstract/document/9764144/

Xu, L., Xu, Y., Xue, T., Zhang, X., & Li, J. (2021). AdImpute: An Imputation Method for Single-Cell RNA-Seq Data Based on Semi-Supervised Autoencoders. *Frontiers in Genetics*, *12*, 739677. doi:10.3389/fgene.2021.739677 PMID:34567089

Xu, X., Lu, Y., Vogel-Heuser, B., & Wang, L. (2021). Industry 4.0 and Industry 5.0—Inception, conception and perception. *Journal of Manufacturing Systems*, *61*, 530–535. doi:10.1016/j.jmsy.2021.10.006

Xu, Z., Elomri, A., El Omri, A., Kerbache, L., & Liu, H. (2021). The Compounded Effects of COVID-19 Pandemic and Desert Locust Outbreak on Food Security and Food Supply Chain. *Sustainability (Basel)*, *13*(3), 1063. doi:10.3390/su13031063

Yang, C. C., Prasher, S. O., Landry, J. A., & Ramaswamy, H. S. (2003). Development of a herbicide application map using artificial neural networks and fuzzy logic. *Agricultural Systems*, *76*(2), 561–574. doi:10.1016/S0308-521X(01)00106-8

Yang, D., & Feng, S. (2022, March). A Collaborative Approach to Integrate Computational Thinking in an Integrated STEM Curriculum. In *2022 IEEE Integrated STEM Education Conference (ISEC)* (pp. 238-240). IEEE. 10.1109/ISEC54952.2022.10025038

Yang, S., Zhu, F., Ling, X., Liu, Q., & Zhao, P. (2021). Intelligent Health Care: Applications of Deep Learning in Computational Medicine. *Frontiers in Genetics*, *12*, 607471. doi:10.3389/fgene.2021.607471 PMID:33912213

Yin, X., Zhu, Y., & Hu, J. (2021). A comprehensive survey of privacy-preserving federated learning: A taxonomy, review, and future directions. [CSUR]. *ACM Computing Surveys*, *54*(6), 1–36. doi:10.1145/3460427

Yuan, D., Zhu, X., Wei, M., & Ma, J. (2019). Collaborative deep learning for medical image analysis with differential privacy. In *2019 IEEE Global Communications Conference (GLOBECOM)*. New York, NY: IEEE. 10.1109/GLOBECOM38437.2019.9014259

Zahid, A., Abbas, H. T., Imran, M. A., Qaraqe, K. A., Alomainy, A., Cumming, D. R. S., & Abbasi, Q. H. (2019). Characterization and water content estimation method of living plant leaves using terahertz waves. *Applied Sciences (Basel, Switzerland)*, *9*(14), 2781. doi:10.3390/app9142781

Zapata-Cáceres, M., Martín-Barroso, E., & Román-González, M. (2020, April). Computational thinking test for beginners: Design and content validation. In *2020 IEEE global engineering education conference (EDUCON)* (pp. 1905-1914). IEEE.

Zhang, X., Wu, C.-W., Fournier-Viger, P., Van, L.-D., & Tseng, Y.-C. (2017). Analyzing students' attention in class using wearable devices. *2017 IEEE 18th International Symposium on A World of Wireless, Mobile and Multimedia Networks (WoWMoM)*, 1-9. 10.1109/WoWMoM.2017.7974306

Zhang, J., Kowsari, K., Harrison, J. H., Lobo, J. M., & Barnes, L. E. (2018). Patient2Vec: A personalized interpretable deep representation of the longitudinal electronic health record. *IEEE Access : Practical Innovations, Open Solutions*, *6*, 65333–65346. doi:10.1109/ACCESS.2018.2875677

Zhang, L. N., & Yang, B. (2014). Research on recognition of maize disease based on mobile internet and support vector machine technique. *Advanced Materials Research*, *905*, 659–662. doi:10.4028/www.scientific.net/AMR.905.659

Zhang, S., Zhou, J., Hu, H., Gong, H., Chen, L., Cheng, C., & Zeng, J. (2016). A deep learning framework for modeling structural features of RNA-binding protein targets. *Nucleic Acids Research*, *44*(4), e32. doi:10.1093/nar/gkv1025 PMID:26467480

Zhang, X., Chen, Z., Bhadani, R., Cao, S., Lu, M., Lytal, N., Chen, Y., & An, L. (2022). NISC: Neural Network-Imputation for Single-Cell RNA Sequencing and Cell Type Clustering. *Frontiers in Genetics*, *13*, 847112. doi:10.3389/fgene.2022.847112 PMID:35591853

Zhang, X., & Davidson, E. A. (2018). *Improving Nitrogen and Water Management in Crop Production on a National Scale*. American Geophysical Union.

Zhao, B., Chen, X., Zhao, X., Jiang, J., & Wei, J. (2018). Real-Time UAV Autonomous Localization Based on Smartphone Sensors. *Sensors (Basel)*, *18*(12), 4161. doi:10.3390/s18124161 PMID:30486422

Zhavoronkov, A., Ivanenkov, Y. A., Aliper, A., Veselov, M. S., Aladinskiy, V. A., Aladinskaya, A. V., Terentiev, V. A., Polykovskiy, D. A., Kuznetsov, M. D., Asadulaev, A., Volkov, Y., Zholus, A., Shayakhmetov, R. R., Zhebrak, A., Minaeva, L. I., Zagribelnyy, B. A., Lee, L. H., Soll, R., Madge, D., ... Aspuru-Guzik, A. (2019). Deep learning enables rapid identification of potent DDR1 kinase inhibitors. *Nature Biotechnology*, *37*(9), 1038–1040. doi:10.1038/s41587-019-0224-x PMID:31477924

Zhou, X., Liang, W., Wang, K. I.-K., Wang, H., Yang, L. T., & Jin, Q. (2020, July). Deep-Learning-Enhanced Human Activity Recognition for Internet of Healthcare Things. *IEEE Internet of Things Journal*, *7*(7), 6429–6438. doi:10.1109/JIOT.2020.2985082

Žižek, S. Š., Mulej, M., & Potočnik, A. (2021). Potoˇcnik, A. The Sustainable Socially Responsible Society: Well-Being Society 6.0. *Sustainability (Basel)*, *13*(16), 9186. doi:10.3390/su13169186

Zrimec, J., Fu, X., Muhammad, A. S., Skrekas, C., Jauniskis, V., Speicher, N. K., Börlin, C. S., Verendel, V., Chehreghani, M. H., Dubhashi, D., Siewers, V., David, F., Nielsen, J., & Zelezniak, A. (2022). Controlling gene expression with deep generative design of regulatory DNA. *Nature Communications*, *13*(1), 5099. doi:10.1038/s41467-022-32818-8 PMID:36042233

About the Contributors

Rijwan Khan received his B. Tech Degree in Computer Science & Engineering from BIT, M. Tech in Computer Science & Engineering from IETE, and Ph.D.in Computer Engineering from Jamia Millia Islamia, New Delhi. He has 17 years of Academic & Research Experience in various areas of Computer Science. He is currently working as Professor and Program Chair of Department Computer Science and Engineering B.Tech 2nd Year at Galgotias University, G.B. Nagar U.P. He is author of three subject books. He published more than 30 research papers in different journals and conferences. He has been nominated in the board of reviewers of various peer-reviewed and refereed Journals. He is Editor of five research books. He is Editor of seven special issues of SCI and Scopus indexed journals. He was session chaired in three International conferences & Keynote speakers in some the national and International conferences

***

Shailendra Singh is presently working as an Assistant Professor at the Department of Computer Science and Engineering, Inderprastha Engineering College, Ghaziabad, U. P., India. He has more than 15 years of experience in academics and teaching GATE/PSUs classes at various renowned institutes like Unacademy, GATEFORUM, ICE GATE, Brilliant Concept, Career Launcher etc. He has earned degrees: M.Tech. (CSE), B.Tech.(IT), UGC NET in CS, GATE(99.24%) in CS, He is NPTEL Gold Medalist (National Topper) in "Theory of Computation"and has authored over 10 research papers in professional journals and conferences. His research areas include Cloud Security, Social Network Security and Machine learning.

Pratyush Bibhakar is Assistant Professor of Sociology currently associated with the School of Liberal Education, Galgotias University, (U.P). He holds a multidisciplinary academic background with a PhD on 'Politics through Art: A Study of Architecture, Sculpture and Political Posters in Russia, 1905-2005' from Jawaharlal Nehru University (JNU) and possesses two M.A Degrees one in Sociology and another in Russian Language, Literature & Culture. His academic interests include areas like Art and Polity, Architecture and society, Education and Social Justice, Literature and Society, Sexuality and Queer studies etc. He has been contributing to various online/offline peer-reviewed journals and other online news media platforms.

Suganya G. is a Ph.D in Management Sciences. Currently she is associated with Kumaraguru College of Liberal Arts and Science, Coimbatore as Assistant Professor. Previously, she has an industrial experience of 4 years in textiles and automobile sector and has been in the field of research and teaching for 5 years. She has been associated as a researcher in a number of collaborative project conducted by University of Leeds (UK) and the Goa Institute of Management (Goa) from 2017 which was funded by: University of Leeds and British Academy in the year 2018 – 2019, AHRC Modern Slavery Policy and Evidence Centre COVID-10 SPF 2020 in the year 2020 – 2021 and UK's Arts and Humanities Research Council (AHRC) in the year 2022. She has organized a number of events and scenario workshops during her tenure. She has been associated as Research Assistant in a project which was sponsored by Indian Council of Social Science Research (ICSSR), Impactful Policy Research in Social Science (IMPRESS).

Pawan Kumar Goel is presently working as an Associate Professor at the Department of Computer Science and Engineering, Raj Kumar Goel Institute of Technology, Ghaziabad, U. P., India. Before that, he worked for several years as an Associate Professor & H.O.D. CSE at Shri Ram Group of Colleges (NAAC A++ accredited institute) .He has more than 17 years of experience in the academics. He has earned degrees: Ph.D (CSE), UGC NET in CS, M.Tech., MBA (HR), and B.Tech (IT) and has authored more than 35 research papers in professional journals and conferences. He has authored and edited more than 5 books with reputed publishers. His research areas include Information Security, Wireless Sensor Networks, Machine learning, information retrieval, semantic web, ontology engineering, data mining, ad hoc networks, sensor networks and network security. He received a International Achiever Award 2023 by Gyan Uday Foundation, Kota Rajasthan, for contribution in the category of "Distinguished Teacher 2023, Educator of the Year 2022 Award by Namaste India Council of Educators on Teacher's Day, 5th September 2022, National Education Excellence Achiever award by Navbharat Rashtriya Gyanpeeth, Pune Maharashtra for excellence in education field on 20/03/2022, Certificate of Appreciation for I2OR National Award 2021 in the category of "Distinguished Teacher" powered by International Institute of Organized Research (I2OR), "Global Outreach Agricultural Award-2020" Established Teacher in Computer Science & Engineering during 5th Global Outreach Conference on Modern Approaches for Smart Agriculture (MASA-2020), He is a member of the Asia Society of Researchers (ASR), the International Association of Engineers (IAENG), IACSIT (International Association of computer science and Information Technology), CSTA(Computer Science Teacher Association), ACM (Association for Computing machinery), UACEE (Universal Association of of Computers & Electronic Engineers), SCIEI, Internet Society ISOC, India, Academy & Industry Research Collaboration Center (AIRCC), IFERP (Institute for Engineering Research and Publication).

Vijay Prakash Gupta has more than 15 years of experience in the fields of education and research. Currently, he is working as an Associate Professor at the Institute of Business Management & Ph.D. Coordinator at GLA University, Mathura. He did his Ph.D. in Management from Uttarakhand Technical University, Dehradun. He has been appreciated and awarded by the Higher Education Department, Government of Uttar Pradesh, for course curriculum design under NEP-2020. He has been given the PERFICIO Awards 2020 for the Best Research and Innovation of the Year 2020 by the BHS Foundation in New Delhi and Campbell University in the United States. He has also been given the Sarvepalli Radhakrishnan Professor Award for Outstanding Contribution in the Fields of Teaching, Research, and Development. He is an editorial member and a reviewer for many reputed international journals. He has published around 20 research papers in reputed national and international journals indexed in ABDC, Scopus, Web of Science, and UGC CARE. He has two patents to his credit. He is regularly invited to serve as the keynote speaker for seminars, conferences and as an expert for Ph.D. Thesis Evaluation and Viva-Voce by institutes of high repute. He has authored five textbooks, edited five books, and contributed chapters to about 10 books. His core competency extends to courses like Business Economics, Managerial Economics, Business Environment, Indian Economy, Marketing Management, Sales Management, and Other General Management Subjects.

Kaushal Kishor received his Ph.D. in Computer science and engineering from AKTU Lucknow, in the domain of Mobile Ad hoc Network. M.Tech & B.Tech in Computer Science & Engineering from UPTU Lucknow. Currently, he is working in ABES Institute of Technology, Ghaziabad as Professor Information Technology and CSE(DS). He has supervised more than 50 projects for graduate and post graduate students. He has more than 18 years of experience of teaching. He has worked for 6 years in IMS Engineering College, Ghaziabad as Assistant Professor of Computer Science & Engineering. He has worked for 4 years in Hi-Tech Engineering College, Ghaziabad as Assistant Professor of Computer Science & Engineering and He has worked for 3 years in G L BAJAJ Engineering College, Greater Noida as Associate Professor of Information technology. He is Gate Qualified 2003 score 94.5 percentile. He has book published and edited (1) "Cloud-based Intelligent Informative Engineering for Society 5.0" pp. 1-234. Chapman and Hall/CRC., 2023, eBook ISBN: 9781003213895 (2) Design and Analysis Algorithms (ISBN NO. 978-93-81695-20-3) (3) Computer Networks (ISBN NO. 978-93-81695-27-2) (4) Compiler Design (ISBN NO. 938169530 x and ISBN-13 9789381695302) (5) Design and Analysis of Algorithms: Techniques and Control Management (ISBN NO. 978-81-8220-516-1) (6) Computer Networks a System Approach (ISBN NO. 978-81-8220-516-3) (7) Compiler Design Principles, Techniques, and Tools (ISBN:978-81-8220-626-7) for various engineering field like B.Tech. and MCA student. He has published 20 papers in peer reviewed international/National journals and conferences. His research interest includes Artificial Intelligence, Computer Networks, Algorithm, Compiler Design Wireless and Sensor Networking.

Indrajeet Kumar received his B Tech in Computer Science & Engineering from Sant Longowal Institute of Engineering and Technology, Punjab in 2009 and M Tech in Computer Engineering with specialization in Computer Network from YMCA University of Science & technology, Faridabad in 2013. He received his PhD on Analysis and Classification of Breast density classification using Mammographic images from GB Pant Institute of Engineering & Technology, Ghurdauri,Uttarakhand, India in 2019. During his PhD he worked on enhancing the potential of most commonly available Breast density classification for differential diagnosis between atypical cases of focal density cases. He served in academia in various reputed organizations like Galgotias Institutions, Greater Noida, U.P, India and graphic era hill University, Dehradun, India, for 6 years. His research interests include application of machine learning, Deep Learning and soft computing techniques for analysis of medical images.

Nitendra Kumar received a Ph. D (Mathematics) from Sharda University, Greater Noida, India and a Master of Science (Mathematics and Statistics) from Dr Ram Manohar Lohia Avadh University, Faizabad, India. Currently, he is working as an Assistant Professor at Amity Business School, Amity University, Noida with interests in the Wavelets and its Variants, Data Mining, Inverse Problems, Epidemic Modeling, Fractional Derivatives Business Analytics, and Statistical Methods. He has more than 10 years of experience in his research areas. Dr Kumar has published many research papers in reputed journals and also published 6 books on engineering mathematics. He contributes to the research community by undertaking various volunteer activities in the capacity of editor for two edited books, Guest Editor of reputed journals of Taylor and Francis.

Noor Mohd received B. Tech degree in Computer Science & Engineering from HNB Garhwal University Srinagar Pauri, Uttarakhand,India, M.Tech degree in Computer Science & Engineering from Uttarakhand Technical University,Dehradun, Uttarakhand, India and Ph.D. degree from Govind Ballabh Pant Engineering Pauri, Uttarakhand, India. Presently working as an Associate Professor in Computer Science Engineering of Graphic Era Deemed to Be University Dehradun Uttarakhand, India. His research interests are in Mobile Adhoc Networks primarily in the area of "An efficient framework for intrusion detection system for mobile adhoc networks".

Sathish Pachiyappan is currently serving as Assistant Professor at CHRIST(deemed), Bannerghatta Road Campus, Bengaluru. He has eight years teaching and three years research experience. He has done his doctoral degree from VIT (deemed), Vellore. He is specialized in Finance and Accounting, doing his research in the same area. He has completed MBA from Anna University and B,Com from SRM University, Chennai. Currently, as a part of his Research work, he had published articles in peer reviewed journals which includes Scopus Indexed Journals; Web of Science Indexed Journals, Australian Business Deans Council (ABDC) listed Journals and EBSCO host Journals. Also, published book chapter in Elsevier, Springer and IGI global. Presented research papers in national and international conferences and also attended Faculty development programmes and workshops which are related to SPSS, AMOS and Econometrics. He went as resource person for various FDP, Guest Lecture, Session Chair for Conference and Workshop to various reputed colleges. He is well versed in handling SPSS,RSTATCRAFT, E-Views (Econometrics) for financial data analysis in research area. He is gold medalist and earned first rank in B.Com at SRM University, Chennai and also earned Class topper in MBA and 39th Rank Holder in Anna University. He received Research Award-2017 for publications in VIT University, Vellore. Recently, received award for "Best Researcher-2020" conducted by Pondicherry Research Society in Pondicherry.

Mritunjay Rai received his Ph.D. from IIT (ISM) in Electrical Engineering with a specialization in Image Processing, his Master of Engineering with distinction from Birla Institute of Technology, Mesra, Ranchi, in Instrumentation and Control, and B. Tech. from Shri Ramswaroop Memorial College of Engineering and Management, Lucknow in Electronics and Communication Engineering. His areas of interest lie in image processing, speech processing, and robotics & automation. Dr. Rai is an active researcher and has published several papers in SCI indexed Q2 journals and at international and national conferences in his domain.

Birendra Saraswat is presently working as Assistant Professor in the Department of Computer Science & Engineering at Raj Kumar Goel Institute of Technology, Ghaziabad. Birendra Kr. Saraswat has received M.Tech (CSE) Degree from Uttar Pradesh Technical University, Lucknow and Pursuing Ph.D (Computer Science & Engineering) from GLA University, Mathura. He has guided B.Tech Project and MCA Project. He had published several research papers in International Journal/Scopus and Conferences of repute and published many patents. His major areas of competencies include Web Technology, C Programming, Python Programming .His research interest area in Machine Learning and Computer Networks. Mr. Saraswat has 16+ years of teaching experience. He has organized many events in the college related to education activity like FDP, Conferences and etc.

Dr. J. Joshua Selvakumar is an Associate Professor at the School of Business and Management at Christ University, Bangalore. He is a marketing management educator and administrator with national and international front-line sales experience with companies such as Mahindra & Mahindra, TTK Healthcare and Excel International in Doha, Qatar. He has 18+ years of teaching in leading Business schools like PSG Institute of Management Coimbatore and others. He served as the Centre Head for PSG Institutions In-House advertising and communication division for over a decade. He has over 42 publications to his credit. He is passionate about integrated marketing communication and sales Management and has published a book titled "Dimensions of Integrated Marketing Communications" with Lambert Publishing, Germany. He has visited several Universities abroad and is currently an adjunct faculty at the University of Toledo, Ohio, USA and a Collaborator with Taylor's University, Malaysia.

Amit Singhal is working as Professor & Head in Department of Computer Science & Engineering in Raj Kumar Goel Institute of Technology, Ghaziabad, UP, India. He received B.E., M.E. & Ph.D. degree in Computer Science & Engineering. More than 40 Research publications in SCI/Scopus indexed, International & National Journals/Conferences. His research interests include Wireless Sensor Networks & networking. He is Senior Member of 'The Institute of Electrical and Electronics Engineers Incorporated' (IEEE) & The Asia Society of Researcher' (ASR) & Life member of The Indian Society for Technical Education' (ISTE)- New Delhi, CSTA- New York, USA, IACSIT- Singapore, IAENG- Hongkong, Internet Society, ICST- Belgium, IFERP, IEEE Computer Society- USA & many other technical & professional bodies. He has been awarded with 'Teacher Innovation Award' for sustainable efforts towards promoting joyful & experimental teaching by 'Sri Aurobindo Society' in Sep. 2019. 'BEST SPOC' by Prof. Bhaskar Ramamurthi, Chairman NPTEL PIC & Director IIT, Madras, "Academic Excellence Award", "Vishistha Shikshak Samman" & "Maulana Abul Kalam Azad Excellence Award of Education".

Padmesh Tripathi is working as an Associate Professor of Mathematics at IIMT College of Engineering, Greater Noida, UP. He received his Ph.D. degree from Sharda University, Greater Noida (India). He completed his Master's and Bachelor's degrees from University of Allahabad, Prayagraj. He has been teaching since last 20 years. He has published many research papers in reputed journals. He has been granted funds from prestigious organizations like: Cambridge University, UK; University of California, USA; University of Eastern Finland, Finland; INRIA, Sophia Antipolis, France, etc. and he visited these organizations. He has been member of Society of Industrial and Applied Mathematics (SIAM), Philadelphia, USA (2011-19). He is life member of Science and Engineering Institute (SCIEI), Los Angeles, USA; Indian Society of Industrial and Applied Mathematics (ISIAM), India; Ramanujan Mathematical Society, India; International Association of Engineers, UK. He is member of EURO working group on continuous optimization (EUROPT), Italy; Society for Foundations of Computational Mathematics, USA. His research interests are inverse problems, signal processing, optimization, data science, etc.

Pratibha Verma is Assistant Professor of Sociology currently associated with the Constituent Government Degree College, Puranpur, India, (U.P). She received the graduation degree from University of Lucknow, UP, India, in 2009 and the post-graduation degree, University of Lucknow, UP in 2011. Currently, she is pursuing PhD degree in the Department of Sociology at the Galgotia University, Noida India. My research interests include medical sociology, agriculture society and rural society.

Index

Printed in the United States
by Baker & Taylor Publisher Services